Thomas King Chambers

A Manual of Diet in Health and Disease

Thomas King Chambers

A Manual of Diet in Health and Disease

ISBN/EAN: 9783337035273

Printed in Europe, USA, Canada, Australia, Japan

Cover: Foto ©berggeist007 / pixelio.de

More available books at **www.hansebooks.com**

A MANUAL

OF

DIET IN HEALTH AND DISEASE,

BY

THOMAS KING CHAMBERS,
M.D. OXON., F.R.C.P., LOND.,

HONORARY PHYSICIAN TO H.R.H. THE PRINCE OF WALES, CONSULTING PHYSICIAN
TO ST. MARY'S AND THE LOCK HOSPITALS, LECTURER ON MEDICINE
AT ST. MARY'S SCHOOL, CORRESPONDING FELLOW OF THE
ACADEMY OF MEDICINE, NEW YORK, ETC.

PHILADELPHIA:
HENRY C. LEA.
1875.

PREFACE.

THE aims of this Handbook are purely practical, and therefore it has not been thought right to increase its size by the addition of the chemical, botanical, and industrial learning which rapidly collects round the nucleus of every article interesting as an eatable. Space has been thus gained for a full discussion of many matters connecting food and drink with the daily current of social life, which the position of the Author as a practicing physician has led him to believe highly important to the present and future of our race.

THOS. K. CHAMBERS.

24 MOUNT STREET, GROSVENOR SQUARE,
      January 1, 1875.

CONTENTS.

PART I.

GENERAL DIETETICS.

CHAPTER		PAGE
I.	Theories of Dietetics,	17
II.	On the Choice of Food,	29
III.	On the Preparation of Food,	87
IV.	On Digestion,	101
V.	Nutrition,	122

PART II.

SPECIAL DIETETICS OF HEALTH.

I.	Regimen of Infancy and Motherhood,	125
II.	Regimen of Childhood and Youth,	134
III.	Commercial Life,	140
IV.	Literary and Professional Life,	144
V.	Noxious Trades,	152
VI.	Athletic Training,	155
VII.	Hints for Healthy Travellers,	169
VIII.	Effects of Climate,	175
IX.	Starvation, Poverty, and Fasting,	184
X.	The Decline of Life,	197
XI.	Alcohol,	200

PART III.

DIETETICS IN SICKNESS.

CHAPTER		PAGE
I.	DIETETICS AND REGIMEN OF ACUTE FEVERS,	231
II.	THE DIET AND REGIMEN OF CERTAIN OTHER INFLAMMATORY STATES,	245
III.	THE DIET AND REGIMEN OF WEAK DIGESTION,	252
IV.	GOUT AND RHEUMATISM,	261
V.	GRAVEL, STONE, ALBUMINURIA, AND DIABETES,	270
VI.	DEFICIENT EVACUATION,	278
VII.	NERVE DISORDERS,	282
VIII.	SCROFULA, RICKETS, AND CONSUMPTION,	294
IX.	DISEASE OF HEART AND ARTERIES,	301

ALPHABETICAL INDEX, ... 307

DIET AND REGIMEN.

PART I.

GENERAL DIETETICS.

CHAPTER I.

THEORIES OF DIETETICS.

WHAT is the natural food of man?

Each animal in a state of nature finds substances suited for its nutrition ready to hand, and within the grasp of the instruments he possesses for their acquisition. And these substances seem, generally, the most proper to sustain the health and strength. So that it has been not irrationally argued, that it would be a useful act of scientific reasoning to infer from the structure of the human organs what kind of food they are most fitted to appropriate, for this would probably prove most conducive to physical well-being.

When, in pursuit of this reasoning, we come to compare man's form with that of other mammalia, his prehensile organs—his teeth, his jaws, and his feet and his nails—do not seem to fit him for grappling with any of the difficulties which the adoption of special kinds of food prepared by nature entails. He can neither tear his prey conveniently, nor crack many nuts, nor grub roots, nor graze. His digestive viscera, in middle life, are too bulky and heavy to qualify him for the rapid movements of the carnivora; and they are not long enough to extract nourishment from raw vegetables. To judge by form and structure, alone, the natural food of an adult man must be pronounced to be *nothing*.

On the other hand, if we read the laws of man's nature by the light of the general consent of the individuals of his race, which is the wisest course,[1] we shall arrive at the opposite conclusion, that his food is *everything* which any other warm-blooded animal can use as nourishment. If we try to construct a universal dietary from the records which each new traveller brings home of what he has beheld habitually eaten, we shall find very few forms of organic matter, capable of supporting mammalian life, which are not appropriated by man also to his own use. By selection and preparation he contrives to remove such parts and such qualities of the substances presented by nature as are noxious to him, and to improve such as suit his purpose; so that as finally swallowed, they are more wholesome to him than to the beasts who eat nothing else. These lists of possible eatables are most interesting to the student of human nature; they lead to inferences as to the action of laws, religions, customs, and associations, in making that abominable to one race which is most highly appreciated by another, and they are an important part of the arguments of those who trace political events and national character to physical causes; but they are not suited to the present volume, which will concern itself with the action on individual health of food generally accessible in the British market. Reference may be occasionally made to a more extended *materia alimentaria*, but it can contribute little to the main arguments proposed.

The power by virtue of which man becomes so truly omnivorous is habit. He can gradually, in time, accustom himself to live on anything containing nourishment, provided he be not limited in quantity, nor restricted in facilities for preparation. The inferior animals could do the same if they only knew how to set about it; for when we bring our reason to bear on their lives, we can effect what at first sight seem most radical changes in their nature, in respect of food; and we can even induce and perpetuate hereditary forms of body suited to the altered circumstances we have brought about. Spallanzani found that pigeons may be fed on flesh, and eagles on bread, by accustoming them to it; the domestic dog grows strong on biscuit, and often suffers in health on being brought back to his native food; our poultry is more

---

[1] " Consensus omnium nationum lex naturæ putanda est."—Cicero *de Legibus*, i, 8.

robust, more fertile, and apparently happier, for being supplied with meat, fat, or soup, and our cats have accommodated themselves to a mixed diet, assimilating their form to that of herbivora, by a considerable increase in the length of their bowels over those owned by their cousins of the mountains. The speechless creatures have not the wits to acquire unaided these new powers; compulsory education is necessary; even for such a simple process as learning to eat turnips, the lamb requires a shepherd to stand over him and forcibly make him chew. Man's chief bodily strength depends on his willingness to submit to the pain of acquiring habits, and on his forcing his domestic stock to submit to it, for the sake of a future advantage.

The solvent actions of the juices of the intestinal canal on food seem to be the same in quality in all classes of animals, and to admit of modification in the proportions of their ingredients according to the diet adopted. Under vegetable food the saliva becomes more copious, under meat there is more gastric juice. The bile of a grazing ox is more watery than that of a man; the bile of a growing boy (who can digest any amount of meat) was found by Gorup-Besanez[1] to contain nearly double the amount of solid contained in that of an old woman (whose age would dispose her to be very little carnivorous).

This shows the importance of what may be called the preparatory or mechanical parts of digestion. The digestive solvents can evidently grow equal to all emergencies of the chemical acts required of them, and the differences in the results of those acts must mainly hang on the mechanical condition of the substances presented to them. Fortunate indeed is it that such is the case, for the mechanical condition of the food is certainly more fully in our power, and more easily influenced by our reason, than the chemical solvency of the secretions. We can choose, according to its hardness, softness, and other external qualities, the sort of victual we put in our mouths; we can prepare it with art, can regulate its bulk and the period of taking it; while the muscles which chew it and swallow it are almost entirely under our direction. But it is only very indirectly that we can influence the saliva, the gastric and pancreatic juices, and the bile.

---

[1] Untersuchungen uber d. Galle, Erlangen, 1846. The relative proportions of solid matter were 17.19 per cent. as against 9.13 per cent.

Assuming, then, that man can easily accommodate himself to a varied and mixed diet—that he has, as a matter of fact, accommodated himself to it—and that, therefore, it will in future, as in the past, best suit his requirements—the next point of interest is the proportion which its several ingredients should bear to one another.

Physiologists have pointed out that in the preparation made for the infant at its entrance into life, and which is a striking instance to the faithful mind of a controlling design in creation, we have a typical instance of what the All-wise considered a suitable dietary. Looking to its qualitative composition, we find milk contains alimentary principles capable of separating themselves, and, in fact, habitually separated for economical purposes, somewhat in the following proportions:

Water,	88 per cent.
Oleaginous matter (cream, butter),	3 "
Nitrogenous matter (cheese and albumen),	4 "
Hydrocarbon (sugar),	4½ "
Saline matter (phosphate of lime, chloride of sodium, iron, etc.),	½ "

This rough average is the best way of stating the facts for physiological purposes; since, as every mother, physician, and farmer knows, the proportions vary considerably in different specimens of even the same species of animal, and are influenced by differences in the mode of living. The argument is, that there or thereabouts, may be found the ratio which there should be in our dietary, in the amounts of the alimentary substances of which the above may be taken as representatives. That is to say, that, supposing a man to consume 200 ounces of victual daily, the contents should be about—

6 quarts of water,
½ a lb. of animal matter, such as cheese, or lean meat, or eggs,
6 ozs. of fat, oil, or butter,
9 ozs. of sugar or starch,
1 oz. of salt, and some small quantity of bone or iron.

A serious flaw in this argument is that while the dietary is prepared for, and truly suits very well, the newly born, we have no evidence that either it is intended for, or would suit better than another, the adult. The milk of our domestic animals so closely resembles that which supported us in infancy, that if we carried the reasoning out to its logical consequences, we should all be feeding together now at the same manger. If the milk represents what the adult ought to make his diet, our bull would require only a little more butter, and our horse only a little less than we do; our goat would want one-third more meaty or nitrogenous matter to be contained in the food than ourselves; and the dog would require five times the proportion of flesh that is laid on his master's table to be afforded him.[1] In point of fact, the life led by the young of all animals is much the same, whereas in adult age they differ widely in their occupations, and in the demand for the sort of viands best adapted to those occupations.

There is greater promise of profit to the dietician in a calculation of the outgoings of matters resulting from the wear and tear of the body, reducing these to ultimate elementary substances, and thus ascertaining in what proportion to one another new supplies of ultimate elementary substances are required, merely to replace those consumed. It is obvious that the food which supplies the demand most accurately will be the most economical in the highest sense. We can measure, for example, the carbon and the nitrogen daily thrown off in the excretions, and then lay down a rule for the minimum quantity of those elements which the daily food must contain to keep up the standard weight. If the diet is such as to make it necessary to eat too much carbon in order to secure a due amount of nitrogen, there is an obvious waste, and the di-

---

[1] The computation of the ingredients of milk is a deduction from the following table of M. Boussingault's analysis:

Milk of	Water, per cent.	Casein and Albumen.	Butter.	Sugar of Milk.	Salts.
Woman, . . .	88.9	3.9	2.6	4.3	0.1
Cow, ...	86.6	4.0	4.0	4.8	0.6
Ass, ...	90.3	1.9	1.0	6.4	0.4
Mare, . . .	90.9	3.3	1.2	4.3	0.5
Goat, ...	84.9	6.0	4.2	4.4	0.5
Sheep, . . .	86.5	4.5	4.2	5.0	0.7
Dog, ...	77.9	15.8	5.1	4.1	1.0

gestive viscera are burdened with a useless load. The same reckoning can be applied to the lime, sulphur, phosphorus, oxygen, and hydrogen, which go towards building up and renewing the tissues of the body. The dietary must contain these, or the body must waste away by the unstayed drain of destructive assimilation; and if it contains any notable excess, not only is it uneconomical, but may be pernicious to the health.

Suppose, for instance, a gang of a hundred average prisoners to excrete in the shape of breathed air, urine, and fæces, daily $71\frac{1}{2}$ lbs. of carbon and $4\frac{1}{4}$ lbs. of nitrogen, which is pretty nearly the actual amount of those elements contained in the dried solids of the secretions, as estimated in current physiological works. Nitrogen and carbon to that extent, at least, must be both supplied. Now, if you fed them on bread and water alone, it would require at least 380 lbs. of bread daily to keep them alive for long; for it takes that weight to yield the $4\frac{1}{4}$ lbs. of nitrogen daily excreted. But in $380\frac{1}{2}$ lbs. of bread there are $128\frac{1}{2}$ lbs. of carbon, which is 57 lbs. above the needful quantity of that substance.[1]

If, on the other hand, you replaced the bread by a purely animal diet, you would have to find 354 lbs. of lean meat in order to give them the needful $71\frac{1}{2}$ lbs. of carbon; and thus there would be wasted 105 lbs. of nitrogen which is contained in the meat, over and above the $4\frac{1}{4}$ lbs. really required to prevent loss of weight.[2]

In the former case, each man would be eating about 4 lbs. of bread, in the latter, $3\frac{1}{2}$ lbs. of meat per diem. If he ate less, he would lose his strength. In the former case, there would be a quantity of starch, and in the latter, a quantity of albuminous matter, which would not be wanted for nutrition, and would burden the system with a useless mass very liable to decompose and become noxious.

---

[1] Dr. Letheby's Analysis gives 8.1 per cent. of nitrogenous matter to bread (Lectures on Food, p. 6). Of this $\frac{1}{4}$ is nitrogen; Boussingault's analysis of gluten giving 14.60 per cent. (Ann. de Chim. et Phys., lxiii, 229). M. Payen makes the proportion of carbon to nitrogen in bread as 30 to 1.

[2] The proportion of nitrogen to carbon in albumen is as 1 to $3\frac{1}{2}$ (15.5 to 53.5 by Mulder's analysis, quoted in Lehmann's Phys. Chemie, i, 343). In red meat there is 74 per cent. of water (ditto, iii, 96).

THEORIES OF DIETETICS.

Now, if a mixed dietary be adopted, 200 lbs. of bread with 56 lbs. of meat would supply all that is required. Besides water,

200 lbs. of bread contains	60 of carbon	2 of nitrogen
60 " meat (including 12½ lbs. of fat upon it),	12 "	2¼ "
	72	4¼

Judged by the above standard, it will be clearly seen that milk does not represent a typical diet for an adult population, the nitrogenous matter being in considerable excess in proportion to the carbonaceous. This is suitable to the young animal, whose main duty consists in growing, that is in appropriating an excess of nitrogenous matter to form an addition to the body daily, but not to the full-grown, who has to develop force, or its equivalent, heat, by the combustion of carbon, and had rather not go on growing.

Calculations such as these, applied to the other numerous, though less bulky constituents of the body, are invaluable. They afford a basis for the administration of food-supply to armies, navies, prisons, and other bodies of men dependent upon us; they enable us to detect the causes of wasteful expenditure, and to distribute limited means in an economical fashion. They tell us why nations which, voluntarily or involuntarily, become dependent on one kind of food for subsistence can never be wealthy, for they devour and waste their substance; and they teach statesmen how to avoid those ruinous revolutions, which, as has been well observed, arise more often from want of food than from want of liberty.

But the calculations must always be open to the correction of continuous observation and experiment. Chemical analysis is much too young an art to be infallible, and hitherto undetected substances and conditions are, year by year, turning up, which modify our conclusions. And a very wide margin must be left for unforeseen contingencies, and a discretionary power be placed in the hands of individuals, or there is a risk lest the administrator should have to regret making too precise a reckoning. He whose income is only just equal to his expenditure, is always on the brink of insolvency.

The most important modification required to be made arises

out of the differences of work demanded. Men may languish in solitary prisons, invalids may lie bedridden, paupers may wait for better times, nations may idle away existence, on a scale of food-supply which is followed by death from starvation when work is demanded. How shall the effect of physical exertion be reckoned? Here the engineers have helped us with their precise and irrefragable science. Joule of Manchester analyzed, about thirty years ago, the relation which the heat used in machinery, as a source of power, bore to the force of motion thus made active. He found means of proving, that raising the temperature of a pound of water one degree Fahrenheit was exactly equivalent to raising 772 lbs. to the height of a foot. And, conversely, that the fall of 772 lbs. might be so applied as to heat a pound of water one degree Fahrenheit. Thus, the mechanical work represented in the lifting 772 lbs. a foot high, or one pound 772 feet high, forms the "dynamic equivalent," the measure of the possible strength, of one degree of temperature as marked by the thermometer. Physiologists seized eagerly on the opportunity which Joule's demonstration seemed to afford them of estimating, in actual numerals, the relation of living bodies to the work they have to do. So much earth, raised on an embankment, represents so much heat developed in the machinery, living or dead, muscle or steel, gang of laborers or steam-engine, which raised it. Both muscle and steel come equally under the great physical laws of the universe which the far-sighted mechanician has expounded. Now, in the animal frame, the supply of heat, and therefore the supply of capacity for work, is that which is developed from latency into energy by the chemical actions, the ceaseless round of unending change, which is an inseparable part of life. The amount of fully digested food, converted through several stages into gaseous, liquid, and solid excretory matters, produces by its chemical changes a definite amount of heat, of which a definite amount escapes, and a definite amount is employed in working the involuntary machinery of the body, and the rest is available for conversion at will into voluntary muscular action. As the mechanician allows for the effect of friction, etc., in making his calculations, so the physiologist allows for the action of diffusion, conduction, imperfect secretion, and so on, in reckoning the quantity of heat available, and allows also for the waste of mechanical power involved in the form and structure of the limbs. To make all these allowances necessitates courses of experiments and calculations which have taken more than a generation, and will probably take more than another generation to complete. But the road seems clear, and already we have gained fruitful information as to the sort of food by which we can expect to get most work out of men and beasts; we have found the cause of many of our failures in distributing victuals; and we have learned how to avoid much cruelty and injustice that our fathers unknowingly perpetrated.

It may be reckoned from experimental calculations, too long to be inserted here, that the expenditure of force in working the machinery of the body—in raising the diaphragm about fifteen times, and contracting the heart about sixty times a minute; in continuously rolling the wave of the intestinal canal; and in various other involuntary and voluntary movements which cannot be avoided even by a mere cumberer of the ground, without doing anything that can be called work—it may be reckoned that the expenditure of force in doing this is equal to that which would raise a man of ten stone 10,000 feet. But a man cannot even pick oakum without expending more force and requiring more to support it. A prisoner on penal diet has half as much again.

There are several reasons for believing that in assigning their physiological functions to the several sorts of food, we should ascribe nearly all the business of giving birth to force to the solid hydrocarbons, starch and fat, by their conversion into carbonic acid, just as we have good grounds for thinking that it is the conversion of the solid hydrocarbon, coal, into carbonic acid, which drives our locomotives. It is not necessary to be acquainted with every step of the process, which, in the body, we confessedly are not, to appreciate the argument. To the nitrogenous aliments seems allotted the task of continuously replacing the wear and tear of the nitrogenous tissues. Flesh food, or that which comes near it in nitrogenous contents, after a few changes replaces the lost flesh which has passed away in excretions; and thus the engineer takes iron ore, makes it into wrought plates or steel, and renews the corroded boiler-plate or worn piston-rod. One of the most cogent of these reasons is that the chief nitrogen-holding excretion, the urea, is little, if at all, increased in quantity by an increase in the work done: whereas the excretion of carbonic acid, in a decided manner, follows the amount of muscular exertion. Now it is very clear that if the supply of power to do work depended on the renewal by food of the nitrogenous tissues, and on their decomposition, the urea would have no escape from being largely augmented in quantity by muscular efforts, and diminished by rest. This is not the case. At first, exercise diminishes the amount of urea (Parkes), and, even when continued, very little increases it (E. Smith, Haughton and several others quoted in Parkes's "Hygiene," p. 383). The very small increase which takes place during the following rest may be attributed fairly to the extra wear of the muscles from extra motion, just as a steam-engine is expected to require more repair than usual when in hard use. But that amount of repair demanded is as nothing, compared with the increase in the tonnage of coal consumed.

To give an example of the mode of working out a problem by this theory: Dr. Frankland ascertains with the calorimeter, which calculates the amount of heat evolved as a thermometer does its degree, the quantity of energy or force, under the form of heat, evolved during the complete oxidation in the laboratory of a given weight of alimentary substance. It was explained before, that heat and mechanical work, being convertible into one another, bear an eternally sure proportion to one another: now, and forever, a definite production of so much heat represents the potentiality of so much motion, used or wasted, according to circumstances. So that from the reading of the calorimeter may be reckoned how many extra pounds ought to be raised a foot high by a man who has eaten an extra pound of the food in question; how many steps a foot high he ought to raise himself (say a weight of ten stone) before he has worked out the value of his victuals. Dr. Frankland has thus estimated the comparative value of foods as bases of muscular exertion, and he has made out a table of the weight and cost of various articles that would require to be consumed in the system to enable a man of ten stone to raise himself 10,000 feet. This is equal to going up a ladder two miles and one-third high—a stiff day's work. Three pounds and a half of lean beef at a cost of at least 3s. 6d. would be wanted; but if little more than half a pound of suet, worth about $5\frac{1}{2}d.$, were substituted, the same effect might be elicited. A liberal three quarts of milk at 5*d*. a quart would do the same thing; but if cabbages or apples happened to be the only available food, 12 lbs. of the former or nearly 8 lbs. of the latter must be swallowed—an intolerable burden even for the *dura messorum ilia*. There is great wisdom, then, in the journeyman tailor who adds suet or bacon to his cabbage, and in the Yorkshireman who puts a slice of cheese in the apple-pie which often serves him for breakfast, dinner, and supper—for the said cheese is very high in the scale of nutrimentary fuel, and 10*d*. will buy as much of it as is equal to the 3½ lbs. of beef. But in wheaten bread we find a true friend, for of it two pounds and a third, costing under 5*d*., supply a food which really might be eaten alone; whereas the consumption of the others as a sole diet is of course theoretical—the average digestive organs cannot bear them. The mere weight, for instance, of 12 lbs. of cabbage would knock a man up, if carried in a vessel so ill adapted to sustain heavy loads as the stomach.

Reverting now to the gang of a hundred prisoners formerly used as an illustration, and supposing we wanted to put them on hard labor involving some exertion equivalent to half Dr. Frankland's unit of ten stone raised 10,000 feet—such, for instance, as carrying 1¼ mile, up ladders, three tons of stone daily—we should find by calculation that the addition of 117 lbs. of bread, or of 58 lbs. of bread with 44 lbs. of lean meat and 63 lbs. of potatoes, to that diet which would keep up their flesh, without labor, would be barely sufficient, and that they would lose a little weight daily. Cases of illness from overwork would be improperly frequent. Give them a draught of milk, or a cup of cocoa and sugar, or some oatmeal-porridge and treacle, or even some green vegetables, and the danger is probably averted. But yet, a wide margin must be allowed for discretionary modification according to circumstances which the immediate administrators of victualling arrangements alone can observe.

Still more necessary is modification according to circumstances required, when we come to deal with individuals instead of masses. When the tailor in Laputa sternly refused to take the usual measurements, and insisted on constructing Captain Gulliver's coat, waistcoat, and breeches on abstract principles, the customer vowed it was the worst suit of clothes he ever had in his life. We should certainly fail in the same way if we did not take the measure of numberless contingencies in the daily life, and numberless peculiarities in the persons of those who consult us about their diet and regimen, or who are (like the inmates of prisons) actually dependent upon us for living at all.

Even healthy persons are not absolutely perfect in the performance of all their functions, and a good deal of the food swallowed escapes digestion and is wasted among the excreta. A considerable margin must be allowed for that. And different forms of food and different preparations of those forms are suited, from their mere mechanical construction, to different constitutions of body. A diet which in an old man may be wasted from lack of solution, and which would therefore be a starvation diet to him, may in a young man be richly ample. On the other hand, the easily soluble, but scanty diet on which the old man might flourish, would not suffice for his active, growing grandson. Then again, very different lives are necessarily led by different men, and even by the same man at various times, and variations in the mere quantity of food, at the instigation of appetite, are not sufficient to accommodate the diet to the work. Brain-work, body-work, indoor work, and outdoor work, all introduce modifications in the daily requirements of the nutritive organs. There are also physiological drains upon the constitution, such as that felt by a nursing mother, and pathological drains, such as that of a purulent discharge, which have to be provided for. And it must be remembered also that many morbid conditions, which yet do not keep a patient from his employment, such as gastric catarrh or tuberculosis, yet interfere with the due solution of the meals. And some, such as gout and diabetes, are aggravated by eatables which are wholesome to a healthy constitution. And there are hereditary tendencies which may be turned into diseases by popular articles of diet, and not a few idiosyncrasies to be allowed for.

With the safeguards of due allowance for the above elements of variation, and a close eye upon current experiment, the theory of demand and supply in dietetics may be turned to the very best account by rational medicine.

CHAPTER II.

ON THE CHOICE OF FOOD.

There is no rule capable of such universal application as that all articles designed for human food should be the best of their kind which can be procured, and that it is false economy to be tempted by a lower price to be satisfied with less eligible wares. The saving in money is always outbalanced by the inferior utility. Economy, when necessary, should be practiced in quantity, not quality, in the sort of food chosen, and not in its degree of perfection.

As to the arrangement of the subject, the naturalist's classification has been that usually adopted by authors. But it seems to me less scientifically appropriate to the matter in hand, and more likely to involve repetition, than one which brings together articles associated in the market, and which, indeed, often combine elements derived from several realms of nature. These articles are here to be considered solely as food for man, and not as independent members in creation.

§ 1. Butcher's Meat.

The lean of butcher's meat should show a deep purplish red tint, with a sort of bloom over it, on the outside of the muscle and a lighter vermilion red with a bare shade of purple in the cut surface. The lean of beef may be a little marbled with fat, but that of mutton should be quite even in hue. When cut in a very fine slice, or stretched in a thin layer over the fat, as naturally, in the ribs, it is semi-transparent and orange-vermilion in uniform streaks.[1]

The surface is quite dry, and even the cut should scarce wet the finger. In substance it is moderately soft, but extremely elastic,

---

[1] The "vermilion" and "purple lake" of the paint-box are the only colors required to copy the lights and shades of healthy meat.

so that no mark is left after pressure. A day or two in the larder should make no difference in this respect.

There is very little odor in a single joint of good meat, and what there is, to most people, is not unpleasant. When made powerful by accumulation, as in a shop, it may be described as refreshing and exhilarating, like a sea breeze.

It should not waste much in cooking. That is to say, it should not contain an excess of water, nor part with it too easily. Dried in the laboratory, healthy muscular fibre, according to Dr. Letheby,[1] does not part with above 70 to 74 per cent. of its weight; whereas, when in an unhealthy condition, it will lose as much as 80 per cent. Cooking is, of course, not the complete chemical desiccation alluded to; but its relation to animal fibre is the same.

When a joint is brought to table roasted, it holds well its gravy, which gushes out when a cut is made, in a rich brown stream, full of appetizing scent and flavor, and called graphically by the chemist "Osmazome."

The tastes of the tissues should be quite distinct from one another. The mutton lean should have no flavor of tallow, nor the beef of suet. Tenderness, of course, is a virtue, but freshness should not be sacrificed to it.

The raw fat of beef, whose base is principally margarin, should be of a light yellow color, like fresh butter; that of mutton, which is mainly stearin, should be very white.

Lamb and veal should have very white and translucent fat; and if the whole carcase can be examined, the fat about the kidneys should be especially observed, as that is not rarely reddish and unwholesome, while the rest is in good condition.

Of these two latter meats, the lean should be pale, but even in tint and free from mottling.

The above-mentioned features of prime meat are still more to be insisted upon when we choose the various internal parts. They decompose quicker, and when decomposed are more unwholesome, than ordinary muscular fibre. They should have a clean bright even color throughout, and be free from spots, speckles, and points of congestion or of bruises.

---

[1] Lectures on Food, p. 235.

Liver, kidneys, and heart become very hard by too great or too prolonged heat. Unless, therefore, a very small dish is required, they should be avoided.

Butchers are fond of palming off the pancreas, or "stomach-bread," in place of the "sweet-bread," which is the thymus gland of the calf. It may be recognized, even when cooked and chopped up, by its large veins and arteries. It is very inferior in digestibility to the more delicate gland.

In roasting, the fat of meat should not run away into the dripping.

If these characteristics are insisted upon, we may be satisfied that our kitchen is provided with the best article, however moderate the price may be; without them we are ill served, however much we pay. They are guarantees of wholesomeness, and therefore worth more than all the negative evidence of absence of unwholesomeness.

The immediate antecedents, and the mode of slaughtering the animal certainly very much affect the nature of its flesh. Chronic loss of health of any kind makes it not only less easy of digestion, but less nutritious in proportion to the quantity eaten. The muscular fibre of a beast in poor condition is pale in color, pinkish, yellowish, or brownish. A quantity of watery fat, of a bad color, is intimately mixed up with the fasciculi. The colors of the two seem to blend so as to present a marbled aspect. Such meat is wet, sodden and inelastic. If it retains the mark of a finger pressed upon it, it is unfit to eat. When left unwiped, it parts with its watery constituents, and soon lies in a pool of blood-stained fluid. When cooked its gravy is pale and mawkish, and it has lost a good deal of its substance. In the museum of the College of Surgeons[1] there is preserved a mutton-chop, the meat of which is converted into a waxy or adipocerous substance, obviously no more fit for food than are those sticks of adhesive matter with which our beaux stiffen their whiskers. This is simply an extremely advanced specimen of a condition which is commencing interstitially in the carcases of all chronically invalided animals. It is that degeneration of muscle with which the classical description of Dr. Quain made all in our profession familiar twenty-five

---

[1] No. 10 in the Catalogue of the Pathological Series.

years ago,[1] as occurring in the human body. It is found not only in organs which have been exhausted by chronic inflammations, but in those which are simply hypertrophied, and with a frequency closely proportioned to the frequency with which the special organs become so enlarged.

It is the more important to notice the effect of chronic disease upon the flesh of adult animals, because it does not in many cases prevent their being fatted up for market and presenting a delusive appearance of prime condition. Thus the "rot," always associated with the presence of fluke-worms in the liver, exhibits itself at an early stage in the sheep by a great tendency to plumpness. The shrewd breeder, Mr. Bakewell, "to whom farmers owe so much" (more indeed than their customers), "used to overflow certain of his pastures, and when the water was run off, turn upon them those of his sheep which he wanted to prepare for the market. They speedily became rotted, and in the early stage of the disease they accumulated flesh and fat with wonderful rapidity. By this manœuvre he used to gain five or six weeks upon his neighbors."[2] But what was the worth of this hypertrophied muscle and adipose tissue? Breeders, if they give a thought to the subject, must be conscious that the heart and arteries do not grow at the same morbid pace with the rest of the body, and the animal, imperfectly supplied with blood, is in a state of extreme anæmia. Indeed, one of the earliest known and best known tests of "rot," is the condition of the eyeball and lid, the inside of which is injected, but pale and tallowy, while the tear-gland is yellow instead of pink.[3] We know the look well in our chlorotic patients.

Not quite so pronounced, but of the same nature, is the anæmia which results from the constant efforts of the farmer to produce a flock which will lay on the greatest weight of flesh in the shortest possible number of weeks. Premature development of size and

---

[1] See the plates in the Medico-chirurgical Transactions, vol. xxxiii, p. 196. The coloring is very accurate.

[2] Youatt, The Sheep, p. 446.

[3] " That dire distemper sometimes may the swain
Though late discern ; when on the lifted lid
Or visual orb, the turgid veins are pale."
          Dyer's Fleece, book i, line 266.

of puberty is in his sight a virtue, both in those destined for the butcher and in those he selects as breeders. It is a saving of time, and time is money. I fear our agricultural societies are not free from the blame of this, inducing competition in bulk by their system of prizes; and I do not see how they can counteract the evil that has been wrought, unless by instituting rewards for prime joints, to be adjudged at the table as well as in the larder.

Good mutton is generally small; indeed large mutton can only be good—these hypertrophied breeds can only be thoroughly prime—by being kept alive till their constitution has grown up to their size. To illustrate the matter by our own race, a school-boy of six foot one never becomes hearty till he is at least one or two and twenty. But after that, the weedy youth may harden and be a fine man. So indeed with these overgrown lambs, if they are kept till four years old, the meat is very choice; but of course the temptation is to bring them to the butcher directly they are as tall and broad as a real sheep; and the farmer looks upon himself as a benefactor to his species, as having made two animals in the time formerly required to make one. But I hear with sorrow of attempts being made to "improve," as it is called, the Welsh breed, and trust they may be unsuccessful. A more promising statement is that Welsh mutton, in the London market, is imitated by a cross between the Southdown and the Scotch.[1] We may pardon the deception, if the meat is as good as its model.

A striking proof of how opposed are the interests of the farmer and of the consumer in the breeding of sheep is, that in the minute experiments and calculations of the advantage of different breeds and various modes of feeding, no attempt is made to reckon the goodness of the dead meat. We are told the weight, but never the quantity of osmazome it contains, though the readiest possible test in the tint of the gravy is very familiar to the eater.

To get good mutton in country places is now a serious problem, and I would suggest to my professional brethren, who are of course permanent residents, that they cannot confer a greater boon on families in the same position of life as most of us are, that is to say, not rich enough to have parks and farms, and yet willing to pay a good price for a good article—I say they cannot confer a

---

[1] Macdonald, Cattle, Sheep, and Deer, p. 483.

greater benefit on these their neighbors, than by inducing them to join in a "mutton club," buying the lambs of a full-sized breed, and keeping them to at least three and a half years old before killing. The price per pound will not be less than charged by the butcher, but it will supply an article twice as good as his.

There has been felt a good deal of alarm, more I think than is justifiable, during the last few years on the subject of the class of parasites which are not uncommon in the flesh of all animals, namely, the *cysticercus* and the *trichina;* and which, when sufficiently numerous to be conspicuous, constitute what is called "measly meat."

The *cysticercus* of the pig is the sort most frequently seen, forming a cyst as large as a hemp-seed. Its commonest habitat is the tongue, on the under surface of which it may be discovered even without cutting into the interior. You may also find the oval holes left by it when dried up in otherwise very perfect hams, and opaque white specks, like seeds, intimately adherent to the muscular fibres, which are its remains dried into calcareous matter. This measle-worm of the pig has been found by the industrious German naturalist Küchenmeister to be the undeveloped embryo of the *Tænia solium*, the tape-worm of most usual occurrence in Great Britain. By keeping them alive in warm milk he was able to watch the development of the animals. The *cysticercus* of the ox is smaller, and is either rarer or seldomer discovered in this country. It becomes the *Tænia mediocanellata*, the species which infests the intestines of Germans, Swiss, and others, producing exactly the same inconveniences as that with which we are familiar. As the measle-worm of the mouse produces the peculiar tape-worm of the cat (*Tænia crassicollis*), as the brain hydatid of the sheep (*Cœnurus cerebralis*) produces the *Tænia cœnurus* in the dog, so the minute larva which infests the flesh of our prey revenges itself on its natural enemy. An old boa-constrictor is always a complete museum of tape-worms, derived from the various living game which it has devoured.

Now, there is no doubt that the boiling temperature entirely destroys the vitality of these creatures. When cooked they can do us no more harm than a baked lion. So that it can very rarely, if ever, happen in civilized countries for them to be transmitted directly to human intestines. Another mode of communication must be thought of, and I think it is not far to seek. Though we do not eat our food raw, dogs very often do, and they distribute far and wide by their excreta all that escapes the solution of the gastric juice. Thus the embryos get spread abroad on the earth, into streams and wells, and especially in our kitchen gardens among the materials of our solids. I had once brought to me a child three years old, with tape-worm, a very rare thing at that age, and thus affording peculiar facilities for detecting its origin. It was the son of a sculptor in the suburbs of London, and all the cooking arrangements of the family seemed perfect, as also their water supply. But in the stone-yard, which had once been a garden, grew a quantity of nasturtiums, and among these the baby used to play, and sometimes ate the flowers and fruit. As the yard was open to the road, it was much frequented by the dogs of the neighborhood, and showed unmistakable signs of their presence. One could not question that here lay the carrier of the little patient's troublesome inmate. The observant and enthusiastic physician of Iceland, Dr. Hjaltelin, also informs me that intestinal parasites are exceedingly common in that country, and that the cause appears to be the distribution of their ova by stray dogs who are always in and out of the kail gardens. Another possible source of *tænia* may be shell-fish, eaten raw, and often containing in their stomachs minute organisms derived from decayed garbage thrown away or used as bait. Only the other day I found in the prehensile organs of a prawn a shred of animal fibre, and saw a man fishing for crabs and prawns at the mouth of the Arun with a piece of paunch. With all these possible sources of infection there is no need to suspect butcher's meat, which is never eaten raw except by some eccentric amateur savage.

The real evil of measly meat consists in its proneness to rapid decomposition in spite of cookery, and to that may fairly be credited cases of illness which are reputed to have followed its consumption. So that it is fair enough that all that is largely infected should be destroyed. In France there used to be appointed to the markets officers called "*langueyeurs*," from their inspecting the tongues of carcasses offered for sale; but their legal authorization, and possible tyranny, seems to have given dissatisfaction, for M. Delpech says they have no lawful authority, but are employed simply as a warrant between buyer and seller.

They have to report to the inspector of markets, usually a skilled veterinarian, who judges if the quantity of measle-worm is sufficient to render the meat unwholesome. The shoulder and the breast are the parts usually examined as tests.[1]

The *Trichina spiralis* is another parasitic inhabitant of live flesh, of a more active character and of a higher grade in creation than that last discussed; for instead of being a solid worm like the tœnia, it is possessed of an intestine.[2] It is sometimes found in human flesh and in pork, appearing as a minute white speck, just visible to the naked eye, which constitutes its nest, in which one or two curled-up specimens are seen, by a microscope, in active movement, but prevented from doing harm by the cyst in which they are imprisoned. The danger consists in its escape and wonderfully rapid multiplication, under special circumstances not very clearly defined; for it exists at most times in considerable numbers without giving rise to any symptoms whatever. But there seems sufficient evidence that in a few instances an epidemic has occurred of its invasion, in overpowering quantities at once, of human bodies, through the food eaten. The symptoms are inflammatory fever and local lesions from the interstitial presence of a mass of quickly increasing foreign bodies. But these instances have been extremely rare. A few years ago a physiologist in this metropolis, having become the fortunate possessor of some specimens of live trichina, instantly invited a crowded conversazione of medical men and others interested in the natural history of our species, and introduced to them by means of the hydro-oxygen microscope his acquisition. Very few of the party, if any, had seen one before; and very few, if any, have seen one since.

The trichina is said to cause degeneration of the muscular fibres in its vicinity, so that the joints infested would not present the healthy appearance described at the beginning of this section. It is killed by the temperature of boiling water, so that if a dish is fairly cooked, it must be quite safe. The cases which have occurred of its proving deleterious have been where the meat has been eaten raw, or imperfectly warmed through and served cold,

---

[1] Delpech, Dict. encycl. des sciences méd., art. "Ladrerie:" an excellent monograph on the subject, date 1868.

[2] Professor Owen has identified the *trichina* as one of the Cœlelmintha, and the College of Physicians has adopted his classification.

with its defects concealed by some enveloping sauce. No decently delicate feeders need be afraid of it. If the leaden dull color of the meat before us has been destroyed, so that it does not look raw when the gravy is run off, and if the peculiar texture of fibre which distinguishes uncooked meat, is removed, we may be sure that the temperature has mounted up to that sufficient to coagulate albumen (150° Fahr.), and that any stray trichina would be killed on the spot or permanently imprisoned in a solid nest.

It may be proper to mention that no form of drying, salting, or even smoking at a low heat, is sufficient to destroy the trichina. So that when travelling in Germany it is wise altogether to avoid the sausages and uncooked ham often served up in thin slices, and which in point of fact, have been proved the sources of trichinous poisoning in the few instances on record.

The whole influence of *fevers and inflammations* upon the flesh of animals thereby affected during life, and whether they should be considered as a reason for its being pronounced absolutely unfit for food, is a moot-point. There is a good deal of hearsay evidence on both sides, but I cannot find any crucial cases recorded as observed by competent witnesses.[1] An enormous quantity of meat is destroyed on this ground, for, according to Mr. Youatt's estimate, one-fifteenth of the whole horned stock of the country die annually of inflammatory fever, milk fever, red-water, hoove, and diarrhœa, and one-tenth of the sheep and lambs are carried off by corresponding ailments.[2] When we add those which perish by accident and accidental sickness, it is obvious that several million tons of meat are thus taken out of human mouths by the law which insists on the destruction of all which bears the marks of disease. Opponents say that if it possess only a fraction of the nutritive power of good meat, it ought not to be wasted, but sold at a lower price, provided always proof can be obtained, that it does not, when eaten, communicate disease. This, as said before,

---

[1] "In no well-ascertained case has it been found that any ill effects have been produced by eating the flesh of diseased animals, although there is abundant evidence that at the outbreak of the distemper in Massachusetts, and before public attention had been directed to its true character, a considerable number of animals, in which the usual premonitory symptoms had appeared, were slaughtered and their flesh sold."—*Second Annual Report of the State Board of Health, Massachusetts*, 1871.

[2] Youatt, Cattle, Preface.

is a moot-point, but yet what is certain is to my mind quite sufficient to justify the exclusiveness of the existing regulations. It is certain that there are some diseases, originating in beasts, which may be communicated to men handling the carcass before it has been submitted to the action of heat, as for example *pustula maligna* and glanders. And some fevers, such as typhus, are common to man and beast, and are indubitably contagious during life, and probably after death, till the flesh has passed through the purification of fire. Now most of those who would buy diseased meat on account of its cheapness, cook their own victuals, and are exposed to all the evils which may accrue from handling a dangerous article. Again, this "braxy" meat, as it is technically called, runs rapidly into decomposition, and becomes a serious nuisance on that ground alone. It is also frequently saturated with the soluble drugs which have been given as medicines by the veterinarian. Ergot of rye, digitalis, opium, tartar emetic, are often administered in enormous quantities. Mr. Youatt advises upwards of half an ounce daily of solid opium for an ox with lock-jaw, which is 240 times the full human dose, and as this is equally distributed to the soft parts by the circulation, in a beast (say) of twenty stone, each pound would contain at least half a grain of the poison. A case is recorded, in which tartar emetic taken by an ox before slaughtering produced serious effects on 107 persons who partook of the meat.[1] One person who died had eaten only half a pound. Tartar emetic was found in the contents of his abdomen.

Acute fevers cause an acute degeneration or interstitial death-in-life of both blood and tissues. Virgil notices that in victims slaughtered during the cattle plague the peristaltic vermicular motions of the intestines, by which the priests told fortunes, are stayed, and that the blood is as the blood of a corpse, scarcely staining the knife—

> Nec responsa potest consultus reddere vates,
> Ac vix suppositi tinguntur sanguine cultri.—*Georg.* iii, 491.

And it may be remarked that the poet, who is a practical farmer as well as the most picturesque of sweet singers, expressly

[1] Quoted by Dr. Pavy from the Central Zeitung für Veterinärmedizin für 1854. Treatise on Food, etc., p. 149.

says that he is here speaking of the early stage of murrain, the fatal later symptoms of which he describes afterwards. The Reports of the Cattle-plague Committee of Privy Council show, that his lines will apply to England as well as to Italy. Now, it can hardly be maintained that it is honest to sell in the market meat thus saturated with natural death, though the animal has been slaughtered before the full declaration of the fever.

The best possible meat may be rendered unwholesome by normal decomposition. The stomach can, indeed, through habit, become used to food in this state; and thus may be accounted for the instances we read in books of travels, of savages, like the Esquimaux, who bury their flesh till it is putrid, or like the Zulus with whom (according to Dr. Colenso) the synonym for heaven is "maggoty meat." Of course, rather than die of starvation, or be reduced to the straits suffered by Hezekiah's army, one would acquire such a habit, and invent a sauce to make it tolerable; but it is scarcely worth while to do so in civilized society. Under ordinary circumstances many cases are recorded in works upon poisons, such as Dr. Christison's, where decayed animal food has produced severe and even fatal diarrhœa, in spite of cookery having concealed some of its repulsiveness. High game has fortunately gone out of fashion, and the most frequent form in which we now meet with decomposing albuminoid matter is that of a fusty egg. Some housekeepers seem to consider this quite good enough for made dishes, and thus spoil material worth ten times what they save by their nasty economy. No egg should be allowed to enter the kitchen, that has the slightest smell of rotten straw. But this seems rather beyond the subject of butcher's meat, and it is time to proceed to another department of the larder. The suitability of different sorts of meat to different constitutions will be considered in a future chapter.

§ 2. POULTRY AND GAME.

Tenderness is the chief virtue in poultry, and is most difficult to find in the winter season. Spring chickens come in with May, but during the five previous months much care is requisite in purchasing this article of the table. A young, and therefore tender bird, may be known before plucking by the comparative largeness of the feet and the leg joints. And when a fowl appears at table with a thin neck and violet-tinted thighs, it is wise to avoid being helped to the leg. These are invariable signs of age and toughness.

The same violet tinge may be noticed in the thighs of old turkeys, which are also distinguished by their hairiness. The age of ducks and geese may be tested by their beaks, the lower part of which breaks away easily in youth.

Besides being tough and indigestible, an old fowl has a rank flavor, like a close hen-house, which arises from the absorption into its flesh of the oil furnished by nature to lubricate the feathers. This is still more perceptible in old ducks and geese. It may often be tasted some time after a meal, and must therefore, like most rank oils, arrest digestion.

Game may roughly be selected by the same rules as poultry. Those who have interested themselves in ornithology may also get some help from observing the undeveloped spurs in young gallinaceous birds, and the pointed long wing-feathers of the young partridge, which become rounded at the tip when he is old.

Poultry should not be too fat. In cooking, the oily adipose tissue becomes rank, and is less digestible than the fat of mammalia.

§ 3. Fish.

The sanitarian has not much advice to give concerning the marketing of raw fish, except that it should be fresh. The guides by which to judge of this are the fulness of the eyeballs, and the bright pink hue of the gills. The sense of smell cannot be trusted to, as it may be deceived by the use of ice. When cooked, the flesh of fresh fish is firm but friable; that which is stale is flabby and stringy, even if preserved by cold from actual putrefaction.

The less salt, and the colder the water is whence our fish comes, the better adapted is it for the table. At Gibraltar it is not hard to distinguish the mullet caught on the Atlantic side of the rock from that which lives in the Mediterranean, a warmer and more concentrated sea; so much is the advantage on the side of the former. An Icelander dining at my house passed by with polite scorn a piece of prime Scarborough cod. Seeing my surprise, he explained that no one who has tasted it at Reikiavik could bear to eat cod in England, and that it was best in the polar circle, braced up by the melting icebergs. The nearly fresh waters of Loch Fyne supply the choicest herrings, and the pure light mountain streams a better trout than our lowland streams, where the atmospheric pressure is greater.

Every sort is best when it is cheapest, for then it is most plentiful and in fullest season. It is a wise plan to contract with your fishmonger to send you so many dishes a week at a fixed sum, and then it becomes his interest to supply that with which the market most abounds. For healthy persons, every kind ordinarily exposed for sale in England is wholesome, provided it be good of its kind, and not spoilt in the cooking. The selection of fish for invalids will be discussed later.

Complete cookery, however, should be insisted upon. The conger-eel, for example, is a very foul feeder, and has been known, if carelessly grilled, to cause diarrhœa, probably from the fetid contents of the stomach saturating the flesh. And at the Pathological Society on May 5, there was shown a specimen of the bunch-headed tapeworm (*Bothriocephalus latus*) which had grown in a person used to eat half-cooked fresh-water fish.

The only sort of reptile of dietetic importance is the *turtle*. It is sometimes viewed as a mere luxury, but is in reality a most digestible and nutritious food, and if more demanded would quickly become more plentiful in the market. The creature grows too slowly for it ever to remunerate artificial culture, but nature supplies it in immense quantities, and its tenacity of life enables it to be brought over alive from the tropics. Fresh turtle is much more costly than it ought to be, but the rival importation of dried turtle fins is reducing the price. This last-named article is of great value, and of moderate price; from it, first-rate real turtle-soup may be made at no more than the expense of mock-turtle, if we deduct the price of the wine used, which to some palates is no improvement. The fins should be soaked for at least twenty-four hours before cooking.

Caviare, the roe of the sturgeon, is best obtained from a fishmonger. The dealer in preserved provisions seems to think it all the better for being preserved, whereas it should be as fresh as can be got, exhibiting its freshness by its softness and light color. That black, hard sort of fish-jam which is sometimes served up, is really unfit for human consumption.

All that has been said above, applies equally to *crustacea* and *shells;* but an additional remark may be made about *crabs.* They should be cleansed scrupulously before cooking, and if that which is removed from their prehensile organs is fetid, they are hardly to be considered safe. The frequency with which crabs disagree unexpectedly with a healthy stomach, may be attributed with reason to the garbage on which the creature lives. And of *oysters* it should be remembered that they are to be eaten raw, or, at most, barely warmed through; for complete boiling makes the flesh tough, so that it is prudent, if they come from near river-mouths, to keep them alive in a shallow dish of clean brine for a day or two, feeding them with meal, and changing the water so as to leave them bare twice a day, in imitation of the tide. They become peculiarly plump and wholesome under this management.

§ 4. Garden Produce.

The commonest fault committed by housekeepers in respect of vegetables, is that they do not supply a sufficient variety, seeming to consider that the meat is the only part of the meal that requires care, and that all the rest is mere garnish, beneath the notice of a Briton, and unfit to sustain his vigorous life. Yet that is not the experience of the observers of mankind. The attention of Herodotus was called to the fact that the Persians, the manliest and most sporting nation in the old world, had at meals not only several dishes, but several courses of vegetable food, preceding a very moderate allowance of solid meat.[1] And Sir Henry Rawlinson describes the diet of this tough race as practically the same now, so that the assumptions of some anthropologists that hunting races are necessarily riotous eaters of flesh, and that carnivoracity strengthens a nation, are not accurate. The Persian gentleman is the spiritual father of the British squire;[2] yet, at

---

[1] Herodotus, Clio, cxxxiii.
[2] He taught his sons "to shoot, to ride, to speak the truth," and then left them to educate themselves; he was devoted to his sovereign to a degree that astonished Herodotus; and he loved a good glass of wine in good company.

many a hospitable board, if a guest does not fancy meat that day, or has eaten enough of it at a previous meal, he will have to fall back upon potatoes, or to solace himself by picking a few bits out of the sauces of made dishes, where the vegetable flavor has been saturated with that of meat and spoilt. Usually, he goes on eating too much nitrogenous food out of sheer idleness.

Another fault is that the vegetables are not sufficiently fresh. Unhappily dead plants do not stink early enough to disgust the nose; but yet, every minute they are kept after their actual death, that is to say, after they have ceased to be capable of growth, renders them in some degree less digestible. Sometimes they are kept too long out of mere carelessness, sometimes from lack of sale, but sometimes also intentionally, to make them look better at table. For a long time, I could not make out why London asparagus so often disagreed with people, till at last I caught a gardener cutting it twenty-four hours before it was wanted, and putting it in a damp warm frame, "to swell," as he said. Cucumbers and broccoli are often spoilt in the same way. The vast wagons of cabbage that one sees coming into London at midnight are often the bearers of two or three days' cutting in small gardens, kept till a full load is accumulated for a single journey; as early travellers by rail may see for themselves. Sprinkled with water they look well, but never regain their fresh character. They ferment in the stomach, and produce flatulence.

Potatoes.—The virtues of a potato are to be mealy and powdery when boiled, and to mash readily into a smooth *purée*. This shows that the starch-granules are in a healthy condition, and that they absorb water and burst the envelopes of glutinous matter which the heat has coagulated. Young potatoes, from not so easily breaking up, require long mastication to render them soluble, and are not then very digestible. But old waxy potatoes are worse, for they seem to unite again into a sticky mass, after being swallowed, and remain for hours undissolved; the worst of all are potatoes affected by the peculiar epidemic called after their name. The diseased part, looking as if it were stained with a drop of ink, remains quite hard in spite of any amount of boiling and digesting: eating it is equivalent to eating so much rotten wood. Potatoes which have begun to sprout, are also indigestible, and frosted potatoes begin to decay immediately a thaw sets in. The best potatoes are "Regents."

Jerusalem Artichokes are largely used in England by people who have gardens, partly because the plant is handsome, and partly because the root is not injured by frost, and so can be allowed to remain in the earth during winter. The dried stem is also convenient for firing. It is a watery vegetable, and though it had the start of the potato in European horticulture, it has never been brought to the same perfection. The fact is it contains no starch, and the "inulin" which replaces that valuable aliment, is only 2 per cent. of its weight, whereas in its successful rival there is a proportion of 16 per cent. It should be eaten only as an occasional change, for the sake of its flavor.

Turnips may have nearly the same things said of them.

Yams and *sweet potatoes* come now into the London market. They are as mealy and wholesome as the commoner tuber, and are sometimes useful to tempt our patients into the use of vegetable diet.

Carrots contain a quantity of pectin, which can be extracted from them in the form of a jelly, and is often used by confectioners to mix with fruit jelly as a diluent. It resides principally in the outer rind, whose thickness therefore in proportion to the pale core is a test of the goodness of the specimen. When soft and friable they are much more nutritious than turnips.

Parsnips may take to themselves the same praise, and ought to be more used, especially with boiled fish. From their sweetness they make excellent fritters, and are liked by children, to whom they are well adapted. However, when old and stringy, they should be avoided.

Salsify is in England considered more a dish for the *gourmet*, than as a food for middle-class tables. This is unjust, for it is nutritious and digestible, and grows easily. It is best eaten alone, fried in a thin coat of batter. It should break readily, and be free from strings.

Leeks make a capital soup and a most digestible side dish. The more white there is in them, and the less green, the softer and better they are. They should have but little smell.

Sea-kale should be perfectly blanched. When colored it is in-

ON THE CHOICE OF FOOD. 45

digestible, as is shown by its being tasted in the mouth after dinner.

Asparagus should be eaten as soon as possible after cutting, and then it is most wholesome. The greenest asparagus is that which contains the greatest amount of the active principles, bitter and resinous, and is therefore to be chosen in preference. I have known timid patients to fear that it must be injurious to the kidneys, because of the peculiar odor communicated by it to the urine. It certainly does no harm, and I doubt almost if it is a diuretic.

In early spring, the fresh young fronds of the male fern make a good imitation of early asparagus, and are decidedly better than the wild asparagus brought to table in the south of Europe. With other substitutes for it I have no acquaintance. The number of them shows how well worth having the real thing is.

Cabbage is the most valuable antiscorbutic we possess. In the slight degree of scorbutus characterized by bleeding of the gums or by purpura, it is eminently successful, and prevents the same thing happening to other members of the household who are wise enough to prefer prevention to cure. It should be soft but crisp before cooking, and show no signs of having been wetted. If it has begun to heat from incipient fermentation, it is most noxious, and generates in the intestinal canal an enormous amount of flatus, consisting not only of the usual carbonic acid, but of sulphuretted hydrogen as well. Fermentation destroys the antiscorbutic qualities of the cabbage, for sour-crout is not nearly so efficacious as the fresh plant.

Sour-crout is prepared by taking advantage of the fermentation as a means of after-preservation. The leaves of the kail are allowed to heat, and then subjected to severe pressure, which arrests the chemical action, and hardens them into a dry mass, which will keep a long time. It requires much soaking, and should not be cooked till free from all sour taste. It should not want chewing, or it is shown to be underdone.

The best sorts are the old white garden cabbage and the summer cauliflower.

Winter greens are of so many sorts that it would be necessary to be a complete gardener to give rules for the selection of each. Their greenness and freshness at a time when all around is brown and decaying is the attraction to them; and it may be said generally, that therefore they ought to be as fresh and green as possible. Under this heading I include savoys and Brussels sprouts, but not broccoli, which should of course be as white as can be got.

I take the opportunity of having to allude to *kohl-rabi* (a new kind of cabbage of which the leaf-stalks are eaten), to say that it is not prudent to recommend to our patients novel varieties of garden produce, unless we are well acquainted with them ourselves. We do not know how to decide if they are good of their sort or not; and much more depends upon that than upon the kind of vegetable.

Cardoons.—Those who grow this delicious thistle, which is seldom brought to market, should take care that the leaf-stalks are at least an inch and a half thick before they are considered fit for cutting.

The artichoke is another thistle, like the last-named, of an ornamental character, and more cultivated in this country. Eaten raw, or only just warmed, as is common in France, it is as indigestible as nuts, which it much resembles; but well boiled till it is quite soft, it may be eaten with impunity even by invalids. After an early dinner it makes a good dish for supper.

Chestnuts are a very good substitute for potatoes with white meat or fowl. They should be thoroughly well boiled, skinned, and served up in a hot dry napkin. Home-grown chestnuts are the best, being more mealy and powdery than those imported. A sweet soup also may be made of chestnuts rubbed through a sieve; but I cannot recommend the polenta cake and bread made of this nut, which are so popular in Italy. It requires a long education to accustom the digestion to them.

Vegetable marrow, squash, elector's cap, and a few other sorts of pumpkin are wholesome diluents, but do not form a substantive diet. The same may be said of the *tomato.* Care should be taken that they are ripe, or they cause colic.

Peas and broad beans should be young, and their skins tender enough to crack in boiling. If they are past the time of life for this to happen, they should be chopped, mashed, or otherwise broken up; for the unbroken skins are very leathery. The longer they are boiled, the harder they get.

Dried peas, split peas, are deprived of their skins already; so that if well boiled, as in soup or pudding, they are very good for food for robust people.

It appears wasteful to throw away the outer shell of the pea. It contains a great deal of nutritious matter, but it is not nice in the commoner sorts. There has been lately introduced a new pea, the shell of which is edible, and seems wholesome.

French beans, from the kidney bean and scarlet runner, are still more required to be young and tender.

White beans are the ripe seeds of the same plants. They are not popular in England, apparently because they do not blend well as an adjunct to meat. But eaten alone with a piquant sauce, they are a most palatable variety of dish, and certainly nutritious.

Lentils, again, are too much neglected. They make a capital soup, resembling pea soup. The peculiar flavor of lentil flour, which is distateful to some persons (reminding them of garden seeds, they say), may be masked by adding to it some sugar and Indian corn flour or fine barley meal. Or if it is wanted for soup, a few bits of celery or asparagus cover the objectionable taste completely. It is sold under the name of revalenta arabica at a higher price considerably than is charged for it as lentil flour at a corn-dealer's.

Mushrooms are best when grown in an open meadow. When forced they are tough and indigestible, and when preserved they are tasteless as well. A meadow mushroom should peel easily, and it should be of a clean pink color inside, like a baby's hand, and have a frill or "curtain" (as botanists call it) attached to the stalk. When the gills are brown they are growing old and dry, and losing their nutritive qualities. The above-described *agaricus campestris* is the queen of its class for cooking purposes in England. It is true there are several other similar fungi eatable, and eaten by experimentalists; but my experience of them is that their flavor is inferior, and that we lose nothing by the safe rule of adhering to the one we know well by sight.

The *Gigantic Puffball* makes excellent ketchup, and can be eaten in the shape of fritters. It must be large and very white, like a great bleached skull. When discolored it is beginning to ripen its spores, and is then poisonous.

The *morelle*, the fan-shaped *chanterelle*, and the black *truffle* should be sweet and fresh. The odor of the last-named when decomposed is so horrible that one can hardly fancy its being tolerated; yet I have known a cook use truffles in this state, and say she thought it was the right smell.

Materials for Salad.—Here again, as in the case of winter greens, the plants used are so many and various that to enumerate them would be as tedious as useless, and to describe their several tests of salubrity would require more horticultural knowledge than I possess. Repelled by the barbarous and barren aspect of a list of their technical names, I was comforted by the recollection of a scene of long ago. A party of young gentlemen and ladies were earnestly disputing about the nomenclature and specific differences of certain plants, and appealed to grandpapa, an elderly Parisian; he settled the matter in a moment—"*Eh, mes chers, ce sont des salades.*" I shall imitate him in condensing them into a class.

Vegetables intended to be used for salad should all be fresh and crisp, and sweet and clean. Their colors should be positive and even, the reds very red, the whites very white, and the greens pure as those in an autumn sunset sky, except in the full-grown leaves, such as watercress. The salad ought to be dressed by one of the daughters of the house, after she has herself dressed for dinner, singing, if not with voice, with her clean cool fingers, sharp silver knife, and wooden spoon,

> "Weaving spiders, come not here;
> Hence, you long-legged spinners, hence:
> Beetles black, approach not near;
> Worm nor snail, do no offence."

The purity of the bowl is more important than that of Titania's bower. So will the guests eat it with light hearts, free from all fears of noxious ingredients.

With a little trouble, not however necessarily attended by expense, a succession may be provided of materials for salad all the year round, so as to have one at table every day. And a great preservative of health I believe it to be for hearty persons. The most difficult season to provide for is the latter end of the winter, and it may be of use to mention that the *dandelion* is then a friend in need. If a pot be placed over the plant as it grows, or the leaves tied up like lettuce, or it be transplanted into a frame, it can be bleached, and thus loses its bitterness. Daisy leaves are also eatable; and thus with a sprig of tarragon, a few cold potatoes, and some ever-constant mustard and cress, giant cress, Australian or curled cress, an olive or two pared thin, or some beetroot and a slice of Madeira onion, a great variety of combinations may be made. Indeed, an inventive lady, with a well-furnished cruet stand, a bottle of Worcester sauce, and some *moutarde de Maille*, might provide a different salad every day of the year. These "scratch" salads are very much improved by a tablespoonful of light white wine. Watercresses rather spoil salad, and are best eaten alone, so as to make a variety when nothing else can be obtained. And the same may be said of radishes, and of endive, which are too strong in flavor to combine well.

Some persons are very fond of tomatoes sliced raw, and eaten as a salad with oil and vinegar. They appear to be quite digestible in this state, if ripe.

Celery and *cucumber* raw are not suitable for eating after a heavy meal. The quantity of woody fibre in them adds an additional load to the stomach, at a time when all its powers are required. With bread and cheese, as a light lunch, they give an agreeable zest, and seem to stimulate the secretion of gastric juice. That is the time for their admirers, and they are many, to enjoy them.

The selection of mere *flavoring herbs*, such as onions, garlic, mint, tarragon, etc., is not a matter for the dietician to discuss. He may, however, say one word in favor of temperance in their employment. An excess makes us unpleasant to our neighbors; and disguising the true flavor of the meat, it leads to our putting up with an inferior article. The object to be aimed at in their use is to promote the secretion of digestive solvents; and the degree in which they attain this object may be judged by the watering of the mouth; a whiff of them excites the flow of saliva, a copious dose runs it dry.

The produce of the kitchen garden, classed according to the main objects which its use serves, may be divided as follows, the order adopted in each class being a rough estimate of the plant's average value as an esculent.

1. *Starchy and sugary plants.*—Potatoes, white and sweet, yams, chestnuts, beans, lentils, peas, Jerusalem artichokes, carrots, parsnips, beet-root, salsify, turnips.

2. *Stimulants.*—Asparagus, wild onions, artichokes, strong onions, garlic, and other substitutes, aromatic herbs and other flavors, mustard, cress, and a few other pungent salad materials.

3. *Antiscorbutics.*—Cabbages, tomatoes, and salad materials in general.

4. *Diluents.*—Cabbages, spinach, turnip-tops, winter greens, broccoli, cauliflower, Brussels sprouts, sorrel, nettle-tops, and in short any leaves sufficiently palatable to eat and soft to swallow, which are green when boiled.

The use of the *first class* is obvious from the powers assigned to starch and sugar by the investigations quoted in the last chapter. Each of these vegetables is a direct food contributing to the force of the body in health. How under certain circumstances some, or all, become unsuitable, will be spoken of in a future part of the volume.

The effect of *stimulating* vegetables is to cause an increased secretion of saliva and gastric juice, thus enabling a greater quantity of food to be dissolved.

Antiscorbutics seem to act by contributing some of the materials of the blood of lesser amount, though of importance to the general vigor of the constitution. Herodotus implicitly attributes the activity and healthiness of the Persian race to the variety of fruit and vegetables consumed by them. And I feel sure that the puniness, infertility, pallor, fetid breath, and bad teeth, which distinguish some of our town populations, is to a great extent due to their inability to get these articles of the table fresh. The watercress seller is one of the saviours of her country. The consumption of lettuce with his tea is an increasing habit worthy of all encouragement in the working man, but he must be warned of the importance of washing the material of his meal.

The last hint is given in view of the frequency of the occurrence of the large "round worm" (*Ascaris lumbricoides*) in the laboring population of some agricultural counties, such as Oxfordshire, for example, where unwashed lettuce is often eaten at this meal. Naturalists will not allow us to think that the creature is a lob-worm, altered by its birth in an abnormal habitat.

But at all events its ova will live for years in moist earth,[1] and may easily be brought in from the garden, which has been manured from all sources.

Diluents contain a large proportion of woody fibre and chlorophyll, which are little, if at all, soluble in the secretions of the stomach, and are not converted into sugar by the saliva, as starch is. And they are not liable to be removed by absorption like water, the most universal diluent. Their use would appear to be to get mixed up with the nitrogenous articles of food, so that the mass may be permeated by the gastric juice and presented gradually to the absorbents. Like gelatin, though apparently not nutritious themselves, they make other things nutritious. Their benefit is made manifest by the improved action of the bowels after their employment.

§ 5. Fruit.

Fruit is hardly ever eaten except because it is nice. A very good reason no doubt, but one that rather removes the consideration of the subject from the scope of the dietician.

The safest time for taking fruit is in the morning or afternoon with stale bread and a draught of water. Thus may be made a very wholesome and digestible lunch. The worst time is after a heavy dinner. Adults often complain that they cannot enjoy fruit as their girls and boys do; the fact is they eat it at a wrong hour of the day.

Grapes, figs, peaches, cherries, oranges, and strawberries, may be considered to be the most digestible; plums, apples, pears, and apricots are less so; while melons, and other cold watery things, are not only indigestible themselves, but prevent digestion.

In selecting oranges, especially for our patients, it is best to take those with the greenish calyx still adhering to them; they are the juiciest and ripest. The fewer pips, the better the orange, or lemon, or grape.

Nuts and almonds have not justice done them by nature. They contain an enormous proportion of a valuable nitrogenous aliment, which in the latter exceeds half its weight, and is called " emul-

---

[1] Davaine, quoted in Reynolds's System of Medicine, vol. iii, p. 194.

sin" from its diffusibility without solution in water; yet this is
so cut off by its natural concentration and hardness from the action of the gastric juice, that it is scarcely digested at all, unless
chewed and cooked much more than usual. When chosen for
food, they should be used in extreme moderation, and at a time
when the stomach is at leisure, and can devote all its powers to
their solution. When pounded and employed as a flavoring, they
are innocent enough. Frying them with butter, salt, and pepper,
makes a tasty and wholesome hot dish for dessert.

Cookery, breaking up the texture of all fruit, makes it much
more easy of digestion.

The Jews, who eat much fruit, assign that habit as a reason
why their people suffered less than others from cholera, during
recent invasions of the epidemic. But it may be remarked that
they sell a good deal of fruit, as well as eat it, and may possibly
be prejudiced in favor of the trade. It can hardly be wise to
consume more fruit than usual, at a time when a chance looseness
of bowels is often the exciting cause of a fatal affection of the
system, excited by the special poison then prevalent.

§ 6. Groceries and Chandlery.

Recent legislation about adulteration has been directed more
against the grocer than against any other of our purveyors. The
stores he sells have all gone through a process of manufacture
which alters their natural aspect, and therefore gives great facility
to fraud. The fraud consists in mixing a cheaper substance with
a more expensive, and disguising the mixture. The disguise may
be deleterious to health, or it may not, or it may even make the
compound more wholesome, but the fraud is the same. In these
pages, however, we have to do only with those adulterations
which render the goods less fit for consumption as food.

The line will be taken of shortly pointing out the characteristics of the best articles, without attempting to enumerate the
sophistications to which they might possibly be subjected, but
against which the possession of good characteristics practically
warrants them. To detect the special method by which bad groceries are made bad, requires the detector to be as acute as the
rogue he is trying to expose; and unless he makes that his sole

ambition in life, he is not likely to succeed. Money-making is a much stronger stimulus to invention than the love of truth. If a customer suspects adulteration, he will act wisely to place the matter in the hands of the legally established analyst of his district, the expense of the proceeding being now made very moderate.[1]

Tea.[2]—The uses of tea are—

1st. To give an agreeable flavor to warm water required as a drink.

2d. To soothe the nervous system when it is in an uncomfortable state from hunger, fatigue, or mental excitement.

The best tea, therefore, is that which is pleasantest to the taste of the educated customer, and which contains most of the characteristic sedative principles. The sedative principles in the leaf consist of an essential oil—which may be smelt strongest in the finest teas, weakest in the inferior sorts, entirely absent in fictitious teas—and of the alkaloid *thein*, which may be demonstrated by heating some tea dry in a silver pot, when the salt will appear as a white bloom on the metal. If there is any bouquet at all, or any *thein* at all, in the specimen examined, it is worth something.

The shortest way to test the *comparative* value of different specimens is to put a teaspoonful of each in one of the little china teapots or cups with covers, here used as ornaments, but originally intended for this very purpose, which has been previously made quite hot. Shake the tea about in the hot pot a few seconds, and then pour on, quite boiling, a small half-cup of water on each. Cover them up quickly, and let them stand by the fire about a

---

[1] Not less than 2*s.* 6*d.*, and not more than 10*s.* 6*d.*

[2] I take the opportunity of the first mention of a purely foreign product, to say that the most interesting way of enlarging our ideas on the subject of food production, is to spend an afternoon now and then at a classified collection of living economic plants, such, for instance, as that at the Botanical Gardens, Regent's Park. It is much pleasanter to think of tea as connected with the pretty little camelia it comes from, than with blue paper packets; and the despised "grounds" will forever after acquire an interest in our minds. Who would have expected pepper, and ginger, and rice, and sugar to look as they do when growing? No consumption of midnight oil over botanical books, gives so much real knowledge as this short hour of healthy observation.

minute. Taste them immediately without milk or sugar, and choose that which has most aroma.

On examination of the contents of the pot after use, there will be found in good specimens very little of the dust or broken leaves. The said dust, in fact, consists of the sweepings of the warehouses, which the Chinese manufacturers make up with rice-starch into pellets, and use to adulterate the real article, under the name of "Lie tea," which expresses its character very well. The hot water dissolves it again into genuine dirt.

As tea is made from more than one variety of plant and from leaves at different periods of maturity, the shape and other characteristics of the foliage are not very distinctive; but I think, as a general rule, that, after infusion, the best leaves are the thickest and pulpiest in texture.

Green tea, normally, contains much more of the essential oil than black; but then its higher price offers a great temptation to frauds, and if it is used, more care is needed in its selection.

Cheap black tea sometimes owes its cheapness to the admixture of leaves damaged by damp, or which have been actually used and redried. This is easily detected by the scent, but as there still remains a quantity of tannin and coloring matter, people will use it, and think they have got an article worth the price it is sold at. However, good hay, or a bunch of wild thyme or mint, would really afford a pleasanter and wholesomer drink. The dried coloring matter is quite insoluble, and the tannin makes the aliment with which it comes in contact insoluble, and indigestible also.

The chemicals used to put a "face," or agreeable aspect, on bad tea are not poisonous, being simply so much inert dirt. They consist of indigo, Prussian blue, whitewash, plaster of Paris, heavy spar, and the like; new things being substituted as the old ones get found out. Their presence, however, shows that the tea is more or less bad, or it would not have been faced.

The finest teas color the water the least. The finest of all, in Europe, the yellow tea which comes overland through Russia, obtainable at Frankfort, and well worth obtaining, communicates only a slight tinge to the infusion. These luxuries are best enjoyed with a slice of lemon in lieu of milk and very little sugar.

In using tea, it must be remembered that the small-leaved and fine-grained tea packs much closer than the coarse, that consequently nearly double the quantity may be contained in a spoonful, and therefore fewer spoonfuls are required.

Coffee contains more of its special exhilarating alkaloid (Caffein) than tea, but somewhat less essential oil. It should not be kept at the boiling-point, or it loses this virtue.

The surest way to have genuine coffee is to purchase it in the bean, with the aromatic scent (which shows that it has been recently roasted) still in it, and grind it yourself. It is easy to add chicory if you think it improves the flavor, but as that root contains no alkaloid, the beverage is weaker in quality. This is desirable under certain circumstances, to be discussed in the second and third part of the volume. A further security is to buy the beans raw, and roast them at home over charcoal; the trouble is repaid by the delicious incense, which alone, among the operations of cookery, it diffuses through the house.

If you have an opportunity of getting it direct from the importer, you will find the best coffee is that from Guatemala. It is probably re-christened "Mocha" in the shops. The smallest and roundest beans are the best. The long oval bean from the West Indies ought to be a good deal cheaper.

Cocoa and Chocolate.—"Cocoa nibs" is the most eligible form in which the plant can be used as a mere beverage, like tea or coffee. They are the seeds merely broken up by rough grinding. But much of the nutriment is wasted in the thick grounds; so that if what is wanted is a supporting food, either these must be well stirred up in the draught, or the extract of the seeds, "chocolate," must be used. This contains a large quantity of fatty matter and may be made a meal of.

Chocolate is an article so disguised in the manufacture that it is impossible to tell its purity or value. Indeed, the makers say it is improved by adulteration, and cannot be sold without. The only safeguard is to buy that which bears the name of a reputable maker.

Sugar.—The baser sort, "moist brown," always contains dirt, sand, and mites. If it is dissolved in warm water, the heavy dirt falls to the bottom, and the mites float on the surface, affording an interesting object for the microscope. Grocers get from handling it *psoriasis palmarum* or "grocer's itch," so it can hardly be a desirable condiment to eat raw. Besides this, a clerical friend of mine was informed by a large and religious grocer in a manufacturing town, that he found it impossible to compete in price with his rivals, without adulterating intentionally the whole of his brown sugar. And he stated (not under the seal of confession) that one of the materials used for coloring, was a mineral of a deleterious nature, but he declined to name it, as, seemingly, in this instance, cunning has advanced ahead of detection.

The obvious moral is always to use loaf sugar or sugar candy, the sophistication of which does not answer.

Treacle.—Common treacle is the waste which drains off from the moulds in which refined sugar is made. It contains a considerable quantity of dirt, acids, extractive matter of doubtful quality, and salts, so that it sometimes acts as a purgative. "Golden drop" is prepared by filtering this stuff through charcoal; it should be clear and light in color, and is then a wholesome article.

Grape sugar is used in England only to adulterate that from the cane; its sale might with propriety be prohibited. But, in Portugal, grape-juice is boiled down with quinces into a sort of jam, the etymological ancestors of all the marmalades—whose name is derived from "*marmelo*," the Portuguese for a quince. Let not the reader be beguiled by a poetical regard for grapes or quinces into eating it. My specimen was a present direct from a country estate in the south, and tasted like gritty molasses and onions. The giver informed me it was very wholesome, but used only by servants and farm-laborers.

Raisins, Sultanas, Currants, Figs, and *Dates,* are the dried fruits preserved by their own uncrystallizable sugar. The *muscatel raisins* are the best, and are prepared by allowing the grape to dry on the vine; the inference from which is that expedients used to hasten the process are inexpedient. The best evidence of the goodness of these articles is their plumpness and softness, combined with the absence of mites, as tested by infusion in water, in the mode recommended for brown sugar. Mites are not known to be poisonous, but they destroy the saccharine constituents, leaving only feculent remains and exuviæ, and converting the remainder into carbonic acid. Now, it is for their saccharine constituents that we employ these dried fruits, both cooked and uncooked. It may be remarked, also, that the skins are very insoluble, those of all sorts of the grape containing a large quantity of white wax, which in fact waterproofs the texture, and prevents its penetration by aqueous fluid. So they should always be split before using in the kitchen. Cakes made of unsplit currants are especially to be avoided, as they are apt to produce pain and purgative effects in the most healthy.

Rice should be as whole and unbroken, and as free from dust and dirt as possible. The presence of weevils in it, constitutes a decidedly damaged article, which ought to be returned. In the future pages, when rice is spoken of in connection with puddings, the Carolina is intended; for curries, or as a vegetable with meat, the Patna is used, since it best retains its form when steamed. Patna rice is also the most eligible thing to eat with jam, or rhubarb, or roast apples, etc., for it has the least laxative action of all cereals, and thus counteracts the inconvenient tendency in that direction of the sweet parts of the dish.

The preparations of *wheat* ordinarily sold by grocers, such as *Semolina*, made from hard wheat rich in gluten, *Macaroni*, *Italian paste*, *Vermicelli*, made of flour from which the starch has been partially removed, are more nutritious than the flours of the corn-dealers; but, at the same time, are less digestible from their being dried up so hard. They are not suitable to be used as vegetable dishes, for they are too nitrogenous for the purpose. But, if they are long boiled till quite soft, they form a substantial meal. The worst sort of macaroni is that stamped in the form of letters, for if it be sufficiently boiled, the shape of the letters is lost, and cooks do not like that.

These preparations are apt to get "weevilly," a state of things usually to be detected by the smell.

Vinegar.—The best vinegar is that made from the acidified white wines of the Loire and Charente. British malt vinegar is deficient in the œnanthic ether which gives a bouquet to the more elegant article, is more apt to become mouldy and to breed worms, and is more often adulterated. As to distilled wood vinegar, although its fundamental composition is identical with that of wine vinegar, yet it has not such a pleasant taste or smell as the latter, for the destructive distillation of the wood gives rise to some empyreumatic products of doubtful wholesomeness, of which traces always are to be found in the product.

There seems nothing gained by scenting and flavoring vinegars. It prevents their being analyzed, and thus excites, perhaps, unmerited suspicion. They smell like lotion, which is unpleasant at dinner.

Vinegar owes its acidity to the acetic acid, which constitutes about a twentieth of its weight in French vinegar of good quality, and in British "proof" vinegar 4.6 per cent. of anhydrous acid.. As to other substances contained in the solution, tartrate and sulphate of potash, tannin, and œnanthic ether appear to improve the flavor without in any way affecting the health of the consumer. But it is not so with sulphuric acid, with which bad vinegar is adulterated. Sulphuric acid—especially if cheap, impure oil of vitriol be employed—cannot be considered harmless if used in the daily food, in the preparation of cabbage, or pickles, or salad, or made-dishes. The more it is cooked, the more concentrated it becomes, for the acetic acid is driven off by the heat, while the mineral remains. The least injury it can do is to corrode the teeth, when present in excess.

To avoid sulphuric acid entirely is, however, not possible, unless you make your own vinegar; and this is really worth the trouble if your consumption is large. For the law allows the manufacturer to introduce sulphuric acid to the extent of one part in a thousand (in France one gramme to one litre), and the article cannot be called adulterated if this amount is not exceeded. The test recommended by the College of Physicians for insuring the goodness of British vinegar used in the preparation of medicine is a solution of 1 part of chloride of barium in 8 parts of water: of this 10 minims will precipitate all the sulphuric acid in an ounce of lawful vinegar. If, after this has settled down, the test solution still continues to form a cloud, the article should not be employed in the preparation of food.

Besides sulphuric acid, cheap dirty vinegar sometimes contains lead and other metals. There was an epidemic of lead poisoning some years ago among the apprentices of a silk-mill at St. Albans, induced by pickled pork. This contamination is provided against in the Pharmacopœia, by ordering the test of sulphuretted hydrogen to be used.

Is it worth while to test or get tested such a mere condiment as vinegar? I think so, for it is really a most useful adjunct to the dietary. It possesses the property of softening and finally dissolving muscular fibre, as you may see by watching its action on a fragment under the microscope; and, in virtue of this solvent action, it is wisely taken with those meats whose fibres are hard, and from their hardness insipid, such for example as boiled beef, fresh pork, brawn, salmon, tunny, sturgeon, eels, lobsters, etc. The resolution also of the albumen in hard-boiled eggs is assisted by vinegar. Acids favor the conversion of cellulose into sugar, which is the first stage of the digestion of the materials for salad —of cabbages, and other green leaves—and their employment in this class of dishes is strictly physiological. On the other hand, to put vinegar on beans is, in the strong language of Monsieur Cyr, "*détestable;*"[1] for it renders insoluble the legumin, which is the most nutritious part of them, constituting, in fact, from a quarter to a third of their substance.[2] Cold boiled beans are sometimes made into a salad, and it is quite true, as M. Cyr says, that the addition of vinegar destroys the flavor, and, probably enough, makes it indigestible. Oil, pepper, mustard and a little white wine, make the best dressing. Beans are a favorite food for persons practicing disciplinary abstinence, and the hint may be of use to them, though not appreciated by the unrestricted world.

Oil.—M. Cyr places olive oil as the highest in order of digestibility of all fatty foods,[3] even above fresh butter. But to merit that praise it must be thoroughly good, quite clear and transparent and free from rancid smell. The paler it is, the better. The white deposit sometimes seen is vegetable albumen, which ought to have been refined out, as it prevents the oil from keeping sweet. Lucca oil, which is the best, has a peculiar, agreeable odor, technically called "nutty." *Olives* gathered young and small are called "French olives," and in this condition are the sort most adapted, by their pleasant piquancy, for eating as a relish. But to use in cookery they are indigestible and tasteless, and inferior to the fruit gathered at a later period of growth, when soft and

---

[1] Cyr, Traité de l'Alimentation, p. 143.
[2] In horse-beans 30.8 per cent., in Windsor beans 29.05 per cent., in haricots blancs 25.5 per cent.—*Payen.*
[3] Traité de l'Alimentation, p. 122.

pulpy with incipient oiliness, and called "Spanish olives." These last, also, are best for salads. In Portugal, they refuse to gather them till they are just beginning to turn purple, and then they are bitter and not so digestible; but I am informed by a Portuguese country gentlemen that they might be just as good as the highest priced French fruit, if the farmers would be persuaded to advance upon the traditions of their grandfathers.[1]

Caviare should be soft, pale in color, and exhibiting the ova of the roe quite distinct. When old and black, and homogeneous in texture, it is out of season and rancid, and arrests digestion. It is wise never to eat it when you see it served carelessly with cold, withy toast. It should be sent up in a toasted-cheese dish.

Pickles.—Grocers appear to consider that the final use of pickles is to ornament shops: so they choose them for the brilliancy of their colors and the elegance of the arrangement of their contrasted forms in the bottle. The consequence is that all sorts of expedients, some of them highly deleterious, are used by manufacturers to enhance "the fatal gift of beauty." In twelve samples of pickles taken indiscriminately and examined in behalf of the State Board of Health of Massachusetts by Mr. Hill, last year, ten were found to contain copper, by the simple process of dipping in them a steel knitting-needle, which in a few hours became coated with the metal. Nine of the samples were also examined for alumina and found to contain it, showing that alum had been used, probably to intensify the lake tints. (Fourth Annual Report, Boston, 1873.)

Black pepper in powder is another article which the conscien-

---

[1] I take this opportunity of alluding to the trade custom of designating peculiar qualities or kinds of food by local names. No fraud is intended, and any legislation which would make the transaction a fraud, would be unjust. The words "Spanish" and "French" do not mean that the olives come from Spain or France, but that they are of the sort made in those countries. A large quantity of "Ostend rabbits" used to come to the London market from near Marseilles, but an importer told me they were bred, fed, killed and skinned like "Ostend rabbits," and therefore to all intents and purposes "Ostend rabbits." So a wine merchant that sells his prime claret as "Château Margaux," is committing no fraud, if his wine is as good as Château Margaux, though it might be proved to have never seen that famous property. Furriers are in the habit of calling tabby-cats' skins "Japanese lynx," and the best "plover's eggs" are usually laid by gulls on the East coast: in neither case is there any intention to deceive.

tious grocer, lately mentioned, declared he could not sell at a profit without increasing its bulk by artificial means. It does not, however, appear that any of the dirt introduced is, to the knowledge of the adulterators, deleterious to health. The simple safeguard is to buy the pepper in corn.

Red pepper when pure is entirely suspended when rubbed in warm water. If a red deposit falls, it is generally red lead, a noxious metal.

Mustard is usually adulterated by the grinders with flour and turmeric, which are not injurious to health, so that the verification of the drug becomes a question for the economist, not for the dietician.

Spices, in general, should not be purchased ground. And it is a prudent proceeding for members of our profession to take the opportunity offered them by the liberality of the Society of Apothecaries, and lay in, at the wholesale market price, a stock of the purest and best spices. They have, thus, test articles which they can compare with those furnished in " our village." It will sometimes be found that the latter, even with the grocer's profit on them, are the cheapest; and then of course they must be adulterated or damaged goods.

Bacon and *Ham*, when properly prepared and not rusty, give us a fat much more digestible and therefore more nutritious than that of fresh pork. The process of salting, and still more that of slowly drying or smoking, removes a great quantity of the water, and coagulates the serum, which tend to make the adipose matter readily run into rancidity. What we have to do in selection is to see that the removal of water is carried as far as possible, and this is accomplished by observing the loss of weight in cooking. Primest bacon, according to Dr. Letheby, should not lose much above a tenth in boiling; and ham wastes much less.

Sausages.—After the sensational descriptions of their manufacture, sent to journals by special correspondents, it may be presumed no one eats sausages, without some acquaintance with the person who has prepared them. There is, or was, in Oxford a large open window through which one could see a stalwart maiden, with her arms glowing from frequent immersion in cold water, chopping and stuffing these savory viands; and I have a lively recollection of the pleasure with which I welcomed them afterwards on the breakfast table. I have often thought since of the scene, when hearing the chopping machine going, in cellars and back premises in London, and have wondered that some honest maker has not taken the course of giving confidence by publicity.

"German" sausages are of unknown composition.

§ 7. Dairy Produce.

Milk.—To people who do not keep their own cows, the purchase of milk is a matter of extreme importance, so much of it is used in every well-managed family, and so sensitive to noxious influences are the junior branches, which are its assiduous consumers. But it must be confessed that to the exclusion of the noxious influences liable to be conveyed in milk, science affords very little help. There are machines for taking the specific gravity, of which the most convenient is the hydrometer sold by surgical instrument-makers as a urinometer. And there are machines for measuring the quantity of cream, of which that generally used is the lactometer, devised by Sir Joseph Banks, a tall test-tube in which the cream rises to the top, and is simply measured.[1] And for more accurate estimation we have Vogel's Galactoscope, in which instrument the quantity of cream is calculated by the degree of opacity caused by the thickness of a layer of milk necessary to obscure the light of a candle. By this, we may guess whether our milkman has put much or little water into the can, whether he has intentionally cheated us by selling the produce of the pump as that of the dairy; but that has not much, if any influence on the appropriateness of the milk for food. If less strongly nutritious, diluted milk is quite as wholesome as undiluted, and more easily digestible. The cases which call for the use of such measures of commercial value are those of public institutions, such as hospitals, prisons, workhouses, where it is our business to make sure that the poor persons towards whom we occupy a paternal position are not defrauded of the limited nutriment supplied to them. If the milk comes from starved animals or from the pump, they ought to have more of it in just proportion.

---

[1] Is it too late to revise the barbarous title of this instrument, and call it (say) the "Cream-reckoner?"

The real poisons whose possible presence throws a dark shadow over the enjoyment of this delicious drink, are quite independent of its richness or the reverse. They are those arising from an unhealthy condition in the cow, or in the human dwellers in the dairy, or from gross carelessness in keeping its produce. After calving, the udder yields a thickish yellow fluid, somewhat stringy and greasier than ordinary milk, slightly reddening litmus-paper.

Till this has ceased to be secreted, whether the period is a day or two, or as many weeks, the milk has a purgative action when raw, and even when boiled and made into a pudding, as is the custom of some parts, is of doubtful wholesomeness. An admixture of it in the dairy produce may be detected by skilful microscopic examination, which shows the presence of colostrum corpuscles. The best way to exhibit them is to mix some water with the colostrum, let it stand a few days till the cream has risen entirely, and then collect the sediment from the bottom of the test-tube, which will be found to consist entirely of them. Their measurement, according to Mandl, is about $\frac{8}{1000}$ of an inch in diameter.[1] Colostrum is apt to coagulate the milk like rennet.

Of more serious importance is the presence of inflammatory or febrile disease in the cow. It is known by the appearance of blood or pus in the milk, but more readily by the sickening odor which exhales from it. If this is accidentally swallowed it disagrees strongly with the stomach, and if the disease is infectious, it may very probably be communicated in this way.

A more likely way, however, for disease to be induced through milk, is its contamination during its transit from the cow to the customer. Where inmates of the farm have been affected by typhoid fever, the infection has been carried to the families in the distant town supplied from that quarter. It is just possible that the germs of the malady may have settled in the milk from the air; but a more probable supposition is that the subsoil water, and so the wells, had become contaminated with the morbid secretions, and had been used in washing the milker's hands, the pails, and pans of the dairy. In a recent notorious case, occurring in London in 1873, this seems to have been fairly proved to be the path of the invasion of typhoid; but it would not have been proved

---

[1] Dr. J. Davy in Med Chir. Transactions, xxviii, 82.

had not Dr. Murchison taken up the investigation with extraordinary energy and perseverance. This danger is unhappily not capable of being warded off by science. The fatal substance which brings typhoid fever into our bodies cannot be distinguished from other organic matters, nor can its existence be made evident by any chemical or microscopic observation. By its works alone, and too late, do we know it. The only possible protection lies in the scrupulous observation of dairy farms by sanitary inspectors, to report on any communication between the drains and wells, and instantly to warn the customers, if the farmers refuse, to stop the sale of dangerous milk as human food. It would be a good plan for the customers of any one establishment to appoint their own inspector.

The evidence of the transmission of scarlatina by milk, is not so conclusive as in the case of typhoid.

There are no other impurities injurious to health known to exist in the milk of our shops: caramel, brown sugar, salt, and carbonate of soda have been detected,[1] but not the chalk, starch, brains, or other substances sometimes asserted to be employed as adulterants.

Milk is sometimes rendered unwholesome in the customer's own house, by the vessels in which it is received not having been properly scoured out with soda. On stale milk, even in minute quantities, there very quickly germinates a blue mould, such as is seen often on cream cheese, and called *Didium lactis*. The mixture of this, adhering to the corners of the can, with the fresh milk, causes it to turn sour, and to give rise to colic and diarrhœa, and, it is not unlikely, also to "thrush" in children, for the crust which forms in the mouth is not a dissimilar form of mould.

The purity of the milk supply is a matter of extreme importance, and fitly forms the subject of legislative interference, provided always that the legislative interference be judicious and does not impede improvement by competition.[2] It is a subject of

---

[1] Fourth Annual Report of the State Board of Health, Massachusetts, 1873, p. 295.

[2] Of such injudicious sort would be, for example, the fixing an absolute standard of cream contents. The standard must be low, or much genuine milk would be condemned; and then when any dealer got milk richer than the standard, he would water it down to the mark, and thus the pump would be more active than ever.

pre-eminent importance to the healthy, above all others. I always feel indignant when I see advertised special milk, in sealed cans or otherwise, for the nursery or for invalids. As if the health of the sick and weakly were more important than that of the strong man, on whose arm those sick and weakly depend for existence. Let us keep our strong men well, and we shall have fewer invalids to attend to. In choosing between two shops, I should always prefer the one that did not advertise a special article.

Cream, when good, is thick, clouty and yellow.

Buttermilk is one of the most wholesome summer drinks possible. It is equally refreshing and nutritious, and to see it given to pigs instead of being distributed to the neighbors makes the philanthropist's heart bleed.

Whey from which the curd has been removed for the purpose of making cheese, is apt to be somewhat sour, from the rennet by which coagulation has been effected. But even then it is a pleasant summer drink, and is certainly very digestible, and rapidly absorbed, for it is in composition more like *serum sanguinis* than anything else. A grate of nutmeg makes it very palatable. In Switzerland it is often drunk as a diet drink, and the inhabitants have such a high opinion of its wholesomeness, that they have founded establishments for the special purpose of receiving patients for the cure of all sorts of diseases by its means. This "Molken-Kur," as it is called, does not however seem to suit English constitutions, unless starvation is required to be the principle of treatment, as is rarely the case.

Junkets and *Curds* are nutritious nitrogenous foods, but they require the stomach to be educated by use for them to be well borne in any valid quantity. Milton's mention of "the junkets" seems to imply that they were formerly more of an ordinary diet in farm-houses than now.

Condensed or "*Swiss*" milk is a device for avoiding the risk of deterioration by shaking. Six-tenths of the water is dried out of it, and sugar is added as a preservative. It certainly is digestible, as is shown by the fact of infants brought up by hand upon it growing fat, and apparently strong, a fact of which most of us have ocular proof. Great care should be taken that only the softest water is used for its solution, and precautions taken against its adulteration. As it is a recent invention it is pure enough at present, but extensive use will probably teach ingenious methods of sophistication.

Clotted cream is simply cream skimmed from heated milk. Great accuracy is necessary to secure the right temperature, yet the union of the offices of cook and dairymaid which it entails does not insure accuracy. If the mistress of the household will make it a few times with her own hands, she will find no difficulty in producing a digestible article by observing the following precautions, and for very shame her example will be imitated.

Clotted cream is simply cream raised by heat, so that a little albumen is partially coagulated along with it. Take a well-scoured stout saucepan (a broad copper one is best) and put a teacupful of quite cold water in it, then pour in the whole milk and heat it over a very slow fire (charcoal is best) till the cream rises; when it does so, take care not to increase the heat, but keep it up. It should never exceed that which the finger can easily bear (about 150°). As the clot rises, divide it down the middle with the finger, and turn it back on itself. Keep doing this till there is no more formed. A *bain-marie* and a thermometer are a refinement upon this method.

Bought clotted cream is apt to make the delicacy unpopular. It is often sour, and adulterated with sugar and flour.

Butter, like milk, is adulterated with an excess of water, which may be detected by boiling; the oily matter floats, and the water underlies it. But neither in that, nor in any other detected falsification is there any source of danger to health. Even when rancid and damaged butter is got up again in order for the market, the processes are such as to make it fit for food, though not so nice as when naturally sweet. The palate is a good guide; but sometimes in autumn and winter, when the grass is scanty, the butter will be flavored with the dead leaves or turnips which stingy farmers will let their cattle eat, and we must not condemn the article as unwholesome because it is nasty at these times.

Cheese.

Classification of Cheeses usually in the Market.

Cream cheeses,	{ Gruelthorpe. { Neuchâtel.
Cheeses made of whole milk rich in cream,	{ Stilton. { Double Gloucester. { Roquefort. { Gorgonzola. { Cheshire. { Cheddar.
Cheeses made of poor or partially skimmed milk,	{ Single Gloucester. { Shropshire. { American. { North Wiltshire.
Cheeses made of skimmed milk,	{ Suffolk. { Parmesan. { Dutch.

Cheese is required for two purposes; one is for eating in small quantities as a fillip to the palate, and the other is in order that it may serve as a substantial food. For the first purpose it may be produced in a rancid, decayed state, and is best when of a rich buttery sort. But to be taken as a meal, to satisfy hunger, newer cheese is better. Poor cheeses, such as the Dutch, are wholesome and digestible if cut in very thin slices and buttered. Toasted cheese is also digestible if it is new and lightly cooked with cream and butter. Tough toasted cheese is about as soluble as leather.

Eggs.—To choose eggs, dissolve one ounce of common salt in half a pint of water, that is, ten fluid ounces measured with a medicine measure glass. In brine of this strength a good egg will sink, a bad egg will float. If held up to a candle a fresh egg will be found to be more transparent than a stale one, or than one with a chicken in it. Fresh eggs are most transparent in the centre, stales ones at the end.

If absolutely necessary, eggs may be preserved some time by rubbing them well with fresh grease when taken warm from the hen-house. But if they acquire a smell of old straw they are unfit for food. Lime gives them a peculiar taste, and prevents the albumen setting.

Rennet.—Rennet is not always to be obtained good in the country. It may be thus prepared for domestic use in making whey, junkets, etc. Take a calf's "bag" with the curd in it (that is the fourth stomach, or abomasus,[1] filled with acid chyle), pick out all the hairs, and wash bag and curd clean. Then, replace the curd in the bag with six or seven ounces of salt, and set them by for a week in a cool dairy. Then, take a strong brine, made of a quart of water to a pound of salt, and pour it cold on the rennet. When it has stood again for a week, the liquor is fit for use.

§ 8. Bread Stuffs.

There is no bread so digestible as that made by an honest, experienced baker. In baking at home you secure the honesty, but you lack the experience. The chance has to be taken of a bad batch through some accident; and then the best must be made of it till it is finished.

If one had to live on bread alone, brown bread would perhaps be preferable, for (as Professor Liebig taught us) it contains in the bran and pollards, which are returned to it after grinding to make it brown, a considerable quantity of phosphate of lime, valuable as nutriment to the bones and other tissues. But the fact is, most of us take in other ways plenty of phosphates in a more digestible form than bran, and the irritating effect which it has on the bowels shows that it is not, in this form, made much use of by them as a nutriment. White bread is generally chosen in preference by shrewd working men who wish to make the money spent on food go as far as it can. It is also far less likely to be withy and tough, and is less often adulterated.

Not but that some admixture of the bran is pleasant, both to the eye and to the palate, as in the flour which is called "seconds," which makes a very good bread, probably in consequence of this flour not being over-ground. Too much friction ruptures the starch-granules, and the dough does not rise well.

Bread should be evenly porous without any large holes, like a fine sponge. The texture should be firm, of which virtue the best test is the being able to cut it up into thin bread and butter. Tough, clammy new bread becomes wadded together into an insoluble mass by chewing, is not penetrated by the saliva or gastric juice, ferments anew, and even in strong persons is apt to produce

---

[1] It is figured in a volume previously referred to, Dalton's Human Physiology, p. 105.

a disagreeable form of heartburn. If from necessity it *must* be eaten, heating and copious buttering, as we heat rolls or crumpets, is the best expedient to make it as digestible as circumstances admit.

A considerable portion of water quickly evaporates from hot bread, causing of course a loss of weight, so that bakers will sometimes try to prevent it by covering up the loaves from the air. The crust is thereby rendered withy, and the crumb is wet and tough, besides which you are buying water at the price of bread.

There is another and more objectionable way in which the loaf is forced to hold water in excess, that is, the addition of boiled rice-flour. It is a sticky gummy paste, which renders the dough more adhesive, and prevents evaporation. So that 21 quartern loaves are made with what ought to make only 20. They may be found out by their sodden bottoms, the water gravitating by standing. To shirk this test, bakers will turn them upside down on the shelves, which always looks suspicious.

I am told by a retailer of glue that bakers buy a good deal of that substance, and the inference is that it is used in the same way to make the dough adhesive.

It is said, also, that alum is added for the same fraudulent purpose, even to good flour. But its object, generally, is to make a damaged article bear a good white color, and stop the excess of fermentation to which it is liable. Alum is easily detected in the laboratory by incinerating the bread suspected to contain it, and our analysts are active in this direction. However, they must guard against being too pedantic, and a distinct line must be drawn between a baker who habitually uses a great quantity of adulterant to dispose of flour which he has wilfully bought in a bad state, and one who now and then rectifies an accidental loss of goodness by the employment of the drug. But certainly, the less alum he uses, the more he is to be trusted.

The best bread grows stale the slowest. "Aerated bread," where the dough is raised according to Dauglish's patent, by forcing pure carbonic acid into it, keeps better than any. It is free from the objectionable presence of the remains of the yeast, not to be avoided otherwise, and is more certain to be wholesome than ordinary bread. It is popular, too, at that test of palatable simplicity, the nursery tea-table.

Yeast is a great difficulty with those who bake at home. Brewer's barm is the best, but it is apt to go dead between one baking and another, and is not easy to be got. Country tradesmen object to sell it, but many retail bakers in London will engage to send regularly by post an ounce or two of "German yeast," which thus arrives quite fresh and active. An orderly cook can keep a ferment in constant action by starting it with some "German yeast" from the regular manufacturers of that article, and feeding the mixture, placed in a cool situation, with some fresh malt and mashed potato or dough, daily.

An acquaintance with the theory of fermentation as explained in all books of physiology, and recently made doubly instructive by the researches of M. Pasteur, will enable an educated person to point out the remedy for all difficulties encountered in the kitchen. There is no more favorable subject for interesting unlettered minds in nature's wonders. I have seen a country congregation quite breathlessly attentive to an account from the pulpit of the recent addition to our knowledge on this head; and I am sure that when their dough has risen well, they have remembered the moral impressed upon them.

Biscuits are too hard for ordinary consumption as a bread-stuff, if made from flour and water only, as "captains" and "ship biscuits" are. They are very useful, however, to travellers, in readiness for those frequent occasions on which the bread is tough, sour, bitter, or otherwise uneatable. They bear well exposure and rough treatment, and if soaked for a few hours in water or milk, they take up several times their own weight of fluid, soften and swell, and with a little cream and sugar make a dainty dish of eminent digestibility. Biscuit powder for infants should be made from this kind.

Fancy biscuits are too numerous to describe, and of various merits. The addition of milk, sugar, eggs, flavors, etc., makes them less digestible than the plainer sort. Those made on a large scale by special manufacturers are the best, because in them the partial raising of the dough is effected by plain carbonic acid, instead of carbonate of ammonia.

When biscuits become damaged, they are often damped and heated anew in the oven. They quickly lose the artificial newness thus acquired, and grow stale and musty. So they are safest purchased in tins, where they are not so likely to be tampered with.

Oatmeal.—The coarsely-ground Scotch oatmeal is the most suitable both for porridge and cakes. If imperfectly boiled, as when prepared in a hurry, or intentionally unboiled as in brose, it is extremely indigestible, and produces the most obstinate cases of pyrosis in the parts of the country where it is habitually used. But when well boiled, and eaten slowly so as to become well mixed with saliva, it is a most wholesome as well as most nutritious food. An oaten diet has bred the Scotch farmer and the English horse, and where will their equals be found?

Emden grits are the best adapted form of oats for gruel.

Barley and *rye* do not appear to possess any distinctive virtues which can give them an interest in the eyes of a medical man. Though useful when other cereals are not to be got, they are inferior to them in solubility and nutritive power, and, at the same time, have not the attractive taste which would cause them to be a temptation against which a warning is necessary.

Maize in various forms is often recommended as a valuable food. It contains a good deal of fattening matter, and on that account is used for fattening geese at Strasbourg, and other domestic animals elsewhere. But its oleaginous constituents incline it readily to grow rancid, when it has a fusty disagreeable smell. When stale, therefore, it is apt to disagree, and in horses often causes a sort of eczema of the skin. To our race, damaged maize, persisted in as a food, is still more deleterious, producing, for example, in Lombardy and the Valtelina, a special endemic cutaneous disease—Pellagra—which is year by year slowly widening its fatal shadow over the finest lands tilled by man.[1]

Maize flour may be refined and made safer by washing out the nutritive portion with alkalies, and in this state professes to constitute "oswego," "maizena," "corn-flour," etc. But the eater should understand that he has before him starch only, and must

---

[1] The Pellagra is the punishment of sin. The farmers, cultivating their lands on the metayer system, the landlord and tenant dividing the crop, are tempted to hide some of the grain in holes and corners, where it gets mouldy. They find it makes their fowls ill, so they eat it themselves. See Lombroso, *Sulla Pellagra*, where figures are given of the peculiar mould to which the author attributes this very serious plague.

not reckon on it for nitrogenous nutriment. The economist will probably think he can buy starch cheaper in the form of rice flour, which, indeed, is often sold under these fancy names, according to the evidence of Dr. Bartlet before the Adulteration Committee this summer.

Other forms of Starch commonly sold.

Arrowroot (West Indian best).
Cassava meal.
Potato starch (uncolored).
Sago (unbleached best).
Sago meal.
Salep.
Tacca starch, or "Otaheite Arrowroot."
Tapioca.
Tapioca meal, or "Brazilian Arrowroot."
Tous-les-mois (West Indian).

The only preference that can be given to one of these over the other depends upon its flavor. All are equally wholesome, and equally suitable for the occasions when a physician wishes to administer starch without admixture.

§ 9. ALCOHOLIC DRINKS.

Wine.—This is a subject terribly overladen with literature, historical, poetical, industrial, scientific, and occasionally nonsensical. So that a simple purchaser who wants to know how to get a good wholesome glass of wine, has no small difficulty in winnowing out the required information from so much chaff.

The first thing a householder should think of before he stocks his cellar, is—what he wants the wine for. Is it to take as a regular beverage, or on festive occasions only? Does he intend to employ it for himself or others as a medicine, or to drink it only because it is nice? Here are the four chief uses of wine, and different wines are suitable for each.

As a regular beverage for a healthy person there is no wine in the English market equal to claret. The intelligence and perseverance of the Bordelais vintager, improving yearly on the traditional experience of centuries, does the best that can be done for a very good grape. Nothing can be more perfectly made than the greater part of the low-priced Bordeaux wine, now brought over direct from the Gironde.

Everybody distinguishes in the grape three parts, viz., the pulp, the stones, the skin. It is on the forms and degrees of pressure exerted to extract the juice, that the presence of these several parts in different proportions in the wine depends. In the pulp is the syrup, which ferments into alcohol, the amount of which constitutes the value of the wine. So as much of that as possible is got out. In the stones are essential oils, which in delicately graduated moderation are wanted to contribute to the formation of vinous ethers as a "bouquet." In the skin and stalks is tannin, necessary to give astringency and preserve the liquid from mouldiness; and there are also coloring matter and extractive, which contribute a distinctive hue, a thickness or "body," and fruitiness. There are, too, in the juice, tartaric acid and its salts in considerable quantity, and citrates and malates in smaller amount; these are rather necessary evils than desirable ingredients, and the owner is glad to see his must deposit the greater part of them in the "*tartar*,"—a product of the vintage for which he expressed his dislike by the bad name he gave it in the old days of strong language.[1] There is also some nitrogenous matter, which in undergoing chemical changes acts as a ferment, and having done its work, ought to disappear, lest it should re-establish fermentation when not required. Now, the perfection of claret, above all other wines, consists in the manufacture being so conducted that each of these elementary constituents of the grape is expressed in exactly the proportion most conducive to the wholesomeness of an alcoholic beverage.

If any of the above-named ingredients, or their products, exhibit themselves conspicuously in the perfected liquor, a peculiar character is given to it, which causes it to be sought out for real or supposed advantages, or avoided for real or supposed evils. We can, thus, classify wines as they appear in the market, in the following groups:

1. *Strong dry wines.*—In these the syrup of the must has been

---

[1] *Sal Tartari* = "hell-salt;" *Cremor Tartari* = "hell-scum," cream of tartar. The history of the word is not known, but Paracelsus found it in use in his day.

in large proportion, and has fermented thoroughly into a large proportion of alcohol.

2. *Strong sweet wines.*—Here the sugar has either existed naturally in such excess, or, more commonly, has been added in such excess, as to stop fermentation and remain sweet.

3. *Aromatic wines*, whose chief feature is a delicate diffusible odor comparable to that of flowers, and thence termed "*bouquet.*" This depends on the union of sundry essential oils with the alcohol in a nascent state, by which œnanthic and other volatile ethers are produced.

4. *Acid wines*, whence the natural acid cannot be eliminated without destroying the flavor. The acid in them is mostly tartaric, and not acetic, as in wines that have turned sour.

5. *Sparkling wines.*—Here the ferment is not allowed entirely to exhaust itself before bottling, so that it goes on slowly under severe pressure, and saturates the wine with carbonic acid, at the same time giving birth to flavors which would not otherwise be produced, and to very exhilarating ethers, without much alcohol.

6. *Perfect wines*, where as many of the above qualities as possible are combined without interfering with one another.

7. *Rough wines*, in which the astringent tannin is predominant.

1. *Strong dry wines* are well represented by sherry, which is the strongest and driest of all; and it is ably supported by Port, Madeira, Marsala, Johannisberg, and a few (very few) others of the products of Northern grapes. These all contain too much alcohol to drink dietetically; much diluent must be taken at the same time to make them wholesome. But for festive use, to take a glass thereof to promote conversation and good fellowship, they are excellent. And, except Marsala, they are well qualified for that post by agreeable flavor.

Constant use of strong wines induces a congested and insensitive state of the gastric mucous membrane, which prevents its glands from secreting freely, and, by reflex action, affects also the salivary glands; these liquors are absorbed slowly, and what sugar remains in them, and much of the alcohol also, becomes converted into acetic acid, which fermentation further causes the oleaginous ingredients in the food to become rancid. Thus is generated "acidity" of stomach, or the presence of an undue amount of nascent acids. A gouty constitution is often thus inaugurated in a previously healthy person, and, what is worse, is capable of being transmitted as an heirloom, just as acquired peculiarities are handed down to their descendants by domestic animals.

2. *Strong sweet wines* are represented in England by Malaga, Alicante, Constantia, Tent, Tokay, Paxarete, Malmsey, the exported Lacryma Christi,[1] Frontignan, Lunel. They are fit to drink only in small quantities, and are best appreciated, with a plain biscuit, when the stomach is not full. Thus taken they are a wholesome substitute for tea.

In order to enhance their flavor and bouquet, the naturally dry vintages are often, either during the manufacture or afterwards, made sweet. The most usual and least objectionable process is checking the fermentation by the addition of boiled grape-juice, and the result of this is an indubitably less wholesome article. By long cellaring the sweetness slowly disappears, and the fine flavor remains; but there is much risk of decomposition, which has to be guarded against by adding an excess of alcohol. Almost all the port now to be had is an artificial wine of this sort. The port drunk by our grandfathers, and up to 1820, was a dry, well-balanced wine, capable of being kept without brandy and without damage for several generations. I have, in Portugal, tasted some perfectly sound, which had been in a private cellar upwards of seventy years. But the summer of 1820 was an extraordinary one, and produced, in the Peninsula, a vintage such as had never been seen before, and the wine, by a preternatural richness of flavor secured a fatal popularity. Since then, the sole ambition of the Oporto merchant is to imitate 1820. He has more or less succeeded, but at the expense of British digestions.

3. *Aromatic wines.*—Several of the before-mentioned have a fine bouquet, but what are intended here are such as are chosen for their aroma almost entirely.

Moselle, the choicer Rhine wines, first quality Chablis, Château Yquem, several Italian wines (such as Orvieto, Monte Pulciano, Capri, d'Asti), Champagne, are of this description. They bear carriage badly, and have to be prepared for the voyage by artifi-

---

[1] That used for home consumption is a badly made wine which will not keep till it is ripe. The choicest Lacryma goes to Augsburg.

cial additions. So that, while wholesome in their native land, they are apt to be much the contrary as found here, and besides have often lost their aroma unless they are brandied.

4. *Acid wines* must be in justice carefully distinguished from wines which have turned sour. The acid of the latter is nascent vinegar, whose presence indicates the probability of other more noxious acompaniments of decomposition being also in the damaged liquid, such as poisonous moulds and fungi, for example; but some wines are really made with an excess of acid, consisting mainly of the tartaric and its supersalts. This happens mostly to the growths of cold countries, such as the Rhine districts, and the environs of Paris. White Burgundy wines (of which Chablis is the best known) and white Bordeaux wines (Sauterne), unless in exceptional years, are more noted for their acid than for their aroma. They are best adapted for taking with rich, greasy dishes, and suit well the dietary of the luxurious districts where they are grown. They are an agreeable adjunct to the usual ingredients of salad dressing.

5. *Sparkling wines.*—Champagne, St. Péray (from the Rhone), Seyssel (or "Swiss Champagne"), Sparkling Moselle, Vino d'Asti, are the best known. Good champagne is by far the wholesomest, and with a minimum of alcohol, possesses remarkable exhilarating power, from the rapid absorption of its vinous ether diffused by the liberated carbonic acid. *Silléry mousseux* contains, according to M. Cyr's table, only from 9 to 11 per cent. of absolute alcohol, but to a sinking fever patient, a glass will give twice the energy that can be obtained from a glass of brandy. The other effervescing wines will be drunk only by those who are reckless or ignorant of consequences.

The test of a sparkling wine is to leave it uncorked. If it be vapid after twenty-four hours, it is bad, and it is good in direct ratio to the length of time it retains its sparkle and aroma. That which roughens the teeth should never be again tasted; it is made of cider and rhubarb stalks; the roughness is from the malic acid it contains.

All these five classes of wine prudence will reserve for festive purposes and occasions; the wise man who wishes to enjoy life, will make them always exceptional, for as idlers have no holidays, so perpetual feasters miss all the pleasures of variety; but I am quite sure that the not infrequent manufacture of occasions for domestic rejoicing, a birthday, a wedding anniversary, a harvest home, a horse sold, the planting of a tree, the calving of a cow, a daughter presented at court, or cutting her first tooth, or any other good stroke of business, is a great promoter, not only of love and happiness, but of personal health. Let the beverages which celebrate the occasion be chosen for their peculiar and exceptional flavors. If they are good of their class, the moderate use will not shorten, but both cheer and lengthen life.

6. *Perfect wines.*—By this term are intended such as possess the virtues derived from the presence of alcohol, of water, of sugar, of ethereal flavors, of fruity extractive, and of acids, without any of them being so predominant as to mask the others, or to require artificial additions for the preservation of soundness and flavor. An enormous acreage is devoted to the production of red wines of this character in the department of the Gironde and other places in France of similar climate. We give the name of "claret" to the whole of them, which is better than pedantically endeavoring to affix geographical distinctions. One of the merchant-princes of Bordeaux, a statesman of the highest integrity, gave me some years ago a hint which I have found of the greatest service in the diagnosis of wine-dealers. "When," said he, " a tradesman offers you, at anything but the very highest price, our wines with the name of an estate attached to them, he is giving currency to a deception. If he uses the terms of First, Second, and Third Quality clarets, it is so-far-forth an evidence of honesty." The wholesale houses and brokers buy up from farmers, many of whom do not make half-a-dozen hogsheads a year, all but the small quantity which is kept to store as "vintage" wines, *vins de luxe*, and indeed much that is quite equal in quality to these speculative articles. The produce is mixed under the superintendence of cellarers, who at Bordeaux form a sort of hereditary caste, handing down their secrets from father to son, and adding fresh knowledge—

"Till old experience doth attain
To something of prophetic strain."

The mixed wine is classed as *première, seconde*, and *troisième qualité*, not from any comparative superiority in wholesomeness, but according to the price it will fetch in the market. Thus, a much better general result is secured than if it were kept separate, as is to a considerable extent done on the Rhine.

If an exporter wishes to pass off some of this blended liquid as the production of some special vineyard, he accomplishes his object very often by adding an artificial scent. If he wants to sell (say) Château-Latour, he uses nuts or almonds, or something which smells like them; if his ambition leads him to aim at Château-Lafitte, he adds a whiff of violets also; if Léoville be his object, violet alone is probably enough. Others seem flavored with cherries. When the wine is originally good, it is not likely that its wholesomeness can be seriously interfered with by this fraud.

But the same cannot be said when imperfect wines, not tempered by admixture, or damaged wines, are cured and fortified for export with flavors, and body, and alcohol. The fault of the two former is not so much that they are actively deleterious, which is not known to be the case,[1] as that they hide the nauseousness of an unwholesome article; and the same might be asserted of the alcohol, if it were inserted in the form of ripe French brandy, but the price of that would diminish the profits too much, and new corn or potato spirit, full of the poisonous oil of grain (amylic ether), is used.

Against the adulteration of claret which does not pretend to be anything else than "first," "second," or "third" quality, we have

[1] The substances sold by adulterators and druggists for flavoring alcoholic liquors are, according to Dr. Hassall, cocculus indicus, grains of paradise, capsicum, ginger, quassia, wormwood, orris root, carraway and coriander seeds, orange powder, liquorice, honey, sulphuric acid, cream of tartar, alum, carbonate of potash, hartshorn shavings, nux vomica, gentian, chamomile, tobacco, opium, juniper berries, angelica root, and bitter almonds. The quantities used of the active drugs are infinitesimally small, and would be nauseous if employed in noxious doses. In a trade circular headed "Important Information for Practical Men," Eichler's Receipts for Liquors teaches how to use also "tincture of green tea," "raisins," "figs," "St. John's bread," "rhatany," "catechu," "elderberry, cherry and huckleberry juice," "brown sugar," "yeast"—and so far we know what we are about. But as much cannot be asserted when we take in hand a mysterious "wine coloring," and still less when the ill-omened name of "body preparation" is given to one of the drugs recommended. (See Report of State Board of Health of Massachusetts, 1873, p. 107).

the valuable safeguard of the enormous quantity that is made, and the small profits which could be got out of the labor and risk of adulterating it.

The commoner Burgundies and the red Rhone wines run our claret very hard in the race for perfection; they err in containing too much fruity extractive, which, except the wine happens to be very strong in alcohol, causes decomposition. They do not keep well, and must be drunk off directly they are ripe, or they become unwholesome; but it is only just to say that they are much improved lately, and there seems little doubt that the scientific advice of M. Pasteur and others is calculated to improve them still further. The cause of unwholesomeness in Burgundy is usually the re-establishment of fermentation, through the formation of mouldiness in the bottle. To detect the presence of this destructive action, cork lightly down about two-thirds of a bottle of the wine, shake it well for half a minute or so, and let it stand: if there is any carbonic acid set free, so as to expel the cork too readily, the Burgundy is unwholesome, and it will, if drunk daily, produce feverishness, tension of the eyeballs, throbbing of the arteries, dry furred tongue, and indigestion.

The grand, old-fashioned vintage Burgundies, such as Chambertin, Clos Vougeot, etc., do not produce these effects, as they are sufficiently alcoholized not to decompose; but their price and their strength fit them only for holidays. They should be treated like port, and taken in a single glass with some exceptionally prime dish, such as venison or a saddle of four-year-old mutton. The primest Burgundies are those which have a peculiar odor of wall-flowers.

Beaujolais was introduced a few years ago from the Mâcon market by means of some choice specimens, but it has not sustained its first reputation. It is apt to turn sour, and at the best has very little bouquet.

The Hungarian wines have been lately widely advertised as superior to claret. Some of them have certainly pleasant fruity flavors, but they do not ripen well in the cellar, and are liable to decomposition. They are inferior to the produce of French vineyards at the same price.

7. *Rough wines* owe their character to the relatively large proportion of tannin which they contain. They have usually a bril-

liant tint, but are deficient in alcohol; and their principal use in the trade is to mix with others to impart color and keeping qualities. The Vin de Cahors and Roussillon come under this category; the latter, being stronger in alcohol than the majority of rough wines, appears in the markets in its own name; at the merchants', and in public houses, as "Burgundy Port." In M. Cyr's table it is stated to contain 16.68 per cent. of absolute alcohol, as against 17.63 per cent. assigned to sherry, and 20 per cent. to port. The ordinary drink of the population in wine countries usually consists of these rough beverages; but, however beneficial they might be, which is questionable, it would be useless to recommend them to those who can procure something more palatable.

Beer.—The only thing to be guarded against in malt liquors is sourness, which needs no comment.

Spirits.—This is, in every respect, the worst form in which alcohol can be habitually consumed. To continue to produce the desired effect it is necessary to continuously increase the strength or the frequency of the dose. Almost all the cases in which injury to physical health has been traced to alcohol, are in reality due to spirit-drinking.

Distilled liquors are by no means to be considered as merely dilutions of their chief ingredient. The products of recent distillation are always deeply saturated with the poisonous amylic ether, which the manufacturers call "oil of grain" or "fusel oil." It is not an adulteration, for gladly would the distillers get rid of it, and would pay largely anybody able to teach them how to do so quickly and cheaply. But it is much more injurious to health than any possible adulteration. I had once an idea that this waste product might be made of economical use, as a medicine or otherwise, and gave it to a considerable number of persons in doses of from one to ten drops. The consequences were invariably feverishness and furred-tongue, and often headache and throbbing of the temples.

After cellaring for a year or so, a great part of this poisonous ingredient disappears by decomposition, leaving only peculiar flavors. And a quicker mode of producing the same effect is to let the spirit drip slowly through the air, at the cost, of course, of much loss by evaporation. But new, unmellowed spirit must be

ON THE CHOICE OF FOOD. 81

absolutely prohibited even from occasional or medicinal use in the dietary.

If "fusel oil" cannot be detected by its peculiar, but not easily described, odor, it may usually be made evident by pouring some boiling water on the spirits or wine, and letting the mixture stand in a small room, or close cupboard, for the night. It is then diffused through the air. It may also be discerned in the breath of the consumer. Chemists have not helped us to any quantitative analysis of this obnoxious substance.

§ 10. WATERS.

It is only exceptionally fortunate people that have a chance of choosing what water they are supplied with. Still, it is of practical use to know what are the good and the bad features of each sort; especially the bad, in order that, if they are exhibited, the source may be avoided altogether, or the dangers provided against.

Distilled water, as it is condensed from steam, is the ideal of perfect purity. It has not even air in it, and therefore tastes flat and metallic. But it makes capital tea, beer, or any other infusion or decoction.

The principal bad feature it can exhibit is derived from its very purity. From the absence of salt, it dissolves and takes up in solution any lead it may come across. So that in ships, or any other places supplied by condensing engines, frequent analysis of the water with sulphuretted hydrogen should be made, lest it should be poisoned by the pipes or cisterns.

Rain water may be, and often is, equally pure. It is, in fact, as it falls, steam condensed in the great condensing apparatus of the sky. But it is better aerated than distilled water. For culinary purposes, for washing, and the like, it is well adapted. But it is apt to pick up dirt on the surfaces where it is collected, and being, like the last, free from salts, also is readily infected by lead.

River water is principally rain-water which has been filtered by passing through the surface soil. A new risk, here, has to be guarded against, namely, that of contamination by refuse organic matter in a state of decay, but not yet sufficiently oxidized into harmlessness. The latter object is naturally secured by the motion and aeration of a flowing stream, and the greater part of the resultant dirt falls to the bottom, while the chloride of sodium and other salts make the water actually more agreeable and digestible. A purified river is the best drinking-water one can have, but unhappily it is not yet quite evident what length of exposure is necessary to secure its purification, and there is the uncomfortable feeling that any organic matter may possibly be of a highly poisonous sort. Water from this source should not be drunk unless it be quite free from taste and smell, either naturally or after filtration; or unless we can trace to an obviously harmless source any taste or smell we find in it. By "harmless" I mean such an impurity as peat, derived from superficial layers of that substance. This is not only harmless, but positively a security, showing that the water has passed through one of the best natural filters in the world; and a slight tinge of it is by no means unpleasant.

Water containing sewage to any appreciable extent gives off a fetid smell just before the boiling temperature, and may be easily detected in this way. The boiling temperature renders it safe from germs capable of communicating infectious disease, but it does not make it clean or wholesome.

The readiest test of the presence of unoxidized organic matter, is to put a drop of "Condy's patent ozonized water for toilet purposes" in a tumbler of water. If the purple-lake hue thus communicated remains for a quarter of an hour, the liquid is safe; if it vanishes, there is more organic matter than there should be. The organic matter may indeed be soap-suds, or some equally innocent portion of our fated peck of dirt; but, on the other hand, it may be the germ of typhoid fever or cholera derived from a source painful to contemplate. As wise persons who have to do with strange dogs always caustic a bite, however free from hydrophobia the animal may appear to be, so it is prudent always to filter river water unless we have tested its absolute purity.

Iron is sometimes spoken of as an impurity which is a recommendation, rather than otherwise, of a water, because iron is given by us as a tonic. But I cannot agree with that opinion. We do not give iron to healthy persons; for if we did, we should often find what I have observed in some who habitually use iron-stained water, viz., deficient nutrition, dyspepsia, and obstinate anæmia. And when we give iron to the sick, we give it as a drug, and not

as a drink, and only in short courses, and moreover we do not
order it to be used in cookery. I should strongly advise iron
springs and streams to be avoided.

Lakes are perfected rivers. The organic matter has first been
oxidized by motion and exposure, and then is deposited by rest.
When Londoners see the happiness and saving due to the bringing a few feet of Loch Katrine through Glasgow, no wonder that
their mouths water for Bala, or some other available lake. Mean-
while, the existing companies do their best to imitate lakes, by
letting their property rest in reservoirs before distribution.

Marshes are, however, in a very different position from lakes.
They are not deep enough to allow the organic and mineral matters to be dropped out of the way, and moreover they are filled
with decaying weeds and insects. It is an observation due to
Hippocrates, that the drinking of marsh water causes enlargement
of the spleen, and many observations have decidedly confirmed
this evidence of the conveyance of ague poison by drinking-water.
Hippocrates remarks also on the unhealthiness of marsh water,
arising out of its frequent change of temperature; it is hot in
summer, and icy in winter, thus tending to produce catarrhs.[1]

Springs are underground streams. Before the surface drainage
which supplies them has got down to their level, it has been most
thoroughly filtered of organic matter, so they are clear and bright.
Moreover, being kept at a considerable barometric pressure, they
hold a good deal of carbonic acid in solution, which renders them
sparkling and exhilarating. But the presence of that carbonic
acid makes them take up also a good deal of lime, iron, and other
mineral constituents of the deep soil. They are "hard," that is
to say their lime forms an insoluble compound with soap, curdles
it and prevents it cleaning your hands. This is a very good test
of the presence of an excess of lime; and another is a deposit taking place in the teakettle after the carbonic acid, which suspends
it, is driven off by heat. Hard waters dry up the mucous membranes just as they do the skin, arrest digestion, and thus cause
gout, gravel, and stone to be prevalent in the districts watered by
them.

Ordinary hard waters owe their objectionable quality to carbo-

---

[1] Hippocrates, vol. i, p. 532-3, edit. Kühn.

nate of lime made soluble by the presence of carbonic acid, and derived from the chalk in the strata through which they have passed. This is in a great measure cured by the means by which it is detected, as above-mentioned, viz., by boiling. But in certain districts, as for example in the Vale of Belvoir, the hard waters contain sulphate of lime, which is not thrown down by heat. So no boiling will rectify them, and in such districts rain or surface water should be employed for drinking and cooking.

Shallow wells have the same virtues and faults as rivers. They are peculiarly liable to be poisoned by the neighborhood of house drains. *Deep wells* have rather the characters of springs. The tests of goodness are applicable accordingly. And here, it may be remarked, that the tests mentioned above are merely hints to excite suspicion, selected as readily applicable without the apparatus of a laboratory. If a doubt arise, no analysis should be trusted but that of a special analyst, for which full instructions are given by Dr. Parkes[1] and others.

The perfections of water are to be—
1. Soft.
2. Clean.
3. With air and carbonic acid in it, to make it refreshing.
4. With salt in it, sufficient to make it tasteless,[2] and to prevent its too ready contamination by lead.

Mineral waters for dietetic, as distinguished from medicinal, use, should have the same virtues. Manufacturers say that "soda water" is always most popular, if it contains a minimum of soda, that is to say if it is simply good drinking-water aerated. And a very delicate beverage is the fashionable "Apollinaris water," the salts in which are in proportions to render it most soft and velvety to the palate, and not in such quantity as to give it any medicinal action. It is as good for health as the water of Loch Katrine.

---

[1] Practical Hygiene, b. i, chap. i, sect. 5.

[2] Tastelessness is sometimes looked upon as an evidence of the purity of water; but this is not strictly true. Distilled water, of absolute purity, has a decided metallic flavor, which is removed by the addition of salt and air. The probability is that by these additions it is more assimilated to the fluids of the body, and therefore is more digestible, more quickly absorbed. For the same reason unboiled albumen is proverbially tasteless.

Potash and *Lithia waters* should be used by invalids only.

I have heard some parents object to their grown-up sons drinking soda water, under the idea that it diminishes their chance of seeing grandchildren around their hearth. I have not been able to trace any such baleful influence.

Toast and water (made by pouring boiling water on a burnt biscuit and two or three cloves) is a wholesome drink, for it secures the neutralization of all organic matter.

Barley water, if well boiled for about a quarter of an hour, is also a good formula for making hard water more digestible. The "pearl barley" of which it is made, should be washed with two waters before using, and about two ounces to the quart is generally found enough to make a drink to quench thirst. Some very thinly shaved lemon rind is the wholesomest flavoring.

Those who drink for pleasure *lemonade, orangeade, ginger beer*, and the like, should always prepare them at home. The small dealers, who brew what is sold, are prone to use the cheaper tartaric, malic (as found in rhubarb), oxalic, and even mineral acids instead of oranges and lemons. And they employ, as flavors, the amylic ethers, or "fruit essences," most deleterious drugs.

Cups of various kinds are wholesome drinks, if not too much fortified with spirituous liquors. Curaçoa should never be allowed to enter into their composition; the peel of a Tangerine orange ground over with a lump of sugar, will give all the flavor without the poison. Borage and cucumber rind are not injurious.

Ice.—In Dr. Bidder's experiments on the gastric juice, he found that low temperature does not exercise any deleterious influence upon it. When absolutely frozen, it dissolved albumen as well as ever, though it was quite spoiled by heat.[1] Again, the secretion of glands is arrested by feverishness of system, or by local elevation of animal warmth above the normal degree, and, in hot weather or hot rooms, it cannot but be beneficial to the stomach to reduce the unusual temperature to which the overheated blood has brought it. Ice, therefore, is one of the most generally useful additions to the dietary of both sick and healthy which the energy of modern trade has made; and the ice-box puts

---

[1] Die Verdauungssäfte, Bidder and Schmidt, Experiments ix, 1, 2, x, 1, xi, 1, xvi, xvii.

a daily supply within reach of us all. The only time when ice is found injurious is, during the exhaustion and real cooling consequent on violent exercise and perspiration. Pond ice, glacier ice, and snow are much inferior to the lake ice with which the English market is supplied. They contain foreign and, sometimes, organic matters, and melt sooner. Ice machines are to be recommended as a means of obtaining cold, when the best ice is out of reach. They are so frequently improved, that it is unadvisable to say which is the best at present.

CHAPTER III.

ON THE PREPARATION OF FOOD.

THE most important element in cooking is, indubitably, the cook. And the most important of a cook's virtues is shown in the selection of food. A good cook is, to a certain extent, born, not made; and if born deficient in necessary faculties, the novice should be made to understand that she has mistaken her mission. The necessary faculties are those of accurate taste and smell—with which, joined to enthusiasm and punctuality, she may become a useful and honored member of society—without which, she is simply an incumbrance. The tests to try her by are plainly cooked eggs, joints, and vegetables. If she regularly sends these up in a state to give a zest to her master's appetite, let him think no trouble or expense wasted in teaching her whatever she desires to learn; if they excite disgust, harm rather than good is done by her technical knowledge of those disguises of inferiority known as "made dishes." Cleanliness may be taught, a variety of receipts may be bought, but a delicate nose is beyond price. Choose a cook young, choose her carefully, and treat her liberally.

It is an old remark that a good cook shows generally a bad temper. There is more truth in it than in most proverbs; for the fact is that in half-educated persons, just indignation is apt to bear the appearance of wrath; and the needful rebukes to over-reaching tradesmen, and the disappointments which one who loves her work must feel, if her enthusiasm is not appreciated, beget a sharpness of tongue and manner difficult to endure. Nevertheless, it is wise to bear and forbear, and to keep an efficient servant when you have got one.

This is not a cookery-book, and therefore, of course, the details of the kitchen could not be entered into, even were I equal to the task. It is not the business of these pages to teach how to make food nice, except so far as that quality indirectly bears upon its wholesomeness. All that can be attempted is, to point out a few particulars in which the preparation of food bears upon what is known of the physiology of digestion and the economics of nutrition. And in a future chapter will be discussed the extent to which a physician should interfere with the cook in the administration of food to his patients, in sickness.

No kitchen is complete without an open range. It is impossible to have a properly roasted joint by any other means, as I learned by visiting the private premises of a "Patent Kitchener" manufacturer, and finding there an old-fashioned fire-place in full operation. He cared too much for his diet to employ his own works. Experience has led me to question even the economy of the closed fire in practical working.

Roasting is the most perfect mode possible of preparing meat for the table. The heat radiated from the open range coagulates the outer layer of albumen, and thus the exit of that which is still fluid is prevented, and it becomes solidified very slowly, if at all. The areolar tissue which unites the muscular fibres is converted by gradual heat into gelatin,[1] and is retained in the centre of the mass in a form ready for solution. At the same time the fibrin and albumen take on, according to Dr. Mulder,[2] a form more highly oxidized and, especially in the case of the former, more capable of solution in water. The fat also is melted out of the fat-cells, and is partially combined with the alkali from the serum of the blood. Thus, the external layer of albumen becomes a sort of box, which keeps together the valuable parts till they shall have undergone the desired changes by slow heat—a box, however, permeable by the oxygen of the free surrounding air, so that most of the empyreumatic oils generated by the charring of the surface are carried off. As these are neither agreeable nor wholesome, the loss is a gain. There is also an acrid volatile product, acrolein, produced by the burning of the fat, which is better removed.

The first part of roasting should therefore be got through rapidly, by close exposure to a bright hot fire, in an open, well-ventilated kitchen. By this, the gravy is retained in the meat,

---

[1] Not, however, the sarcolemma, which an experiment of Professor Kölliker's seems to remove from the class of substances yielding gelatin. See Kölliker's Mikros. Anat., vol. ii, p. 250.

[2] Quoted in Moleschott's Diätetik, p. 450.

till, at the first incision, it flows out of a reddish color. After the complete coagulation of the albumen on the surface, the joint should be removed a little further off from the fire, so as to roast gradually through.

The whole time of roasting depends partly on the weight of the joint, partly on the sort of meat.

Brown meats, such as beef, and mutton, and goose, require a quarter of an hour for each pound.

Veal and pork, the same, with five or six minutes added at the end to make sure of absence of red.

White-fleshed birds take somewhat less; for example, a turkey of 8½ lbs. (according to M. Gouffé) only an hour and three quarters, a capon of 4 lbs. fifty minutes, a fowl of 3 lbs. half an hour, a pheasant thirty-five minutes, a partridge fifteen minutes.

The fire should be thoroughly lighted before commencing, and kept up evenly; two gills of broth put in the pan, and the larger roasts basted with it five or six times, the smaller three times, during roasting. Before removal from the spit, some thick fleshy part should be pressed with the finger, to ascertain that it is soft. Uncooked parts retain their elasticity. All these times have been calculated on the understanding that there is no draught to lower the temperature between the fire and screen, for, however airy and well-ventilated the kitchen should be, such an irregular distribution of heat is most noxious, and overthrows all calculations for the due roasting of meat.

M. Brillat-Savarin pronounced *on est né rôtisseur;* in this he does not show his usual wisdom, for an eye on the clock will supply the lack of an instinctive knowledge of time.

Roasting properly conducted is the most scientific and wholesome, and on that score the most economical mode of dressing meat.

Baking naked meat at a high temperature is a feeble imitation; and the way cooks have of baking first, and then browning the outside, so completely reverses the needful order of the required processes that it may be designated a fraud. Baked meat is ill-flavored and indigestible from the saturation of the substance with empyreuma; but it is not so when the temperature is moderate, and when the materials are further defended from it by a layer of some bad conductor of heat, such as a thick pie-dish, or a crust, or a coat of clay (as practiced by the gipsies). No empyreuma, or product of charring, is then formed; and the fat and gravy which ooze out, assist in the cooking. The process goes on even after the dish is taken out of the oven, if it is kept hot by being enveloped in a thick flannel, or placed in one of Silver's Norwegian cooking boxes. The "Cornish pasty" is the most perfect dinner that a laborer, or sportsman, or artist, can have brought to his midday rendezvous. Meat or fish, and potatoes, or anything in short that the taste or purse dictates, are enveloped in a thick solid crust, baked slowly, and then packed in several layers of woollen. The basketful will keep warm for hours, and is the *ne plus ultra* of outdoor refreshment.

Vegetables and fruit demand the same slow treatment. For a large party, apple or gooseberry pie should be baked all night in an old-fashioned red dairy pan.

Eggs should not be used, or at any rate very sparingly, in bakes; for submitted to heat for a long period, their albumen becomes more and more tough and insoluble.

Rapid boiling has, in a minor degree, the same case-hardening effect on the meat as roasting; but the interior albumen seems, by this process, more hardened and less digestible. In boiling a joint, the heat should be kept up for five or six minutes. Then, it should be cooled down by the addition of three pints of cold water for each gallon of boiling water, and retained at that heat.[1]

In boiling fish, the addition of salt to the water, or the use of sea-water, makes the flesh firmer, and retains the flavor in the interior; but in making stock for *souchées*, the softer the water the better.

Mutton is best boiled in hard water, for the same reason as fish.

Slow boiling makes, it is true, a nourishing soup, but converts the muscular fibre into a mass of hard strings, which, eaten or not eaten, are in nine cases out of ten equally wasted. They are to

---

[1] That is to say, reduced with water at 50° from 212° Fahrenheit to 170°, or to the extent of 42°. The formula in the text is given as a specimen of the best, perhaps the only, mode of directing cooks how to reduce temperature. The female mind abhors meters of all kinds, and degrees Fahrenheit mentioned in our orders would infallibly entail their rejection, as only fit for hospital nurses. Medical men cannot be too cautious to avoid introducing anything reputed "chemical" into the kitchen.

be found in the fæces, exhibiting all the beautiful transverse striæ of their original state, quite unaffected by their intestinal journey.

The utility of *Soups* and *Broths* depends on several circumstances which modify the advantages accruing from their liquid state. In the first place, heat seems to have an effect in some degree proportioned to the period of application to albumen, rendering it more or less insoluble, at the same time that, to a delicate palate, there is a decided loss of flavor. Thus soups and stews which are kept too hot, are wholesome enough during the first few hours, *may* be digested at a railway refreshment room for some hours after, but on the second or third day give the rash stranger, beguiled into a cheap French dinner, an almost certain diarrhœa. Though finely divided, the minute fragments of muscular fibre seem to be, individually, rendered insoluble by continued heat.

Then, again, a high temperature, too long continued, extracts from the meat all its gelatin—an innutritious material, which envelops the fragments of fibrin in the stomach, and prevents their being acted upon. And this is all the more likely, when an overanxious cook tries to make the soup what she calls "good" (that is, strong, stiff, and glucy) for invalids.

Again, if the soup is, by straining, made clear and ornamental, a great deal of the most valuable part of it is removed: the *bouilli*, if not over-boiled, contains the chief constituents wanted as nourishment.

This subject will be reverted to when discussing cookery for the sick.

Soup is rendered more wholesome and nutritious for healthy persons by the addition of vegetables. Thus, the "Administration de l'Assistance publique" in France, adopted, by the advice of a commission of physiologists and physicians, a formula for the preparation of bouillon embracing this addition. Reduced to approximate English measures the recipe is as follows:

Water,	4 pints.
Meat (with bone),	2 lbs.
Carrots, turnips, and other vegetables,	6 ozs.
Salt,	¾ oz.
Roast onions,	¼ oz.[1]

---

[1] Cyr, De l'Alimentation, p. 49.

It may be safely commended for adoption in "soup-kitchens."

Boiling is a form of cookery peculiarly adapted to vegetables. Dr. Paris[1] remarks that it deprives them of a considerable portion of contained air, which is injurious to digestion when in excess. Potatoes should be steamed or boiled in their skins, and not so long as for them to fall to pieces from the breaking of the starch-granules; when skinned, they ought to retain their shape. On the other hand, the cabbage tribe and carrots can hardly be boiled too long. It is essential that soft water should be employed, and it is the securing this, that makes steaming such an advisable form of boiling, for steam is of course, soft water.

Particular care should be taken that vegetables are thoroughly boiled soft all the way through, and dried on a cullender.

A certain quantity of oleaginous matter renders vegetables, in which there is much combined water, less massive in the stomach. Thus, roast potatoes are better for the addition of some fresh butter, and mashed potatoes for a little cream beaten up with the *purée* after it has been passed through a sieve; and, again, milky rice pudding does not collect into a lump as plain rice is apt to do. In making the latter dish up for baking, eggs should never be used. Baked white of egg is the most insoluble form of albumen possible.

Plain boiled rice should always have a little fresh cold butter mixed up with it. In that way it will serve as an accompaniment of meat at dinner.

Stewing has the advantage over dry baking, that there is no charring or formation of empyreumatic gases; and the heat is not too great. The principal objection to it is that the meat often gets saturated with the fat gravy, and is, thus, too rich for some persons.

It is, however, not nearly so liable to this objection as *Frying*. Unless conducted with great skill, this process coats each particle of the food with a medium difficult of penetration by the gastric juice, for it is oily, whereas the secretion of the stomach is watery. Butyric acid and other rancid and empyreumatic educts are formed, and disturb digestion, producing, not rarely, flatulence and heartburn. The art consists in frying "lightly," as cooks phrase it,

---

[1] Cyclopædia of Practical Medicine, i, 576.

that is to say, quickly and evenly, and with constant motion, so that the high heat required does not char any part. For those who do not dislike the flavor of oil, it is a more manageable medium than butter, and generally turns out a lighter dish. Good Lucca or Provence oil is also less likely to be sophisticated than bought butter.

Hashing is not to be encouraged. From the two processes that are gone through, the animal fibre is too much hardened to be readily digested. Cold meat is wholesomer, and may be made as palatable with mayonnaise or some such sauce. If it must be done, at least let a water-bath (*bain marie*) be used.

Marinating, that is, baking in vinegar and water with layers of bay-leaves and pepper-corn, is suitable for the more oily kinds of fish.[1]

Broiling imparts a peculiar tenderness to the meat, by the rapid hardening or browning of the surface, preventing the evaporation of the juice. It is, in fact, roasting applied to small portions of meat. Tradition commends it as a suitable cookery for the meat of persons in training; but, perhaps, in no sort of dish does it shine so much as in fish.

In the preparation of *mixed dishes* and in *seasoning* it should be a general and almost universal rule that the different ingredients should be as far as possible cooked separately. The reason is obvious,—each article, from its texture, requires, for its perfection, to be submitted to heat for a different period. Too long exposure destroys its flavor and solubility. To take familiar examples,—an egg if made into a custard, or just coagulated, is tasty and wholesome; but if baked for half an hour in a pudding it becomes useless as an article of food. Spices lose nearly all their flavor, while retaining all their irritating qualities, if mixed in a dish before boiling; yet if heated up separately and for a shorter time, they retain it, and will suffice in much smaller quantity. Soup, on the other hand, requires less boiling than the vegetables which are usually put in it. If boiled together, the latter are therefore sure to be underdone. If baked in a tartlet, jam loses

---

[1] Another mode of wholesomely cooking oily fish, such as sprats, pilchards, or herrings, is to stew or bake them in a deep dish in layers, with a layer of breadcrumbs between each.

all fruity odor and taste, and sinks into the paste. It should be only barely heated after the paste is done. Just warm an oyster, it is sapid and digestible; bake it in a beef-steak pie, it is leathery and insoluble. It ought to be put in cold, just before the dish comes to table. Onions require long cooking; but the other seasoning herbs usually used in broths should not be sprinkled in till a late stage of the boiling.

The effects of *salting* are often misunderstood. It is well known that living exclusively on salted meat produces scurvy; and it is imagined that the injury to the system arises from the saline matter thus introduced into it. Such, however, is not the case. Chloride of sodium is such a large constituent of our blood that it cannot possibly be noxious. The unwholesomeness of salt meat, as an article of diet, depends on its deficiencies, not on its excesses: it has lost, according to Baron Liebig's calculation, half its nutritive value by the removal of its fluids and salts by the brine; and the dried-up remnant is difficult of solution. Soaking in water may soften and remove the salt, but it does not restore its nutritive value.

Smoking and *Drying* are not quite so injurious to the texture of the flesh as salting. It would seem that in the latter the hardening process goes on continuously; in fact, the salt remains and continuously extracts the aqueous constituents. But the drying takes place once for all, and the article gets no worse when it is once prepared, until decomposition occurs. These modes of preparation seem peculiarly adapted for fish, which naturally perish so readily, and which are more injured than even meat by salting.

Tinned meat cannot make any claim to a recommendation except on economical grounds. When heated up for the table it is too much cooked to be digestible or pleasant. It is best eaten cold, with some of its own jelly, and salad or mayonnaise sauce.

The process of "tinning" meat is well known to consist in the expulsion of air from the material by means of heat; and it is to be feared that so long as this procedure alone is employed, the injury from the excess of temperature is unavoidable. The aim of the inventor should be, to effect the same object by the aid of air-pumps in the cold.

Ice seems to promise more favorable results than the last-named device, for preserving meat during its conveyance over long distances—from the overstocked producer to the hungry consumer. The attempt has hitherto failed from several avoidable causes; but the Committee of the Society of Arts, who have taken up the matter, are sanguine of final success in putting the Australian fresh meat butcher in direct communication with the English laborer. It should be remembered that meat which has been thus preserved requires immediate cooking, as it quickly goes bad from the change into a higher temperature of air. It should, if possible, be brought home in an ice-box, and kept there till wanted.

Vegetables preserved by drying undergo an interstitial hardening of the tissues which renders them insoluble in the saliva, as may be observed by their want of flavor. They are inferior to fresh vegetables for use in preventing scurvy—indeed they seem inferior to lime-juice for that purpose—so that they cannot be wholesome for the healthy. It is a fact, not explained as yet, that long transport, especially by sea, however rapid, has a somewhat similar deleterious effect on vegetables; so that we do not get the advantage which might have been anticipated from the increased rapidity of communication with distant supplies. A notable instance is that of Algerian peas.

The preservation of food by excluding the air by means of oil is a subject which requires further experimenting upon. Delicate fish, such as sardines for example, can be kept in this way for a long time; and from time immemorial the method has been adopted for wine in Italy. A thin layer of fresh oil in the neck of the flask obviates the necessity of a cork in wine not intended to travel, and is much more efficient, preventing even the lightest and most perishable liquors from turning sour. Potted meats and sausages are often judiciously preserved by a layer of lard, which seems effectual in retaining moisture and preventing decomposition. Why do we not apply the same principle to the storing of jams, gooseberries, plums, etc., for winter use?

Preservation with sugar has the disadvantage of introducing an ingredient which is cloying to the appetite, a mask to the natural flavor of the vegetable, and, moreover, apt to generate an excess of acid in the stomach.

As a rule, forced vegetables, and fruit out of season, are not to be recommended. The natural period of its perfection is long enough for us to enjoy each; and, then, a change is as wholesome as it is pleasant. To bring them to table sooner gratifies merely a vulgar ostentation or impatient gluttony, and receives its just punishment in a premature weariness.

Certain articles of diet yield their savors best when their valuable ingredients are got out in the form of an *Extract*. An illustration of this may be found in the *Guatemala mode of preparing Coffee and Chocolate*,[1] which is as follows:

"Coffee berries are used that have been stored dry one year. Taking enough for one or two days' consumption, they rub them in a linen cloth, and lay them in the sun before roasting. This operation is always performed by the lady of the house, over a quick charcoal fire, in an iron cylinder, which she keeps turning for ten or fifteen minutes, till the berries are roasted on all sides. The most esteemed bean is the 'peaberry' (which bears the round monocotyledonous form, instead of the dicotyledonous), in consequence of its browning more evenly. It is not a different species of coffee, but an accidental variety, more common on some trees than others. At the critical moment of perfection, the berries are emptied into a basket and stirred round to prevent their further concoction. They are then ground, and a very strong liquid extract is made from them by infusion in hot water. A common percolating coffee-pot will serve this purpose, and save straining. A dessert-spoonful of this essence, with either boiling water or boiling milk poured on it, forms a cup which is the invariable *bonne bouche* of every meal.

"Cacao trees grow wild, and are also cultivated round some Indian villages; the berries are dried in the sun; and an Indian woman generally comes to your hacienda to make them into chocolate. Kneeling before her mill, consisting of a smooth curved surface with a heavy stone roller, she grinds the kernels up with an equal weight of sugar, and flavors with cinnamon or vanilla, the latter a wild product of the forest. It is ground three times over, and, when in a smooth paste, is cut into oblong square pieces, each enough for one cup. It is then dried on a plate over a charcoal fire. When required for use, it is dissolved in boiling water or milk, and a thick froth worked on the top of each cup with a

---

[1] Dictated by Mrs. Osbert Salvin.

swizzle-stick or whisk. The cup is often filled up with milk-cheese, cut into small dies."

A principle of rational cookery, much overlooked by professors of the culinary art, is that each article of diet should be so prepared that its own natural flavors and other characteristics should be enhanced, instead of being masked or destroyed. Condiments and sauces should be so moderately used, as never to be a prominent feature; and then, should be so blended and balanced, as to make it difficult to identify them. There seems to me nothing utopian in the idea of a universal sauce, adapted to all sorts of animal food, making them all more savory and more wholesome at the same time.[1] There is ample scope for individual taste in the selection of the variety suited to the user's palate.

One is almost ashamed to mention cleanliness as an essential in cookery, the idea of the contrary condition in eatables is so repulsive; but cooks do not seem aware how often their dishes are unpalatable, and, therefore, unwholesome, solely from being prepared in a vessel which has a disagreeable flavor remaining in it. Soap is sometimes employed in washing pots, instead of soda; and the taste of the rank train-oil seems to adhere to the metal, and to infect a succession of otherwise excellent material; and so adheres the odor of onions, and of several other condiments, to a steel knife. The use of printed paper to stand glass and crockery upon, in cupboards, is also objectionable, as the oily effluvium from fresh printing-ink is very rank, and acrid, and penetrating, as our patients who suffer from hæmorrhoids well know to their cost.

Fish is an article very often spoilt by injudicious endeavors to make it palatable. "Melted butter," in reality, requires the hand of a first-rate artist, whereas every kitchen maid thinks she can concoct it. Unless M. Gouffé's instructions have been followed to the letter, it is best avoided altogether. A few drops of Chili vinegar, or black pepper vinegar, or elder vinegar, or Worcester

---

[1] This is an old ambition, see Pepys's Diary, February, 10, 1661-9. The Duke of York "did mightily magnify his sauce, which he did then eat with everything, and said it was the best universal sauce in the world, it being taught him by the Spanish ambassador; made of some parsley and a dry toast, beat in a mortar together with vinegar, salt, and a little pepper; he eats it with flesh, or fowl, or fish. . . . By and by did taste it, and liked it mightily."

sauce, or a slice of lemon, or a sauce made by boiling pepper and salt in plain water with a few favorite herbs, assist the digestion of the fish, whereas greasy sauces impede the process, chemically and mechanically.

Grilling does for fish that which roasting or boiling does for meat, and is a commendable mode of preparation, especially as it obviates the temptation to take sauce.

Those professional lectures are usually the most instructive which take up some universally known and simple matter relating to a subject, and use it in illustration of the principles involved in the art to be taught. To exemplify the elements of sanatory cookery, perhaps nothing could be more fitly chosen by a lecturer than the cooking of an egg. First, for the name—our great-grandmothers (if proverb register language truly) talked of "roasting" an egg: we call the same process "boiling," marking the fact that there is no essential difference between the two, the end of both being the same, namely, the bringing albumen into a mechanical condition more suitable for the digestive viscera than when raw. Then, an egg, more clearly than any other meat, exhibits the *virtues of freshness*, and the vices of defects in that quality. Boil an egg warm from the nest, and you recognize it by its creamy, lightly coagulated, eminently digestible form of albumen. Even by next day it is less perfect, and steadily degenerates in value, till it becomes the most hateful of poisons. Here, may be pointed out, the *importance of selection* according to external obvious qualities.

Next, should come a scientific picture of the *coagulation of albumen* by heat. Albumen begins to coagulate at $140°$[1] very slowly, but does not form a solid mass with rapidity under the temperature of nearly $212°$. Till it has formed a solid mass it is easily permeable by heat, and the central parts are solidified, therefore, equally with the exterior; whereas, coagulated albumen is a bad conductor of high temperature; and if the outside sets quickly and firmly, the inside remains light and semi-solid. Boil an egg at a slow heat, and it is not cooked till hard all through; put it into quite boiling water, and the white sets soon, and leaves the albumen of the yolk soft. This illustrates the rules given a

---

[1] Simon's Chemistry (Day's translation), vol. i, p. 16.

few pages back, about applying *at first a high heat* in roasting and boiling, and afterwards moderating it, when the hardened surface has inclosed the deeper parts as in a box. The *shell of the egg* may be a text to demonstrate, how in baking and frying, the external media, the crust and the oil, retain the whole of the substance, so that there is no loss during these economical culinary operations, though the meat is less wholesome.

Then, that this loss of wholesomeness is due to the saturation of the article of diet with the products of dry distillation, and, also, that enhancement of natural savor, solubility, and wholesomeness run in parallel lines, may be illustrated from the observations of Beaumont on the different times occupied in the gastric digestion of eggs in several conditions. He found, for example,

				Hours.	Minutes.
Eggs, whipped and diluted,	occupied in digestion,			1	30
" fresh raw	"	"		2	0
" fresh roasted	"	"		2	15
" soft boiled (or poached)	"	"		3	0
" hard boiled	"	"		3	30
" fried	"	"		3	30

It may be observed that this is just the order in which they are tasty; that is to say, the degree in which they come with facility into contact with the sensory nerves distributed through the mucous membrane; so that pleasure and duty here, as usually in natural operations, become one.

General rules for the *preservation of food* are somewhat deceptive. They lead to their being too much practiced for the display of the pride of ingenuity. The fact is that food is always the worse for storing in respect of its wholesomeness, even if its taste is not injuriously affected.

But economical reasons exist for the restricted exercise of this art. The principle consists in excluding the evil oxidizing influences of air and moisture. In dry goods this is done by keeping them dry and warm and closely covered up. Starch, rice, tapioca, sago, macaroni, vermicelli, sugar, sweetmeats, jams, salt, and dried and salted meats, tea, coffee, etc., require this treatment. And they should be kept in a different cupboard from odorous goods, such as candles and soap, or they will catch the objectionable flavor. But with most fresh organic substances a different treatment is necessary to the attainment of the same end. They contain in their texture itself sufficient moisture and air to oxidize them into decomposition, and the more stagnant these are the more surely do the chemical actions result. It is necessary, therefore, to let them have free ventilation; their external surface should be frequently wiped, or at least blown over by a current of air, so as to let the old moisture escape and fresh be absorbed. Thus, meat should hang exposed in an open larder, and be often dried. Lemons should be purchased in the summer and suspended in nets for use at the time when they are dear. Onions and garlic should be strung up in an outhouse (not the larder). Parsley, thyme, mint, and other herbs should be dried in the wind, out of the sun, put each into a separate paper bag, and hung up in the kitchen. Where apples and pears and chestnuts are stored, the window should be left open, and the fruit frequently turned. Too much draught makes vegetables withy; so they should be laid on a stone floor behind the door. Potatoes are best stacked in dry sand.

The date when each article is stored should be written down.

Eggs are an exception to the usual rule respecting organic substances. They cannot be treated in the same way by reason of their structure, yet it is impossible to avoid keeping them for culinary purposes. They are best preserved by being washed over with a solution of gum and packed in a square box of bran, which is to be turned over a quarter of a turn every day.

CHAPTER IV.

ON DIGESTION.

THE most recent systematic teacher of physiology defines digestion as "the process by which food is reduced to a form in which it can be absorbed by the intestines and taken up by the blood-vessels."[1] The process consists in the aliments being passed along a canal, thence called the "alimentary canal," running through the body, where it comes in contact with fluids oozing out from various glands, which mixing with it dissolve it, and reduce it to a homogeneous juice or "chyme."

The more perfectly and quickly this end is attained, the more healthy is the digestion.

These fluids resemble one another in being watery, saline, and albuminous, like the serum of the blood, so that they are well adapted for passing readily through the mucous membranes by endosmose. They pass in, as they passed out, freely, and carry in with them the nutriment prepared in the alimentary canal. But there the resemblance ceases; and the differences of the fluids poured from the different glands are very marked, and their actions on the various constituents of the diet are strikingly specialized.

1. *The saliva*, secreted from the glands of the mouth, is alkaline, glairy, and adhesive, and possesses the power of converting very rapidly the unabsorbable starch into soluble and absorbable sugar.

2. *The gastric juice*, secreted in the thin layer of gland which lines the cavity of the stomach, is acid, and can dissolve, so long as it is acid, the solid albumen and fibrin of flesh food.

---

[1] Dalton's Human Physiology, fifth edit. chap. vi., where will be found a complete résumé of the scattered contributions of others up to the present time, often illustrated and criticized by the light of the author's original investigations on this subject. An old résumé of mine, entitled Digestion and its Derangements (London, 1856), is out of date, as well as out of print.

3. *The pancreatic juice* is again alkaline,[1] very full of albuminous organic matter, and capable of exerting a peculiar influence on fatty matters. It disintegrates them, and reduces them to a state of emulsion, so that the mixture of fat in the watery fluid it floats in, is white and opaque. In this creamy state, it is capable of soaking through the mucous membrane and passing into the blood, by the same way as the starch and fibrin.

4. *The bile* is very different from any of the above fluids in chemical reaction and contents, and it does not appear to possess the power of dissolving starch by turning it into sugar, or of dissolving fibrin, or of emulsioning fat in any degree equal to pancreatic juice, though it is itself of a soapy nature, as is shown by the commercial use of ox-bile by carpet cleaners to remove greasespots. It is neither acid nor alkaline, but it possesses considerable bleaching powers, and, also, arrests decomposition in animal substances. In this latter capacity, it is highly conducive to our comfort by obviating that intolerable odor which distinguishes the excretions of patients in whom the bile is deficient. But, with all this, it does not clearly appear how any of its known physical properties can aid in the constructive assimilation of nutriment. Yet aid it does, most decidedly. If, in consequence of accident or disease, the flow of bile into the alimentary canal is wholly cut off, the animal rapidly emaciates, and dies starved. Even when the supply of bile is only partially diminished, as in the instance of some of our patients, there is a marked deficiency of nutrition, shown in loss of weight and anæmia. Now, this would not happen if bile were a mere excrementitious fluid containing ingredients resulting from the disintegration of the tissues, and designed merely for their removal from the body. It has assimilative functions to perform beyond the simple drainage of the blood, though what those functions are can be expressed only in the vaguest terms at present, and the most prudent and most recent physiologists decline to give a definite opinion on the uses of the bile.

5. *The intestinal juice* is secreted alkaline, and possesses the power, like the saliva, of converting starch into sugar; but, unlike

---

[1] Dr. Dobell questions its alkalinity. Perhaps it varies with the state of the body.

ON DIGESTION. 103

that fluid, does not lose its power in presence of an acid. It, also, has a certain solvent power over albumen, inferior, indeed, to that of the stomach, but stronger in this respect, that it is not checked by the bile or pancreatic juice.

Thus, we see that a meal, as it passes downwards, is irrigated first by a watery fluid which, as it dilutes and adds to its bulk, dissolves or fits for absorption a great portion of its starchy constituents. Then, it is further irrigated by a still more dilute fluid, which dissolves its meaty part. Then, the fat is washed out of it by the stream of pancreatic juice. And, simultaneously, the bile is poured on it in a continuous stream, which makes it, in some unexplained way, more easy to be taken up as nutriment. Afterwards, the intestinal juice, oozing out in small quantities throughout a long canal, seems fitted to make up the deficiencies of any of the previous solvent acts.

The daily quantities of these fluids, as estimated, mainly, from the results of Drs. Bidder and Schmidt's experiments, may be reckoned, at least, to equal the following:

Of saliva, ... 3¾ pints,
Of gastric juice, ... 12 pints,
Of bile, ... 8¾ pints,
Of pancreatic juice, ... 1½ pints,
Of intestinal juice, ... ½ pint,

making in all nearly three gallons. Of this, ninety-six per cent. is water, of which only so much passes away in the stools as prevents them from being inconveniently solid. The rest, therefore, that is to say two gallons and a half, is restored to the blood by absorption.

"The clearest notion we can gain of the business performed by all this two dozen pints of water which exude on the mucous membrane of the intestinal canal, and are by the same membrane taken up again, is by viewing them as a *circulation*. It is constantly going its rounds like an endless chain, finding and taking up inside the solid structure of the body substances which ought to come out and be got rid of, finding outside nutriment which the body wants, and conveying it in.

"Truly, when a man contemplates with the eye of the reason this unceasing journey, this great current so entirely removed from the cognizance of our senses, he is at first confounded with the novelty of the ideas it excites (*ingenti motu stupefactus aquarum*) and almost refuses to receive them. It is highly important, therefore, to bring it frequently before the mind till it becomes habitual, for there is no view of living phenomena so practically weighty for the medical man."[1]

In addition to this, there is a quantity of fluid introduced as beverage. Water passes straight through the mucous membranes unchanged, and alcohol, with a change of no weighty importance; and dissolved in them are numerous minor substances not strictly dietetic, though valuable by acting pleasurably or medicinally on the nervous system.

Rightly, then, has digestion been likened by an old chemist to a process of "rinsing;"[2] all that is required is washed out of the alimentary substances, and the remains passed on to be got rid of along with the waste products of chemical life.

Of the reason why these various secretions are able to digest the various constituents of our food, we know absolutely nothing. We find an albuminous matter in saliva, and we call it "*ptyalin;*" we find an albuminous matter in gastric juice, and we call it "*pepsin;*" we find an albuminous matter in the pancreatic juice and we call it "*pancreatin;*" but the only distinguishing point about each is that it acts upon starch, or acts upon flesh-meat, or acts upon fat, just as the fluid from whence it was concentrated acts. Ultimate chemical analysis merely shows their resemblances, and not their differences. The nearest approach to a similar proceeding in nature is the operation of malting, where, aided by heat and moisture, starch is turned into sugar by the presence of "*diastase;*" and it is remarkable enough that when albumen is artificially dissolved by gastric juice in the laboratory, a peculiar odor somewhat like that of malting is given off. But here our information ceases; there are no more points in diastase, to distinguish it chemically from ordinary albumen, than there are in pepsin or pancreatin.

In bringing to bear upon dietetics the observations of physiologists, the main things the physician has to consider are the mechanical condition in which the food should be brought, the influence

---

[1] Digestion and its Derangements, p. 31.
[2] Letheby On Food, p. 48.

of its several solvents, and the times when they are ready to receive it.

Chewing is the first provision made for securing a due mechanical condition. The perfection consists in so breaking up the mouthful that it should be as completely as possible permeated by the saliva. The object of this is, in the case of meat, to soften it in preparation for swallowing and for future solution in the stomach; in the case of starchy matter, to convert it into "glucose" or sugar.

With regard to the amount of chewing required by flesh food, there is a good deal of popular misconception. Persons with bad, false, or tender teeth are often found to fancy that a vegetable diet is more suited to their imperfect power of mastication than an animal one; and we not unfrequently see mothers instructing their children carefully to chew meat, and neglecting the same precaution in respect of vegetables. Physiology teaches an opposite caution. It is desirable, indeed, that the jaws should break up muscular fibre, lest it should perchance stick in the gullet, and be certainly difficult of penetration by the gastric juice in the stomach; but to a vegetable aliment the performance is owing of more important functions. It is still more indispensable that it should be broken up, for it has to be immediately acted upon; and it is indispensable also that it should be detained in the mouth till enough saliva to convert its starch into glucose has been secreted. Complete mastication, therefore, important for both, is still more important for vegetable than for animal food; and the leisurely performance of the operation cannot be prudently omitted by a mixed eater.

It may serve to remind us of this, to reflect that while lions and tigers and wild dogs bolt their food, cows not only spend the greater part of the day over their nibbled meals, but give it a second chewing when in repose.

Doubts have been thrown (Dalton's Human Physiology, page 116) upon the importance of the action in the mouth to the conversion of starch into sugar; but the following easy experiment seems sufficiently convincing. Take some boiled starch, say in the shape of arrowroot, and heat it with potassio-tartrate of copper. There is no change in the blue color of the salt. Now, put some in the mouth, and hold it a few moments only. When it is again heated with potassio-tartrate of copper, the metal is precipitated, and shows by its brilliant yellow color an abundant quantity of sugar.

The saliva, then, begins to convert starch into sugar immediately; and it is not slow to extend its operation to the whole mass submitted to it. A protraction of the foregoing experiment will show this. A mouthful of boiled arrowroot held in a healthy mouth for five minutes, will show afterwards scarce a trace of starch remaining.

But it is true that the morsel is hardly ever allowed to remain in the mouth long enough for its complete conversion; hardly ever is it sufficiently boiled and chewed for the saliva to affect the whole of it. Much free starch and free saliva must be carried down the œsophagus. During its passage the action goes on, and, doubtless, as much saccharine transformation takes place in the latter as in the former locality. But in a minute or two it must arrive at the stomach, and there the acidity of the viscus is said to put a stop to the saccharization. On this arrest of the salivary action by the presence of acid, Dr. Dalton's opinion of its slight influence in digestive solution is grounded. Nevertheless, once that the mass has passed through the pylorus, its acidity is neutralized, the action of the remaining saliva recommences on the starch yet unconverted, and this action is reinforced by the intestinal juice.

By the unconverted starch, I mean not only that which was unchanged on arriving at the stomach, but also a good deal set free since that stage of digestion. For, besides the saliva, there practically comes into play in the solution of starch that which I have described as temporarily arresting it, to wit, the gastric juice. Cookery, even when most efficient, rarely ruptures the whole of the granules. Many escape in the best, and in bad cookery the majority escape. They cannot, therefore, be affected by the saliva, till their albuminous envelope has been dissolved by the gastric juice. Then, the amylaceous matter may be converted into sugar, either, rapidly, by the saliva present, or, more slowly, by the pancreatic and other intestinal secretions.

For the reduction of starch, therefore, so as to bring vegetable food into a condition capable of easy digestion, the first point is that the salivary glands should secrete a sufficiency of fluid; and this not merely at the time of mastication, but that they should go on supplying it as long as any starch remains unconverted. Then, it appears extremely probable that the gastric glands aid the future carrying on of the process, though the acidity of the stomach prevent its continuance at the time of its prevalence.

Now, the salivary glands in the healthiest persons are liable to derangement from purely external circumstances acting on the nervous system. Temporary emotion affects them temporarily, and chronic emotions affects them chronically. We all are familiar with the dry lips of the coward, the lover, the pitiful, and how the tongue cleaves to the roof of the mouth when pain is endured, or when bad news is brought. "Bread eaten in sorrow" can hardly be swallowed, so long it takes to moisten the morsel. Again, bodily exertion parches the throat. It cannot be expected that meals of mixed food, swallowed when the body is under the influence of the circumstances quoted in illustration, should be dissolved, or nourish the tissues as they ought. And there is nothing surprising in the fermentation of the undigested vegetables, and the formation of flatulence by the carbonic acid which results.

Under the same circumstances, a portion of the solid meat remains undissolved, and is often thrown away unaltered by vomiting or diarrhœa; for the stomach is influenced contemporaneously with the mouth, as is clearly shown by the proverbial loss of appetite from mental causes.

It is in this arrest of secretion that the sedative action of alcohol comes in useful. It is an anæsthetic, and prevents the effect of the nervous system upon the alimentary canal from being so deleterious as it has been shown naturally to be. A few teaspoonfuls of good strong wine or dilute spirit will often restore the lost power of taking food, and is an instinctive indulgence, as a protective against the sundry blows inflicted on digestion by the exciting nature of social life in the present regimen of the world. It is possible to imagine a state of society, as among the Pitcairn islanders, for example, where everybody was apparently the better for taking no alcohol in any form, but even in that instance, the abstinence does not seem to have lengthened life, and it is certain that in Europe, it would shorten it for many of our most active and useful citizens.

Equally important is the absorption of the sugar thus formed from the starch. In health, a very great part is absorbed in the mouth and gullet, sometimes all, for chemists have great difficulty in finding it in the stomach, unless it is swallowed in excess. Some of it is probably converted into the lactic acid, which aids the solution of flesh food, and the rest taken up as sugar by the intestines. Still, even in health, a good deal of both starch and sugar escape, and appear in the fæces. But, in the catarrhal state, the mucus which lines the membrane is an almost impermeable impediment to osmosis, from its insolubility in water, and arrests absorption in proportion to its quantity. All mucus is a degree of disease, and every Briton knows how easily it is formed by very slight external influences.

It is clear, then, that for the easy digestion of starch the whole of the alimentary canal must be in a normal condition, and the nervous system not exhausted by recent excessive use.

The Mechanics of Digestion refer to the provision which is made for the food being duly brought into the sphere of action of the solvents described above. First comes *Chewing*, the importance of which, as a means of saturating the mass with saliva, has been already insisted upon. Here may be mentioned, further, the importance of its completeness, for the sake of reducing muscular fibre to a fine pulp, so that it may be quickly infiltrated by the gastric juice on its arrival in the stomach.

"In the human subject, the teeth combine the characters of the carnivora and the herbivora. The incisors, four in number in each jaw, have, as in other instances, a cutting edge running from side to side. The canines, which are situated immediately behind the former, are much less prominent and pointed than in the carnivora, and differ less in form from the incisors on the one hand, and the first molars on the other. The molars again are thick and strong, and have comparatively flat surfaces, like those of the herbivora; but, instead of presenting curvilinear ridges, are covered with more or less conical eminences, like those of the carnivora. In the human subject, therefore, the teeth are evidently adapted for a mixed diet, consisting of both animal and vegetable food. Mastication is here as perfect as it is in the herbivora, though less prolonged and laborious; for the vegetable substances used by man, as already remarked, are previously separated, to a great extent, from their impurities, and softened by cooking, so that they do not require for their mastication so extensive and powerful a triturating apparatus. Finally, animal substances are more completely masticated in the human subject than they are in the carnivora, and their digestion is accordingly completed with greater rapidity."[1]

However much natural selection may have rendered stronger the surviving species of other animals, it has, in the case of the human teeth, proved injurious to the perfection of our race. Artists have, unhappily, taught us to see loveliness in button mouths, bud-like lips, round dimpled chins, tiny pearly teeth, and to recognize aristocracy in hatchet faces. You take up a skull in one of the bone-houses near old fields of battle (say at Hythe, where Pict and Briton, and others besides, grin so grimly at us), and you cannot enough admire the evenness, the firmness, the completeness of the set of half-worn, yet quite sound teeth; but you are fain to confess that to have a wife, or a son, or a daughter with a prominent square jowl like that, would be a severe trial to your æsthetic feelings. You are incurably perverted by the mediæval association of moral purity and intellectual refinement with pitifully weak jawbones; and a big-mouthed broad-nosed Helen would never have been the mother of your children—yet would she have saved you many a dentist's fee. Narrow jaws can hold but few teeth; if the natural numbers come, some must be extracted, or else they crowd together, and decay from pressure; and there is no feature which is so markedly hereditary as narrow jaws; as the mother is, so is the offspring.

When, by the grinding machinery above described, the food has been reduced to a pulp, it is easily embraced by the tubular muscles of the pharynx and œsophagus, and passed by their steady, wavelike motion downwards to the stomach; the passage is opened before it, and its return is prevented by the closing of the tube behind as it goes onwards. The sensibility of the œsophagus is so very slight, that we do not, in ordinary circumstances, feel the morsel going down; but if it is peculiar in shape and nature, we become aware how slowly and steadily it proceeds, and ought to proceed. For during this passage, much of the sugar and soluble salts, and the watery part, is taken up by absorption, and if

---

[1] Dalton's Physiology, chap. vi, p. 109.

it is hurried by bolting mouthful after mouthful in rapid succession, scant justice is done to the victuals, and a risk of indigestion is incurred. The healthy circulation of saliva in this round is very great; we have seen lately how many pints are poured out daily in the mouth, and of this but little finds its way to the intestines, the rest being taken up in the œsophagus and stomach, principally in the former. So, the importance of not interfering with its due action can be easily understood.

Once in the stomach, the mass of food acquires a rotation from the wavelike movement kept up by the peristaltic muscles of the stomach. Their alternate contractions and relaxations press on the mass, much as the undulations of a serpent carry it over the ground—only in the latter case the undulating body is free, in the former it is fixed. The surface, being the part subjected to the moving power, moves quicker than the centre, and thus the whole contents of the stomach are rotated as a uniform mass, from left to right, and continually irrigated by the gastric juice along the depending, lower, and larger part of the sac. As it passes the opening of the pylorus at the other end, the narrowing of the sac squeezes out, with a somewhat quicker motion, such portion as is dissolved into creamy chyme, and it oozes on into the duodenum, leaving the still undissolved substances to flow along the shorter and upper curve of the stomach back to the starting-point. Thus, a slow, rotatory movement of the whole mass is sustained till all is dissolved, or, at least, so far reduced in size as to get through the pylorus.

The rotation seems to be continued in the intestines, if one may judge fairly of the action of their peristaltic fibres by the movements seen, immediately after death, in the intestines of a vigorous animal slaughtered for food. And the biliary ducts appear, also, to have a peristaltic action, rolling out gradually and regulating the flow of bile into the duodenum. Now, over all these involuntary and unfelt but constant wavelike movements, the nervous system presides; and they are, without doubt, seriously affected by all that affects the nervous system, notably by mental emotion and bodily exhaustion. The œsophagus is sometimes so paralyzed by a sudden shock during a meal, that it does not close behind the victuals swallowed, and they are thrown up by a sort of regurgitation. Even some hours after a meal, the arrest of the stomach's action by emotion may cause vomiting. And many instances are on record of a mental emotion so arresting the biliary ducts as to produce jaundice. The late Dr. Macleod, of St. George's Hospital, used to relate a scene which he saw in his own practice: a young lady with distended abdomen was charged with being privately married, and her eyes and skin had got bright yellow before he left the room. And, in the wild times, when people really did get into a passion sometimes, they are stated by bystanders to have often become jaundiced. Few of us have ever seen an adult in a rage of the old sort, so that we must not question the accuracy of the statement.

Whether absorption of the dissolved alimentary substances is affected by similar causes is not clear. I see no reason to suppose that it is, as endosmosis is such a purely physical act, that the nervous system can control it only very indirectly. But, as the supply is cut off, it cannot, of course, be active during the prevalence of such interferences.

The subject of the impediments to digestion arising out of disease of various kinds, is of too great weight to be cited merely as an illustration, and will be considered more fitly in a future part of the volume. Here is the place for a few reflections on the suitability of different articles of food.

Aliments may be called digestible when they yield readily all their nutritious particles to the fluids destined for their reduction to chyme. It is clear from what has gone before, that a comparative estimate of their digestibility cannot be made merely by reckoning, as Dr. Beaumont did, the duration of their sojourn in the stomach. That well-known physician is familiar to all, from having made a series of observations on a Canadian voyageur who had a permanent opening into his stomach, a healed gunshot wound. This was easily looked into, and the time consumed before the victuals passed out noted. But the results, which are often taken as an infallible guide by the dietician, are not truly in accordance with what general experience teaches us on this matter. For example, salt tripe or pig's trotters, by this theory, take only one third of the time demanded by roast beef, the former dish being dissolved in the stomach in one hour, and the latter requiring three, while veal took four hours and a half. Sour-crout seemed to be twice as digestible as soup made of beef and vegetables. Mutton was three quarters of an hour longer in digesting than beef, and so on. Green vegetables, again, are hardly at all touched by the gastric juice—are we therefore to conclude that they are useless?—surely, the general observation that they are very digestible by a healthy man is more in accordance with fact.

I do not lay much stress on the objection to these experiments arising out of individual peculiarity of constitution; for in fact the subject of observation was a remarkably robust man, little affected by external circumstances, and able to carry on, in spite of the hole in his side, the laborious duty of conducting timber-rafts down the American rapids.

In order to give observations like Dr. Beaumont's their full value, they should be corrected by a set of similar tests conducted upon the duodenum and ilia, for which a good opportunity has not yet occurred. Experiments on animals have been made, it is true; and they establish the fact that the intestinal juice is of great importance, and that the pancreas, if not the liver, aids in the solution of food;[1] but the alimentary canal, and the habitual diet, of the lower creatures are so different from ours, that a comparison of the behavior of different articles of food in their case, affords no practical experience for human dietetics.

Something in the way of a comparative estimate might be made by laboratory experiments on the juices removed from the body, separately and mixed. Pending advances in this direction, the following general conclusions are, perhaps, all the rules our knowledge enables us to lay down:

The degree of Cohesion has an important influence upon digestibility.—Tough articles, incapable of being ground up by the teeth, remain unused by the alimentary organs, while fluids and semifluids lead the van of digestibles. The tissues of young vegetables and young animals are, for this reason, more digestible than those of old specimens of the same class. And emasculated beasts, having softer muscles, are better suited for the table than perfect males. It is desirable also that the post-mortem rigidity, which lasts several days in some animals, should have merged in softness before the meat is cooked. But this object, usually at-

---

[1] See the plates illustrative of Dr. Dalton's experiments, Human Phys., p. 146.

tained by hanging in the larder, may be arrived at equally well by *immediate* cooking, before the rigidity has set in. In warm climates and exceptionally warm weather, the latter course is the preferable of the two.

That culinary preparation is the most efficacious, which most breaks up the natural cohesion of the viands. And it may be observed that the force of cohesion acts in all directions; and it is of no use for a viand to be laterally friable, if it remains in longitudinal strings.

Fat interposed between the component parts of food diminishes its digestibility.—It is the interstitial layer of fat between the bundles of fibres of beef that makes it less digestible than mutton, and that causes larded meats often to disagree.

Dilution favors digestibility.—Yet it may be carried too far. Even water will sometimes run through the alimentary canal, carrying on too rapidly the matters dissolved in it. And inert substances, such as woody fibre, if mixed up in too great quantity with food, dilute it so much that the central parts of the mass do not come at the digestive mucous membrane. Gelatin, probably in the same way, retards the digestion of food if too concentrated.

Too high a temperature retards digestion.—It is not merely that the gastric, and presumably the other digestive solvents, are decomposed by heat above that of the body, but the amount of secretion and the muscular acts necessary to forward solution are arrested by its local action. This applies principally to starchy foods, which are often taken hotter than at all suits the salivary glands. Meat has time to cool before it gets down to the laboratory prepared for it in the stomach.

The application of these rules to practice is not difficult; but it is obviously impossible to compare articles of diet unless they are of a definite quality; and, therefore, it will be understood that all those spoken of are supposed to be of the very best sort, and dressed in the way best adapted for securing their virtues.

Some years ago, I printed what I called a "Ladder of meat-diet for invalids,"[1] which I will repeat in a future part of this volume, when we come to consider the regimen of the sick. I

---

[1] The Indigestions, p. 101, 3d edit., Philadelphia, 1870.

there conclude with roast joint of mutton, which is the "promised land" of the convalescent; so, that dish may fairly here begin the list, which may form a sort of skeleton framework—from its actual or possible position in which may be judged roughly the comparative time which each article is likely to require for its digestion. When time and strength have to be economized, or where a full quantity is required for purposes of nutrition, it is wise to adhere to the leading names on the list. Where moderation can be calculated upon, and full leisure secured, a healthy man cannot be called imprudent for indulging his appetite for variety by descending to the very bottom. Indeed, to do so, within reasonable bounds, contributes to high health; for it is certain that an important element in making the most of food is variety. It is not enough to supply in proper amount the proximate dietetic substance; both in our own race and in domestic animals there is risk of a falling off in condition unless different substances of the same class are employed in rotation. The very strongest, perhaps, can bear uniformity without injury, but to the average man or beast it is as finally noxious as it is distasteful. Dr. Parkes suggests that the good effect of variety is probably on primary digestion, improving the appetite and so causing more food to be taken by counteracting the cloying result of sameness.[1] But I think it goes further than that, for few can fail to have noticed in their own personal experience, if once the attention be called to it, how often a most indigestible dish, when partaken of as an occasional luxury, has seemed to sit easy on the stomach, and to nourish well though its quantity has been spare. The great art is to give it time and space, and to be moderate.

Dr. Parkes makes the further very practical suggestion, that where variety of sort of food is unattainable, variety of cookery, to a certain extent, fulfils the same object.

Table of Precedence in Digestibility of some Articles of Animal Food.

Sweet-bread, and lamb's trotters.
Boiled chicken.
Venison.
Lightly boiled eggs, new toasted cheese.
Roast fowl, turkey, partridge, and pheasant.

---

[1] Parkes, Practical Hygiene, p. 186, 4th edit.

Lamb, wild duck.
Oysters, periwinkles.
Omelette (?), tripe (?).
Boiled sole, haddock, skate, trout, perch.
Tripe and chitterlings.[1]
Roast beef.
Boiled beef.
Rump steak.
Roast veal.
Boiled veal, rabbit.
Salmon, mackerel, herring, pilchard, sprat.
Hard-boiled and fried eggs.
Wood pigeon, hare.
Tame pigeon, tame duck, goose.
Fried fish.
Roast and boiled pork.
Heart, liver, lights, milt, and kidneys of ox, swine, and sheep.
Lobsters, shrimps, prawns.
Smoked, dried, salt, and pickled fish.
Crab.
Ripe old cheese.
Caviare.

The comparative digestibility of various vegetable dishes is easier to estimate than that of animal food; it is in a direct ratio to the facility with which they are reducible into a homogeneous mass, by mechanical means, from their natural form. And they are more readily digested if this reduction takes place through chewing in the mouth, rather than by mashing in the kitchen, as they in the first way become permeated more thoroughly by the saliva. If, on the score of defective teeth or other reasons, a preference is given to artificially broken-up vegetables, they should be retained in the mouth longer than is required by the mere preparation for swallowing.

The influence of the mind over digestion must not be forgotten. "Bread eaten in sorrow" remains unabsorbed, and it is not without reason that, even in the earliest times and among the most barbarous tribes, companionship during meals has always been sought. It is not only painful reflections which disturb the digestion; any concentrated thought is equally injurious, and injurious in a close proportion to the intellectual powers of the indi-

---

[1] The "tripe" as made in America would seem from Dr. Beaumont's account to be more digestible than the rich dish we prepare from it here.

vidual. The only people fit to feed alone are those fluttering butterflies whose intellects do not dispose them to concentrate their thoughts, and whose good luck exempts them from the need of trying. And even these instinctively seek society. To the brain-worker and the body-worker, cheerful distraction at mealtimes is a rule of imperious necessity, the habitual neglect of which entails chronic disease and the early failure of vital powers as a certain punishment.

The adjuncts of family meals should be studiously made as agreeable as possible. A change of clothes, clean hands, and courteous manners, should not be reserved for company, but enforced as a daily habit. If allowed to be omitted, it becomes a labor instead of a matter of course. Table decking is an elegant art, capable of exhibiting the good sense, as well as the good taste, of the artist, and highly promotive of ease of mind in the company, however small, or however familiar. If flowers are lacking, there are always leaves to be had, and I have seen toadstools with mosses and lichens so arranged as to form a centrepiece that Cellini must have praised. I do not see why we should not have music and singing at domestic meals, as well as at city feasts. All are not eating at once, and a change of performers might be kept as long as required. The cook, also, should be encouraged to make the dishes which are exposed to the eye as pleasant to look at as possible, not so much by adornment, which is apt to be vulgar, as by concealing all that is untidy and suggestive of painful idea. The forms of animals, in fact anything which makes us remember that the food has been a living animal at all, should never be conspicuously displayed, but rather covered with such vegetable garnish as is capable of harmonizing with the character of the dish.

Ease of body, as well as of mind, is requisite for complete digestion. Muscular exertion should be avoided immediately before and immediately after all substantial meals. The repose previous need not be long; a nap of forty winks, dressing, and washing, are usually enough to prepare even an exhausted pedestrian or hard rider for a good dinner. The best test of due preparation is a healthy appetite without any feeling of faintness or squeam. The rest after meals requires rather more judgment and self-control. Instinct induces us to take it, but does not tell us to avoid excess. Now, it is certain that to a healthy person excess is very possible. Sleep, for example, after dinner retards digestion, and allows the distended stomach to act injuriously on the circulation of the brain. It is proper only for very aged persons or invalids.

I have heard it argued by persons more ready at observing the facts of natural history than at reasoning upon them, or perhaps still readier at finding an excuse for laziness, that dogs and other carnivorous animals naturally betake themselves to sleep after a repast. That is true, but, then, it must be remembered, that in the wild state their chance of a meal comes but seldom; they must take the food when they can get it, and to guard against starvation overload their stomachs. The lethargy which follows is a necessity, and there is no evidence that it prolongs their lives. Those breeds of dogs which live most in the company of man, and feed on the mixed and cooked diet of their masters, usually give up the practice of sleeping after meals along with their gluttony, and to all appearance seem to suffer less frequently from indigestion than their cousins at the kennels, who require very careful treatment to preserve their health.

The best employment after a hearty meal is frivolous conversation, accompanied by such gentle sauntering movements as are encouraged by a well-ventilated drawing-room or garden. Then is the time, also, for those true games where luck and skill are so combined as to have the character of game and not of business.

I have ventured so far to go beyond what might seem the limit of a dietician's tether, because it is upon these social considerations that depends the determination of the best times of meals for healthy persons. For the heaviest repasts, those hours should be selected when we can secure to the fullest extent leisure of mind and body, and the opportunity of applying the aids mentioned above as tending to promote them. It is useless to prescribe the times for meals, or even their number, unless with a regard to the disposal of the remainder of the day, whether that is regulated by choice or necessity.

The intervals between meals, also, depend on the occupations which fill them up. Sleep retards digestion, and therefore a considerably longer period may be allowed to elapse between the last food at night and the first in the morning than is suitable during the day. Violent exercise of mind or body also retards digestion, and therefore, when this is practiced, food is not called for so soon as on a day of rest. I have often observed that, in spite of a late breakfast, a keen appetite often comes on Sundays before one o'clock to lawyers and merchants, who on working days do not care to eat till two or three, nor even then. Whereas, busy medical men whose work is more continuous, though less severe, adopt the early Sunday dinner hour only with reluctance.

There can hardly be any exceptions to the rule, that after the night's sleep, and the long fast which has emptied the digestive organs, food should be taken before any of the material business of the day is taken in hand. Work done before breakfast is more tiring, and, with due deference to certain well-meaning enthusiasts for early rising, is not done so well as after the stomach has been fortified with what it must require, if in a healthy state. The hour of rising must, therefore, regulate the hour of breakfast. It is no proof of health or vigor to forego it without inconvenience, nay rather the contrary; but it is proof of health and vigor to be able to lay in then a solid foundation for the day's labor. The natural appetite for food should be fully and completely appeased; and if there is a desire for meat, there is no reason for declining it; indeed, where mental work has to be done, it had better not be omitted. For mere bodily toil, a breakfast merely farinaceous, such as porridge, bread, milk, and butter, is most adapted, to which the usual additions *de luxe* may be made according to choice.

Not more than five hours should elapse before food is again taken. To some persons, from habitual neglect, the appetite does not arrive so soon; and then if they sit down to an unaccustomed luncheon, they feel stupefied by it, and quote this experience as an evidence that a midday meal is unwholesome for them. The stomach requires a gradual education after it has got into bad habits; so, beginning with a biscuit and a little milk, the patient should advance in quantity, till he arrives at the amount which is shown to be the proper amount by his sitting down to his subsequent dinner hungry and unexhausted. The proper amount varies much in different persons and different circumstances, and the only general description that can be applied to it is "moderate."

Instead of a light intercalary meal, some families prefer to take a substantial high dinner in the middle of the day. This seems to suit idle people and children, but if hard work has to be resumed immediately afterwards, very frequently indigestion is the result. It is the cause of that form of congestive dyspepsia to which the middle classes in Germany are extremely subject, and which drives them instinctively to eliminative mineral waters, like swallows to the South.

The best time for an adult's largest meal is when the business of the day is done, say, somewhere between five and eight. If it is taken earlier, there is time to get hungry again before bed-time; and if later, sleep comes too soon on the top of it. For light eaters the later hours, for heavy eaters earlier hours are most suitable. It may be observed that the court dinners of the city companies and the merry-makings of Greenwich ichthyophagists, where the guests meet to eat largely, are usually early.

It is superfluous for a healthy person, not influenced by any of the peculiar circumstances which will be hereafter considered, to devote special attention to considerations concerning the wholesomeness or digestibility of his dinner. He is apt to leave off this and leave off that, under the impression that they have once disagreed with him, till his bill of fare becomes most injuriously restricted. Variety in diet is of essential importance to health, and a succession of several imperfect or even unwholesome kinds of food is better than a monotonous repetition of a perfect aliment. Occasional feasts and occasional fasts constitute the natural mode of life for an intellectual and social animal. This paragraph applies to all meals, but I have inserted it apropos of dinner as being the principal.

By variety is implied, not a great number of dishes at once, which is confusing and oppressive, and destructive of the object aimed at, but a frequent (why not daily?) difference in the principal dish, to which the few other accessory dishes are harmonized. Some of the most appetizing dinners one has ever eaten have really consisted of one article, novel and unexpected. The famous Mrs. Poyser sagely remarked that a man's stomach likes to be surprised, and no surprise is possible if the same monotonous superfluity is repeated day after day.

In the course of the evening a cup of tea seems to give a fresh fillip to digestion, and supplies liquid which is required for solution of the viands. Some persons are afraid of its keeping them awake, and will find a good substitute in extempore lemonade—a cup of hot water poured on a slice of lemon, some chips of the rind, and a lump of sugar. But tea of an afternoon is by no means to be recommended. The habit of taking it began as a sort of fashionable whim about a dozen years ago, but spread so far downwards through the middle classes that medical men ought to exert themselves to stop it. If the dinner hour is so late that too long a time interposes between it and luncheon, let the latter be moved onwards, and, if necessary, breakfast also. For the dilution and washing away of the gastric secretion weakens its power of digesting the subsequent dinner, improperly blunts the appetite, and not unfrequently generates flatulence and dyspepsia. A biscuit, and an orange or an ice is a much less injurious indulgence at the same hour.

A man in health ought to be satisfied with three meals a day, and should educate his stomach to take enough at them to supply his requirements. The practice of constantly nibbling at odd times induces a flow of saliva almost continuous, like that of herbivorous animals, and neutralizes the gastric juice, so that meat is not fully digested.

The last meal should be sufficiently late for the whole not to be absorbed before retiring to rest. Going to bed hungry is liable to induce a habit of restlessness at night. If business or pleasure keep you up much longer than usual, it is better to take a light farinaceous supper, which, in this case, induces sleep. This, however, is a very different thing from sleep in a state of repletion, which (as was before observed) disturbs the circulation in the brain, producing painful dreams, unrefreshing rest, and feverishness.

An average adult may consider that he is taking enough to supply the ordinary requirements of healthy activity, if he eats in twenty-four hours the equivalents of a pound of meat and two pounds of bread. The English soldier, on home service, receives from government $\frac{3}{4}$ lb. of meat and 1 lb. of bread, and he buys about $\frac{1}{2}$ lb. additional bread and 1 lb., or so, of other vegetable food.[1] Dr. Parkes calculates that this quantity of nitrogenous ali-

---

[1] Parkes, Practical Hygiene for Medical Officers of the Army, p. 523.

ment is somewhat deficient for the maintenance of high vigor, so that I have ventured to add ¼ lb. of animal food, reducing by a few ounces the supply of vegetables.

The nearer a man approximates to this allowance the better, if he has no individual peculiarities of size, temperament, occupation, climate, or state of health to allow for. To all but exceptional cases, anything beyond this is excess—not hurtful necessarily, perhaps even beneficial as an occasional change, but still an excess. To continue such excess as a daily habit, puts a person in a position more prone to ill health than he would be naturally. To err by defect is equally injurious if persisted in; but it has this advantage, that it oftener obtains a compensation which to a considerable extent rectifies the balance. Thus, the soldier at home, above quoted in illustration, gets frequent little treats, partly out of his own and partly out of others' pockets, which fill up the corners left by the government ration. And, perhaps, the most perfect diet is one just within the limit proposed, with an occasional transgression as a change.

It is possible, certainly, for a high liver to make an equivalent change by abstinence; and if he is lucky enough to be ill and go to a doctor, he is perhaps advised to do so; but I have never yet known one to fast voluntarily as a preservative to health, though he must have often felt the beneficial effects of its being forced upon him. It is forced upon him, sometimes, as a direct consequence of the habit of overeating; a surfeit of accumulation comes on, or he brings it on by going a little further than usual, and then nausea and loss of appetite perform the good office of making a change by cutting off the supplies. Unless relieved in this way, the high liver gets torpid and stupid by day, restless by night, there is a sluggish circulation in the veins from the overloading of the blood with useless material, there are congestions and engorgements of the internal organs, and dark dirty discolorations of the skin, thick urine, flying unaccountable pains, neuralgia, rheumatism, gout, obesity.

The evils of too great restriction of food will be considered in a future chapter, when we come to the subject of poverty as a modifying circumstance in dietetics.

CHAPTER V.

NUTRITION.

WITH the passage of the liquefied alimentary substances through the mucous membrane lining the tube, digestion, strictly so-called, ends. Anything which afterwards occurs to the incoming matter may more properly be classed with nutrition.

Prominent among the needful changes in the material for building up the tissues, stands that which takes place in the liver. The biliary secretion from this organ has been spoken of in the previous chapter as taking a part, though not a leading part, in the solution of fat in water, as staying the too rapid decomposition of albuminous food and excretory matter, and as supplying some of the liquid vehicle in which the aliment is conveyed through the walls of the digestive tube. In all these the liver is subordinate, and its place may be readily supplied. But the anæmic degeneration of the blood, the gradual pining emaciation, and ulceration of the tissues, and finally death, which follow its suppression—all of which point to some deficiency of assimilative capacity and not merely to the stoppage of an excretion—show that it performs in the circle of life some special duty which cannot be spared or replaced by another gland. The blood going into the liver is altered before it comes out of the liver much further than is involved in the mere formation of bile.

During the last thirty years, the changes, necessary to life, wrought on the blood by the liver have engaged a large share of the attention of physiologists. I have been comparing the account which I printed in 1856[1] of what had been done up to that time with those of the present day, and I find a progressive advance towards the light, which seems to promise our sons not only interesting, but practical knowledge. It was then established that "a copious amount of sugar is formed in the cells, and is forwarded into the hepatic vein, but not into the bile-ducts," and "that the sugar is formed in consequence of taking food."[2] It has been

---
[1] Digestion and its Derangements, chapter vii.   [2] Ib., p. 160.

now ascertained that it is not only an immediate consequence, that the sugar does not pass from the food into the liver, but that the sugar-making in the cells of the gland continues for a long time after the alimentary canal is emptied. It appears, moreover, that the sugar, though more copious even than had been supposed, is a very transitory stage in the circle of chemical changes. It is so momentary that it can be made evident only by an artificial arrest, just as electrical action in the earth, though so vast in its results, touches our senses only through the partial derangement which shows itself in the lightning's flash.

But let it not be reckoned as of small moment because of its latency. It is an observation as old as Thucydides, that the influential forces in the world are the most active when most unnoticed; for, indeed, perturbed crises are usually the results of their disturbance. We may see in diabetes the serious evil of the further conversion of the sugar being interfered with, and of its remaining in the blood only to run off by the urine, and we can hardly doubt that the breach of a previous link in the chain, the formation of sugar, may cause equal discomfort and explain at a future time some obscure morbid states.

Those who take delight in the glimmering through the mist of future light, will find much to interest them in Dr. Dalton's sketch of this subject, and the details of the experiments repeated by himself, with ingenious precautions to avoid error, in the ninth chapter of his "Human Physiology."

That the spleen and lymphatic glands take also a part in nutrition, or blood-making, is pretty clear, but which part is entirely unknown.

The defectiveness of our information about the stages and machinery of nutrition does not, however, prevent us from having a clue to a good deal of practical knowledge of the comparative NUTRITIOUSNESS, for different purposes, of different kinds of diet.

By nutrition, two ends have to be accomplished, the growth or repair of the body, and the production of motion or force. The first indication is more or less fulfilled in proportion to the quantity of digestible nitrogenous material, the latter in proportion to the quantity of digestible carbohydrates which the aliments contain. It is very clear that the digestibility must be insisted upon, otherwise we should make serious mistakes in our valuations of food. For instance, according to the well-known table of M. Payen, a pound of chestnuts and a pound of milk contain very nearly the same quantity of nitrogen; yet, to expect that a baby would grow as well upon one as the other would be criminal folly. When, then, circumstances require us to foster growth, to increase the vigor of the muscles and nerves for short temporary exertions, to replace preternatural wear and tear, meat is valuable in the direct ratio of its solubility in the stomach. When the regular performance of a daily round of moderate exertion alone has to be provided for, carbohydrates in the form of farinaceous and oleaginous food may with advantage constitute the chief of the diet. So that before giving any general rules for the selection of a dietary which will best perform its duties, it will be needful to review the special circumstances for which it is required, which I propose to do in respect of normal conditions in the second part, and in respect of morbid conditions in the third part of this volume.

PART II.
SPECIAL DIETETICS OF HEALTH.

CHAPTER I.

REGIMEN OF INFANCY AND MOTHERHOOD.

It may be presumed no Englishman doubts that the best food for a new-born infant is a mother's milk.[1] Even deviations from the normal condition of the general system, or of the breast, should not be allowed too readily to deter a mother from suckling, till there is evidence that the secretion is disagreeing with the child. Unless diarrhœa or thrush occur, it may be taken for granted that proper nutriment is afforded, and if proper nutriment is afforded, we may be sure a woman's health is not affected by the inconveniences which she may be enduring.

I have known a woman, unable to feed herself from severe rheumatic fever, have her child held to her breast and be nourished only from thence, without any harm following. And in several instances slight inflammation of the breast has seemed to be benefited by the flow of milk induced by suckling. Indeed, I have once seen a breast with an abscess in it supply healthy nutri-

---

[1] Strange to say the opinion is not universal. Dr. Brouzet (Sur l'Éducation Médicinale des Enfants, i, p. 165) expresses a wish that the state should interfere and prevent mothers from suckling their children, lest they should communicate disease and vice! A still more determined pessimist was the famous chemist Van Helmont, who thought life is reduced to its present shortness by our instinctive infantile propensity, and proposed to substitute bread boiled in beer and honey for milk, which he calls "brute's food" (in the chapter "Infantis nutritio ad vitam longam").

ment to a child, the mother feeling sure that the abscess did not communicate with the lacteal tubes. I fancy no medical man would sanction a persistence in the latter risk, but still it were to be wished that our accoucheurs would be somewhat less hasty than they generally are in debarring mothers from suckling on slight grounds. A certain injustice is inflicted on the child, and problematical benefit conferred on the parent.

During the months of suckling it should be the object of the mother first to provide herself with an appetite, and secondly, to provide herself with proper food. The appetite often fails simply from want of fresh air, especially in those who are used to enjoy it, the remedy for which state of things is sufficiently obvious. Sometimes the disrelish for food is a symptom of the exhaustion induced by the labor, and then small doses of sal volatile or a light bitter, such as gentian, will remove it. Sometimes it is a direct gastric anæmia, arising from going too long without food. The patient should eat directly she begins to feel hungry, and not wait to feel very hungry. But at the same time she should be careful not to overload the stomach; in fact, though she eats often, she should not eat more than when in ordinary healthy exercise. A great mistake is often made by endeavoring to supply the wants of strength and appetite by an extra supply of wine or malt liquor. The nurse should never take more than she is accustomed to; if she does, it makes her eat less and digest less, though she does not feel the debility which is the consequence of the innutrition. Beer increases the quantity of the milk, just as it increases the quantity of the urine, but it also renders it thin and watery in the one case as in the other. Indeed less than she is accustomed to is the more rational rule of diet, for the happy peaceful circumstances of her situation usually exempt her from the mental wear and tear, and the exhaustions of the nervous system incidental to social life, which it is the special purpose of alcohol to compensate.

The most proper food is cow's milk, fresh and unskimmed. It can be taken at all times, in a great variety of forms, and nobody has ever been known to take too much. If it turns sour, limewater mixed with it not only corrects its acescence, but also supplies a valuable aid to the growing bones of the infant. In the solid dietary again, milk may fairly be taken as the type of the due admixture of alimentary principles, because not individual growth, not the production of force, but the secretion of that very substance is the object of the selection of diet. So that we cannot do better than take the proportions of nitrogenous, carbonaceous, and aqueous constituents of the lacteal secretion as a guide to the proportions of these principles in the diet of nursing mothers. Analyses of milk are to be found in all physiological works, and if it be reckoned roughly that in food as presented for our consumption, there is 50 per cent. of combined water, I think it will be found that the following scale of diet corresponds pretty closely to the proportion of the several constituents there enumerated.

Supposing the full diet to consist of three pounds of solid food, that will require six pints extra of uncombined aqueous fluid to make it as fluid as milk. And the three pounds of solid food should consist of

> 14½ oz. of meat,
> 13 oz. of fat, butter, sugar,
> 20 oz. of farinaceous food and vegetables,
> ½ an oz. of salt, lime, etc.

Small women and small eaters, especially if they have small children to bring up, will require less; but let the reduction be proportionate in each of the several classes of alimentary substances. And at first from the exhaustion of parturition, from the want of exercise and of fresh air, the appetite turns against meat. Let then milk, especially boiled milk with arrowroot or the like, chicken broth, egg-custards, fill up the deficiency. Only insist that enough is taken.

The observations by Dr. Barker, of New York, on this subject are so much to the point that I cannot forbear quoting them. He says, "Give the puerperal woman as good nutritious food as she has an appetite for, and can easily digest and assimilate. You will at first find many nurses who will not accept these views, and they may fail to carry out your directions in this particular; but my experience has been that after a time the intelligent ones become enthusiastic converts to this course. . . . Your patients rest and sleep better, and their functions are established with less disturbance than they would be with a spare or insufficient diet.

Since I have adopted this measure with my puerperal women, am very sure I have much less frequently met with those annoy ing and troublesome nervous phenomena that so commonly follo parturition, as the nervous system is then apt to be in a conditio of exalted susceptibility. The function of lactation is thus ger erally established without that disturbance of the system calle milk fever, formerly so common."[1]

It may be noticed that the Professor says nothing about wir and malt liquors. They are conspicuous by their absence in h dietary. And in truth the less a nursing mother takes of ther the better, so that her temper and digestion do not obviously suffe from the restriction.

The child should be put to the mother's breast as soon as sl wakes from her first sound sleep after its birth. The waiting fe three or four days is an old-fashioned relic of the days of drug ging, when it was considered wrong that the young bowels shoul be relaxed by the colostrum of the first milk, but right that the should be griped with castor-oil. Not to use the first milk wasteful and injurious. The best substitute for it is cow's mil diluted and sweetened as hereinafter described.

The education of the infant must begin immediately after birtl In the first place it has to be taught to suck, for which ever monthly nurse has her own device, and will only laugh at an male who should presume to interfere. Next it has to be taugl not to be always sucking, whenever the whim takes it or tl mother comes in sight. Regular definite times, the intervals be tween which are gradually lengthened as the child's strength an growth allow, give a rest both to the stomach of the receiver an the breast of the giver, which conduces to the due digestion of tl nourishment. As a general rule the daily allowance of milk re quired by a healthy infant is on the first day very small indeed on the second day it takes about a quarter of a pint; on the thir day two-thirds of a pint; on the fourth or fifth day it will cor sume a full pint. And this quantity augments gradually till b the sixth month you must not calculate on less than two pini

---

[1] "The Puerperal Diseases," Clinical Lectures delivered at Bellevue Ho pital, by Fordyce Barker, M.D., p. 27.

being wanted. The distribution should be in an inverse ratio to the quantity. During the first two months the child should have the breast eight or nine times daily, if the quantity yielded is small, and six or seven times if it is large. After that a gradual reduction may be begun, which before weaning should have arrived at the number of four meals daily, which is the most proper for the digestion of mixed diet.

If a mother, with or without reasonable cause, deputes her duties to a wet nurse, she ought thoroughly to understand that the expedient is not without drawbacks. All the best accoucheurs agree that in choosing a woman for the office, observation of the figure, the complexion, the color, the teeth, or even the shape and development of the breasts, and the analysis of their secretion, are all unimportant compared with a knowledge of the regularity of the catamenia. In this respect, it stands to reason we must take the applicant's own character of herself, a serious temptation to dishonesty. An unmarried woman may not improbably have a concealed constitutional taint which is communicable through the milk, and at the best is an unpleasant inmate in the family. A poor married woman, however respectable, is removed from a starving home to sudden abundance, and invariably overeats herself, and it is fortunate if she does not overdrink herself too. She pines and grows anxious about her own child if it is alive, and insists upon having her troublesome husband to see her openly or secretly, on the pretence (a fallacious one) that his visit increases the flow of milk. Moreover, a rich mother cannot but feel some compunction in purchasing for her own offspring what is stolen from another, who is sometimes seriously affected by the fraud, and retires disgusted from this false world.

At all events, a trial ought to be first made, under the superintendence of a medical man, of fresh cow's milk or goat's milk, and of Swiss condensed milk.

Cow's milk should at first be mixed with half its bulk of soft, pure, tepid water, in each pint of which has been suspended a drachm of " sugar of milk " (which is procurable at any chemist's, being used for grinding up powders), and two grains of phosphate of lime finely powdered. If the milk has been partially skimmed, as is often the case in cities, then a good tablespoonful of cream should be added to each pint, to make the mixture equal to human. If it has not been skimmed, a couple of teaspoonfuls of cream is sufficient.[1]

The advantage of using goat's milk, is that the animal can be brought up to the very nursery, even in cities, and will supply nourishment directly to its little master's lips if called upon. Children do not seem to dislike the peculiar taste.

Swiss milk has been already alluded to in the second chapter of the first part (p. 65). No inconvenience has as yet been proved to arise from its use, but, at the same time, no superiority to fresh cow's milk has been confirmed. As it is already sweetened in the preparation, no additional sugar is required, but care must be taken to dilute it sufficiently to make it resemble not ordinary milk, but milk and water.

Laputa never devised anything more preposterous than "Liebig's food for infants." It is composed of malt flour, wheaten flour, bicarbonate of potass, water, and cow's milk, and the following is recommended as the simplest mode of cooking it:

> Take of wheaten flour, ½ oz.
> Malt flour, ½ oz.
> Bicarbonate of potass, 7¼ grs.
> Water, ½ oz.
> Mix well, and add of cow's milk, 5 oz.

Warm the mixture, constantly stirred, over a very slow fire till it gets thick. Then remove the vessel from the fire, stir again for five minutes, put it back on the fire, take it off as soon as it gets thick, and finally let it boil well. Before use, strain through a muslin sieve.

The object of all this is to convert the starch of the flour into dextrin and sugar, and to elaborate a product which after all is

---

[1] In the above recipe household measurements are used as the nearest possible approximation to the following formula:

Whole cow's milk,	600 parts.
Cream,	13 "
Sugar of milk,	15 "
Phosphate of lime,	1½ "
Water,	339½ "
	1000

(Dict. Encycl. des Sciences Méd., art. "Lait," 1868.)

not nearly so like the natural sustenance as may be made with cow's milk, water, and a little additional sugar of milk. It is needless to say what a number of unnecessary risks are incurred of some one of the numerous articles employed being adulterated, of inaccurate measurement, of dirt, and of careless preparation by tired servants, who, it will be observed, have to keep the fire low for the first part of the process, and then to coal it up, making no small smoke, for the final boil. Sensible parents will be content to leave the recipe for some coming race who may prefer art to nature.

It is advisable to nourish the infant directly from the breast; and where this cannot be done, as soon as possible after the milk is drawn. Our senses tell us of a peculiar aroma given off by fresh milk which quickly exhales, and appearances seem to warrant the conclusion that this contributes to soothe the sensitive nervous system of the suckling, and so assists digestion.

The best diet for an infant during the first six months, is milk alone. It is true, man is a tough animal, and can stand with impunity much rough usage, and therefore a vigorous baby often seems none the worse for a certain quantity of farinaceous food; but the first appearance of flatulence, gripes, screaming, ill temper, or other ways infants have of complaining of dyspepsia, should make the nurse desist from these attempts to hurry on natural development.

It is only when the coming teeth are on their road to the front that the parotid glands secrete sufficient saliva to digest farinaceous food. When dribbling begins, then is the time to begin with the various preparations of these substances bountifully supplied by nature and art. Till then, anything but milk given to a healthy baby must be tentative, and considered in the light of a means of education to its future dietary, and must not take the place of milk.

Among the various means of education, I would select as most generally applicable broth or beef tea, at first pure and then thickened with a little tapioca or arrowroot. Chicken soup, made with a little cream and sugar, serves as a change. Baked flour, biscuit powder, and tops and bottoms should all have their turn; in fact, change is necessary, or the child is apt to get too fond of its soup, and to neglect the really essential nutriment of milk, and to wean itself prematurely.

The consequences of premature weaning are most disastrous, but insidious. The child continues to present the external aspect of health, its muscles are strong and elastic, but the bones do not grow in equal proportion. It is active and anxious to walk, but the limbs give way and become distorted. If it is ill enough to be taken to a medical man, he calls the condition "rickets," but there are crowds of poor creatures affected in this way whose parents refuse to see that anything is wrong till the malady has gone too far for cure. The suspicion that rickets was due to this cause has long been prevalent in the profession, but it is to M. Jules Guérin that we owe the proof derived from direct experiment. This pathologist found not only that rickety children had almost invariably been prematurely weaned, but that the disease was capable of artificial and intentional induction. He took young puppies and young pigs, specimens respectively of carnivorous and herbivorous animals, and he produced a rickety softening of the bones of each by removing them early from the mother, and giving the one set meat and the other set vegetables before the natural period. Professor Trousseau has backed the deductions of M. Guérin with his valuable *imprimatur*.

The time for weaning should be fixed, partly by the almanac, partly by the growth of the teeth. The troubles to which children are liable at this crisis are usually gastric, such as are induced by hot weather; so that in summer it should be postponed, and in winter hurried forward. The first group of teeth nine times out of ten consist of the lower central front teeth, which excite no wonder in any but very young parents by appearing any time during the sixth and seventh month. The mother may then begin to diminish the number of suckling times; and by a month she can have reduced them to twice a day, so as to be ready when the second group makes its way through the upper front gums to cut off the supply altogether. The third group, the lateral incisives and first grinders, usually after the first anniversary of birth, give notice that solid food can be chewed. But I think it is prudent to let milk, though not mother's milk, form a considerable portion of the diet till the eye-teeth are cut, which seldom occurs till the eighteenth or twentieth month. At this period even very strong children are liable to diarrhœa, convulsions, irritation of the brain, rashes, and febrile catarrhs. In these cases the resumption of complete milk diet is advisable, and sometimes a child's life has been saved by its reapplication to the breast. Now these means are the most readily feasible when the patient is accustomed to milk; indeed, if he be not so, the latter expedient is hardly possible.

CHAPTER II.

REGIMEN OF CHILDHOOD AND YOUTH.

The diet of childhood requires from its rational guardians as much attention as that of infancy. The passions at this age overpower the instinct, and reason has not yet asserted its throne. Children should have four meals a day, but meat only at one, or at most two; the latter when only a small portion at once is allowed. When in health they should have no wine or beer except as a festive treat, no coffee, strong tea, or other exciting drink. Once-cooked succulent meat without sauces or condiments, eggs, plenty of farinaceous pudding, mealy potatoes, carrots, spinach, French beans, rice, bread, fresh butter, porridge, roast apples, oranges, should form the staple of the nursery commissariat.

As to quantity, we may take the full diets of hospitals as fairly representing what should be the minimum proper for those who are not restricted by fortune in the matter of food, and the mean amount upon which a growing child can continue to flourish. For it may be observed that while all extravagance is herein avoided, the food is intended mainly for those who are convalescent from acute disease, and have not only to remain well, but to recover flesh.

The following estimates are taken from the published tables of diet. The quantities named are those which may be fairly put before each child of each article, and by selecting the larger quantities we may give a very full allowance. If the smaller amounts of one are eaten, then care should be taken that the fuller weights of others are chosen.

And let it always be understood that food is not to be dispensed with pedantic accuracy as if it were a pharmaceutical prescription. Even in hospitals considerable latitude is allowed, and still more in private nurseries should we avoid making life a toil by too much interference. It is only in cases of prominent and persistent excess in one direction or the other, that we should bring our adult reason to bear on infantile instinct.

Full Dietaries for Children at various Hospitals.

Age.	Bread.	Butter.	Milk.	Meat.	Vegetables.	Pudding.	Hospital.
Under 7	Unlimited.	1 oz.	¼ pint.	2 oz.	4 oz.	Unlimited.	St. George's.
Under 7	12 oz.	?	½ pint.	2 oz.	8 oz.	Twice a week.	London.
Under 8	5½ oz.	about ⅔ oz.	½ pint.	2 oz.	4 oz.	¼ pint.	Children's Hospital, Great Ormond Street, and Evelina Hospital.
Under 8	8 oz.	about 1 oz.	?	2 oz.	4 oz.	0	Leeds Infirmary.
Above 8	8 oz.	about 1 oz.	¾ pint.[1]	3 oz. broth ½ pnt	6 oz.	Gruel, ⅓ pint.	Children's; Great Ormond Street, and Evelina Hospital.
Under 9	6 oz.	?	1 pint.	2 oz.	6 oz.	½ pint gruel or broth.	Birmingham General Hospital.
Under 9	7 oz.	½ oz.	1 pint.	4 oz.	4 oz.	To order.	St. Bartholomew's.
Under 10	12 oz.	¾ oz.	1 pint.	2 oz.	4 oz.	6 oz.	St. Thomas's.
Under 10	6 oz.	?	1½ pint.	2 eggs.	?	8 oz.	King's College.

Extreme monotony should be avoided. It is a great inconvenience to young persons in after-life to have been brought up in such a narrow round of indubitably wholesome victuals that they cannot eat this or that. They should especially be guarded against family whims; and if the parents are conscious of prejudices against any of the ordinary foods of mankind, they should educate their descendants to take these as a matter of course. For it is astonishing how ingrained some of these acquired idiosyncrasies become, and indeed after full manhood they may be concealed but are never quite overcome. Yet few of the minor thorns in the rose-bed are so vexatious to oneself and others. I shall not soon forget the annoyance of taking a young man to a Greenwich dinner, and finding that he never ate anything which swam in the waters. Thus occasional abstinence, in the shape of no meat or the substitution of fish, and occasional festivities, consisting of food given deliberately because it is nice, are not out of place in the

---

[1] Including what is put in mashed potatoes.

nursery. Most kind fathers and mothers act on this principle, but they sometimes needlessly let the indulgence trouble their conscience.

The articles of diet should be as good and as clean as can be obtained, but no criticism should be permitted to those who sit at table. A boy or girl should be ready to eat anything which is set before them, and not refuse even badly cooked or strange meats; for in roughing it through the world, whatever position they are in, the choice often lies between that and going without.

The plan adopted at many schools of working before breakfast is not conducive to health. If it is inconvenient for the household to prepare the meal immediately the pupils are dressed, the most that should be exacted is the repetition of some light task prepared overnight; but better than that is to let them have half an hour's run in the play-ground. Violent exertion also of mind or body before and after other meals should be discouraged by a suitable arrangement of the hours of work and play. Æsthetic pursuits, drawing, dancing, singing, may be made so to combine relaxation and amusement as to leave the powers of digestion unexhausted, and may be practiced up to the time of meals.

To the full development of the digestive organs, muscular exertion in the open air is essential, and it is doubly valuable when it is of a pleasurable character. Proper exercise always involves a rational style of dress; for ill fitted and uncomfortable clothing is soon rejected by those who rejoice in the natural movement of the limbs.

It is even more necessary for girls than for boys that a sufficient playground should be attached to places of education, for they cannot be allowed to wander about the country like their brothers, and the funeral processions falsely called exercise are almost useless. In town, gymnastics or riding on horseback may be made substitutes for games; but the money required for these would be much more profitable if expended in the rent of a field or lawn.

For families who are so fortunate as to be near a river or lake there is no exercise for girls so good as rowing a light oar or sculling. It opens the chest, throws back the shoulders, straightens the back, and insures the shoulder straps of the dress not impeding movement, so that the liver and stomach gain space to act.

Many a sculpturesque figure will acknowledge her debt to her boat for her beauty; and a few weeks' instruction in swimming at Dieppe or Trouville takes away all sense of danger from the amusement.

Up to the period of puberty the daily use of wine should be allowed only during illness and by the express advice of a medical man. Its habitual consumption by healthy children hastens forward that crisis in their lives, checks growth, and so habituates them to the artificial sensation induced by alcohol that they can scarcely ever leave it off when they wish. This restriction does not exclude occasional festivities, and boys in active exercise seem to digest well a glass of well-brewed beer at dinner.

Between puberty and full growth the principal thing we have to guard young people against is overloading the stomach. Their meals should be sufficiently frequent to avoid this, otherwise the stomach from habitual distension becomes larger than is appropriate for the size of the trunk, and there is in after-life a tendency to gastric flatulence. Lads sent to learn a business in the city are often much neglected in the matter of a midday meal, and have to make up for it by gorging themselves in the evening. This spoils their breakfast next morning, and they really get starved from over-repletion. The best luncheon a growing young man can have is a dish of roast potatoes, well buttered and peppered, and a draught of milk. Or the same vegetable with a little bacon or fish may be made into a Cornish pastry, which if wrapped up in flannel will keep hot for several hours. In the summer boiled beans and bacon, or bread and cheese and lettuce, with a glass of claret or a draught of bitter beer, may take its place. But let the repast be confined to one dish, and then they will not eat too much. Red meat in the middle of the day is too heating during active life, so that if the conventional form of a sandwich is the only convenient lunch found practicable, let it be made of eggs, or fowl, or cold fish, flavored with a little salad dressing, or the like.

Youth is the time when gluttonous habits are acquired. The commencement of them is easily detected, and they should unsparingly be made as disgraceful as they really are. Ridicule is not always a wise engine to employ in education—it is too powerful—but against gluttony it may fairly be used. Let it not, however, be supposed that excess in gratifying the palate is at all a laughing matter. It is a vice just as truly as sexual excess is a vice; and there is the less excuse for its becoming an habitual vice, in the fact that the temptation to acquire it is strongest in youth, and becomes weaker as full growth is attained. That it is regarded as a serious vice by the highest authority is shown to all time by that wonderful history of the civilization of a specially favored race preserved in the commonest of books. Kibroth-Hataavah—"The Graves of the Greedy"—remained for future generations as a standing memorial of Heaven's wrath, and of the natural punishment of sins against natural law. The worldly Lord Chesterfield is equally explicit in denouncing the vice, when writing to his son at school, and though his outspoken sentences are couched in language too old-fashioned for quotation, they are well worthy of the attention of both parents and children.

The gorging themselves with pastry and sweetstuff at the confectioner's, as practiced habitually by school-boys, and often by girls when they get a chance, lays the foundation not only for indigestion in after years, which is its least evil, but also for a habit of indulgence which is a curse through life. A schoolmaster who should effectually check this without needless restriction of liberty, and make greediness unfashionable among his pupils, I would rank far above the most finished scholar in Europe. An important step towards it is to give the boys enough to eat at regular mealtimes.

Yet are asceticism and hypocrisy to be equally eschewed with gluttony. The hearty enjoyment of what is pleasant to the taste at proper times is quite consistent with, indeed usually goes along with, habitual temperance; and one of the most practical lessons knowledge of the world can teach is that all pleasure is enhanced by self-restraint.

Young people should not be brought up to the habit of taking physic. As a rule, the British mother is very fond of dabbling in doctoring, and apt to try her first experiments on her own family. If there is any definite disease discoverable, a professional man is called in, but if a child is only weakly, or troubled now and then with unimportant ailments, she tries this and that which has been recommended by her friends, without suspecting the probable truth that the cause of the imperfect condition lies in some irrational regimen pursued. She cannot make out what is the matter;

surely it would be wiser to consult some one who can, or at all events, who knows that he cannot, and will not act till he does. The consequence of frequent drugging is sometimes real illness, generally a debilitated state of the digestion, and almost always a disposition to fly to drugs for the immediate relief of petty inconveniences, which in reality impedes their cure by more far-sighted means. Boys get laughed out of this at school, but girls are seldom so fortunate, and grow up with the idea that something which calls itself physic is a necessity of human life. Now, in all the pharmacopœias there is not a single active article which, joined to its virtues, has not the vice of deranging more or less gastric digestion. It is that which makes it a medicine and not a food. Assuredly, its secondary or final effect in suitable cases, is to restore digestion, but when taken needlessly, it cannot but be injurious even to such a tough animal as a boy. The proper place for the family medicine chest is, not the bed-room or the boudoir, but the store-room, where there is some little trouble in getting at it, and where it should be locked up along with a stomach-pump, and other provisions for emergencies to be applied by skilful hands.

CHAPTER III.

COMMERCIAL LIFE.

The continually increasing numbers who devote themselves to commercial pursuits, and the still larger numbers whom they influence as dependants as they grow in importance, make the habits of the class a matter of serious social consideration.

The commercial man measures his usefulness in the world by his success in rapidly accumulating honest wealth. Honesty, therefore, being presupposed, the most conscientious is always liable to the temptation of wishing to compress two days' work into one, so as to be rich in half the time taken by his neighbors. To speak of this as a "struggle for life," is silly; of those who labor hardest in our cities, there are very few who would not acknowledge that one-tenth of their anxious toil would supply the daily needs of themselves and families. They are in reality egged on by ambitious rivalry, which uses for its purposes that insatiable hunger for hard work innate in the British breast. The haste to be rich is most unwise, and not only often defeats its own purpose by prematurely incapacitating the haster from further struggles, but if it is successful, it surely deprives middle life, or at least old age, of its occupation.

A man whose unusual exertions have made him rich rapidly, is sure to have been too much engrossed by his business to take an interest in other things. He may have kept himself, as a duty, acquainted with the pursuits and sympathies of his fellows, but he is incapable of making them the occupation of his thoughts. He is driven to look to the past only for the genuine interest of life.

Much more often the health suddenly breaks down before the desired object is attained, and the power is wanting to engage in other pursuits, to take the place of business which is perforce given up. The expenditure of strength, in the hurry to grasp wealth, has resulted only in weakness and poverty.

It was a piece of shrewd advice administered by an old merchant to a young one—"If you want to die rich, live as long as you can."

The most important rule for one engaged in any business which involves headwork or responsibility to lay down for himself is to strictly confine his business to its own times and places. Retail shopkeeping of all employments allows the greatest number of hours to be occupied in attention to its interests; for, if fairly prosperous, it does not exhaust the brain, and yet offers the gentle stimulus of movement and conversation. However, the principals of many large concerns of this kind are more heavily weighted, and if they want to enjoy health must draw a strict line between the hours devoted to money-making and those devoted to living, just as much as if they were merchants or manufacturers.

The result of a neglect of this rule, of bringing the counting-house into the dining-room and bed-room, is indigestion and sleeplessness.

The principal meals should be breakfast and dinner, breakfast before and dinner after the work of the day. But a break in the middle for luncheon is very important, indeed is imperative for all but exceptional cases. At breakfast and dinner, animal food is necessary to a hard worker; but it is not required at luncheon, and often causes heaviness and feverishness during the afternoon. Any large quantity of fat or butter also seems heating, especially if it is cooked, as in pastry. Farinaceous food, vegetables, fruit, should be the staple of the midday meal, with only so much of anything else as is wanted for a relish, the less the better.

Many commercial men give up vegetables because they find that taken at a mixed meal, along with meat, they cause flatulence. If they will separate the two classes of food, which require the digestive powers of different and somewhat opposite solvents, the saliva and the gastric juice, if they will take vegetables at one meal and meat at another, they will often find the difficulty overcome, and full quantities of both digested without fermentation.

The habitual use of stimulants in the middle of the day is to be deprecated; nevertheless where an unusual amount of cerebral exertion has been gone through, a cheerful glass of wine or beer will often prevent over-fatigue—let not, however, the demand or the supply grow a daily habit.

The daily use at dinner of a moderate amount of alcohol in some form contributes, I am sure, to the health of brain-workers. Light perfect wine is the best form, next beer, next strong wine and water, last spirituous liquors.

Commercial work can be done only in the town, and it must be confessed that town air and influences are not the most favorable to health. On this score many nowadays spend their nights at long distances from their place of business, so that no more time than is absolutely essential should be spent at a disadvantage. The success attendant upon this plan of residence in the country is closely proportioned to the earliness of the time at which business can be left. Unless an hour or two can be given to relaxation in the purer air before dinner, I do not think the labor of rushing backwards and forwards is compensated for. It is pleasant, doubtless, to see the junior branches of the family flourishing among green fields, but not when the bloom is gained by the exhaustion of the bread-winner's strength. Those who can afford it, will do better to fix their permanent residence near their work, and live temporarily in the country for a few months during the long days.

Besides the reason mentioned above another may be given for the long hours borne by retailers, namely, that their shops are better ventilated and lighted than most of even the wealthiest merchants' counting-houses. To pass from the magnificent dwelling of his wife and daughters to the dull stuffy den of many a prince of commerce, recalls the image of Samson grinding in the dark through the treason of his money-loving spouse. Things were not so bad when the family lived over the offices, and a softening female influence civilized the whole house. But now work and life seem to be seeking a divorce from one another, and the place of business is growing more and more gruesome, and, like another ill-omened locality, is not to be alluded to in polite society. A ladies' mission for the improvement of these dwellings is urgently called for. Unlike other missions, it could dispense with promoters, secretaries, speeches, committees, subscriptions, and collectors. Or rather, all these agencies united could embark in the family conveyance, or even in a one-horse fly, and begin operations at once with a builder and decorator as assessor. The scale of expenditure should be proportioned to that of the other home; it will probably add very little to the yearly bills, nothing in comparison to silk gowns and spring bonnets.

The healthiest exercise for a commercial man is riding when it is possible. It diverts the thoughts, especially if the nag be skittish, prevents the stagnation of the abdominal bloodvessels, and promotes a due flow of bile. But the outside of an omnibus is better than nothing at all, and is within the means of every one. Much walking is usually found too fatiguing, and if adopted as a duty, is apt to be monotonous. Boating and cricket are suitable for the younger members of the commercial world, but they occupy more time than can often be spared, and have to be kept for holidays. The more violent athletic sports are open to still more objection, and if it is attempted to pursue them at the same time that the thoughts are occupied in business, they exhaust the vital powers, and weakness is the result.

CHAPTER IV.

LITERARY AND PROFESSIONAL LIFE.

THAT dogmatic expression of Büchner's, "No thinking without phosphorus,"[1] has gained an unhappy notoriety. Strictly taken it is a groundless assumption, for it is impossible for us to have any evidence that intellectual being may not exist joined to any form of matter, or quite independent of matter at all. We certainly do not know enough of the subject to lay down a negative statement. And if it be held to mean that the amount of phosphorus passing through the nervous system bears a proportion to the intensity of thought, it is simply a misstatement. A captive lion, tiger, or leopard, or hare, who can have wonderfully little to think about, assimilates and parts with a greater quantity of phosphorus than a professor of chemistry working hard in his laboratory; while a beaver, who always seems to be contriving something, excretes so little phosphorus, at least in his urine, that chemical analysis cannot detect it.[2] All that the physiologist is justified in stating is that for the mind to energize in a living body, that body must be kept living up to a certain standard, and that for this continuous renewal of life a supply of phosphatic salts is required. The same may be said with equal justice of water, fat, nitrogen, chloride of sodium, oxygen, etc. The phosphates are wanted indeed, but wanted by pinches, whilst water must be pouring in by pailfuls. One might go on thinking for weeks without phosphates, but without water a few days, and without oxygen a few minutes, would terminate the train of self-consciousness. The practical points taught us by physiology are that for the integrity of thought the integrity of the nervous system is requisite; and for the integrity of the nervous system a due quantity of such food as contains digestible phosphatic salts.

---

[1] Ohne Phosphor kein Gedanke, Kraft und Stoff, sec. 122.
[2] See the analyses of the several kinds of urine in Simon's Chemistry, vol. ii, pp. 144, 342, and 350.

For the intellectual direction of the nervous system it is at the same time essential that it should not be oppressed by physical and mechanical difficulties. The presence in the stomach or blood of imperfectly assimilated nutriment impedes its functions in close proportion to their amount; so that not only the chemical constituents but the mode of administering food must come into the calculation.

The most perfect regimen for the healthy exercise of thought is such as would be advised for a growing boy, frequent small supplies of easily soluble mixed food, so as to supply the greatest quantity of nutriment without overloading the stomach or running the risk of generating morbid half-assimilated products.

The physiology of the action of alcohol has a very practical bearing on the physical regimen of the mental functions. Alcohol has the power of curbing, arresting, and suspending all the phenomena connected with the nervous system. We feel its influence on our thoughts as soon as on any other part of the man. Sometimes it brings them more completely under our command, controls and steadies them; sometimes it confuses or disconnects them; then breaks off our power and the action of the senses altogether. The first effect is desirable, the others to be avoided. When a man has tired himself with intellectual exertion, a moderate quantity of alcohol taken with food acts as an anæsthetic, stays the wear of the system which is going on, and allows the nervous force to be diverted to the due digestion of the meal. But it must be followed by rest from mental labor, and is in fact a part of the same regimen which enforces rest—it is an artificial rest. To continue to labor and at the same time to take the anæsthetic is an inconsistency. It merely blunts the painful feeing of weariness, and prevents it from acting as a warning. I very much doubt the quickening or brightening of the wits which bacchanalian poets have conventionally attributed to alcohol. An abstainer in a party of even moderate topers finds their jokes dull and their anecdotes pointless, and his principal amusement consists in his observation of their curious bluntness to the absurdity of their merriment.

There is no more fatal habit to a literary man than that of using alcohol as a stimulant between meals. The vital powers go on getting worn out more and more without their cry for help.

being perceived, and in the end break down suddenly and ofte irrevocably. The temptation is greater perhaps to a literary ma than to any other in the same social position, especially if he ha been induced by avarice or ambition to work wastefully against time; and if he cannot resist it he had better abjure the use o alcohol altogether.

As to quantity, the appetite for solid food is the best guide.] a better dinner or supper is eaten for a certain amount of fermente liquor accompanying it, that is the amount most suitable. If worse, then it may be concluded that an excess is committed, how ever small the cup may be.

Although nothing can take the place of alcohol in this article o diet, yet fermented drinks are not suited to the nervous system a all in proportion to their alcoholic contents. The fruity ether and aromas evolved in the process of fermentation, and which d not seem capable of existing in a digestible form without alcoho are even more powerful in repairing the waste of the nerve powe Burgundy has acquired a special fame as food for the brain, an claret runs it hard, while good, sound, unadulterated beer is homely creature little inferior to them. All of these are superio to sherry and port, and to spirits, for reasons given in a forme chapter of this volume on the choice of food.

Mental activity certainly renders the brain less capable o bearing an amount of alcohol which in seasons of rest and relaxa tion does not injuriously affect it. When any extraordinary toi is temporarily imposed, extreme temperance or even total absti nence should be the rule. Much to the point is the experience o Byron's Sardanapalus:

"The goblet I reserve for hours of ease,
I war on water."

The posture of the body usually adopted by literary and man; professional persons engaged in writing is a matter worth consider ation. Chamber counsel are notoriously subject to piles an venous congestion of the rectum; women who sit much with thei work in front of them get also congestion and irregularity of th uterine organs; cold feet from sluggish arterial circulation ar frequently complained of by otherwise hearty sedentary workers The ill health which these symptoms indicate may often be pre vented by the use of a high desk at which the work may be done standing for a time now and then; and a further change of posture may be obtained by an easy chair which will allow of thinking with the body thrown back and by occasional walks about the room.

Athletic sports are scarcely consistent with steady, hard brainwork. Probably only the most muscular try to persist in them, and they acknowledge that their intellects are readiest and strongest when they are taking quite moderate exercise, and not when their muscles are corky and their limbs light. There is a peculiar state of health into which those are apt to fall who, having for a long period kept themselves in training for boat-racing or other muscular exertion, afterwards adopt a life which involves mental labor and responsibility, even though they get a fair amount of bodily relaxation. The leading symptoms are emaciation, weariness, depression of spirits, and an unnaturally high specific gravity in the urine, which is, however, abundant and full-colored, thus showing an excess of destructive assimilation which cannot but be very injurious.

Fresh air and relaxation of mind are of more importance than exercise, which last is indeed mainly valuable as securing them. The limits of weariness should not be transgressed. The attempt to compensate for excessive literary toil by excessive bodily toil is based on a false conception of the relations of matter and spirit, worthier of an ancient Gnostic than of a modern philosopher, which has more than once led to fatal results. I had for some years as a patient a literary lady who wrote much and well in magazines. She would go straight from her study to her garden and glebe, dig furiously and mow with a scythe, despising or rather luxuriating in fatigue. Gradually paralysis came on, showing itself first as "writer's cramp," and then creeping over the whole body. The mind and senses were as perfect as ever, and so long as she was able to move the tongue she dictated lively monthly articles, and at last died apparently of sorrow at being unable to communicate her thoughts.

Tobacco should not be indulged in during working hours. Whatever physiological effect it has is sedative, and so obstructs mental operations. But as a relaxation afterwards it is in moderation beneficial. As a calmative before retiring to rest it has the sanction of a vigorous brain laborer, John Milton, whose supper, we are told, consisted of bread, water, olives, and a pipe of tobacco. There is a flavor about the fare of the happy days he had passed with an elegant literary circle in Italy.

The daily habits of Robert Southey, a man who more than any other made literature a healthy profession and a successful profession, are thus described by his son in his "Life," vol. iii, 2, and vi, 6: "Breakfast was at nine, after a little reading, dinner at four, tea at six, supper at half-past nine, and the intervals filled up with reading or writing; except that he regularly walked between two and four, and took a short sleep before tea, the outline of his day when he was in full work will have been given. After supper, when the business of the day seemed to be over, though he generally took a book, he remained with his family, and was open to enter into conversation, to amuse and to be amused."

"My actions," he writes about this time to a friend, "are as regular as those of St. Dunstan's quarter-boys. Three pages of history after breakfast (equivalent to five in small quarto printing); then to transcribe and copy for the press, or to make my selections and biographies, or what else suits my humor, till dinner-time; from dinner till tea I read, write letters, see the newspaper, and very often indulge in a siesta;—for sleep agrees with me, and I have a good substantial theory to prove that it must; for if a man who walks much requires to sit down and rest himself, so does the brain, if it be the part most worked, require its repose. Well, after tea I go to poetry, and correct and re-write and copy till I am tired, and then turn to anything else till supper; and this is my life—which, if it be not a merry one, is yet as happy as heart could wish."

And a very rational mode of living it was, deserving of its reward. The country air and quiet among the lakes and mountains, the association with kindred and loving spirits, the old-fashioned dinner-hour excluding uncongenial society, the regular exercise, and the sound night's rest, with temperance, soberness, and chastity, preserved his mental powers fresh and vigorous in old age, to leave to future generations undying memorials of sympathy with all that is best in humanity.

Milton describes himself as "with useful and generous labors preserving the body's health and hardiness, to render lightsome, clear, and not lumpish obedience to the mind, to the cause of religion and our country's liberty, when it shall require firm hearts in sound bodies to stand and cover their stations."

His blindness probably interfered with the activity of his muscular discipline in later years, for he was a martyr to gout towards the end of his life.

Samuel Johnson is another type of the literary man pure and simple. Scrofulous, awkward, hypochondriacal, and corpulent, he was averse naturally to bodily exertion, yet he walked a good deal, and worked steadily and patiently without bursts of industry or idleness. Passionately fond of company and of eating and drinking, he restrained himself, and indulged only when the labor of the day was over. His knowledge of physiology and medicine kept him from quackery, and his medical advisers were the most rational physicians of the day. After middle life, his own observation of his health led him to abstain entirely from wine; I have heard my grandmother describe the air of dignified patience with which he passed the bottle which she often pressed upon him at her father's table. He sat up late at night indeed, yet that was not for work, but to rest the mind with sportive and varied conversation. He had his reward in the retention of his mind, even when its material organ had broken down.

Shelley was a vegetarian, and, perhaps, his peculiar way of living, combined with the fact of not writing for a livelihood or to please others, estranged his sympathies from human kind. But at all events, his temperance did not weaken his exuberance of thought and diction. What would have happened had he consumed more phosphorus, it is impossible to say; but he could hardly have been a more rapid composer or stronger wielder of words.

Walter Scott passed a genial sociable existence, took much exercise, dissuaded his younger friends from substituting gig-driving for riding, and always insisted on having seven or eight hours of utter unconsciousness in bed. Had he passed his whole life in his study, he would have written probably worse and certainly less, for he would have had a shorter life to write in.

It is true that Byron assumes in his poetry the character of a débauché, and says he wrote "Don Juan" under the inspiration of gin and water. But much of that sort of talk is merely for stage effect, and we see how industrious he was, and read of his training vigorously to reduce corpulence, and of his being such an exceptionally experienced swimmer as to rival Leander in crossing the Hellespont.

It is especially when the mind of genius is overshadowed by the dark cloud of threatened insanity, of hypochondriasis, or of hysteria, that a rational regimen preserves it to the glory of God and the advantage of man. Nothing but daily exercise, temperate meals, and a punctual observance of regular hours of study and rest, could have kept burning the flickering candle of reason in poor suicidal Cowper. Most rarely and faintly do his writings exhibit a trace of the gloom which made life to him, as he described it in his last words, "unutterable misery."

On the other hand, the keen poison of his own genius slew in youth Kirke White, when he surrendered himself to its exclusive cultivation:

> That eagle's fate and his were one,
> Who on the shaft that made him die
> Beheld a feather of his own,
> Wherewith he wont to soar so high.

The elegant appreciater of nature, the author of "The Seasons," faded away from lazy and self-indulgent habits. The great all-loving soul of Burns produced so little because it was drenched in drink and idleness, not excessive indeed, but sufficient to ruin his usefulness.

Apropos of this last matter, we may give to some people the same caution which Swift administers in a letter to Pope: "The least transgression of yours, if it be only two bits and a sup more than your stint, is a great debauch, for which you will certainly pay more than those sots who are carried dead drunk to bed." The machinery of sensitive souls is as delicate as it is valuable, and cannot bear the rough usage which coarse customs inflict upon it. It is broken to pieces by blows which common natures laugh at. The literary man, with his highly cultivated, tightly strung sensations, is often more susceptible of the noxious and less susceptible of the beneficial results of alcohol and other indulgences than others. His mind is easier to cloud, and there is a deeper responsibility in clouding it.

Equally when we descend into the lower regions of Parnassus, the abodes of talent and cleverness and the supply of periodical literary requirements, we find the due care of the body absolutely essential to the continued usefulness of the intellect. The first things to which one entering the profession of literature must make up his mind, are to be respectable and healthy.

What noble fragments one finds in Savage and in Poe! and how sad to know that they are fragments instead of stately structures, solely because the builders had not the wisdom to live regular lives!

CHAPTER V.

NOXIOUS TRADES.

The digestive organs are liable to suffer from the position assumed at work by certain handicraftsmen, and the discomfort hence arising leads to the adoption of an unwholesome dietary, which in the end intensifies the evil.

Shoemakers contract a peculiar sort of gastralgia, partly from the pressure of the last against the epigastrium, and partly from the constriction of the abdominal viscera, especially the stomach, by leaning so far forward to work. The use of the upright bench, in which the last is held firm by a stirrup, and an erect posture always preserved, is the best remedy for the evil, and the thanks of the country are due to Mr. Sparkes Hall for his advocacy of this method of getting over a difficulty as old as the Pharaohs at least.

Against the constipation and hæmorrhoids which the same posture induces the best preservative is the free use of fresh butter, a cold tub every morning, and an occasional dandelion pill.

The discomfort which they cause in a sensitive condition of the stomach causes vegetables to be avoided by many shoemakers. They can hardly bear to take sufficient to sustain health. So long as this is the case, they should eat as many oranges and lemons as they can, or in default of them, fresh rhubarb, and try the plan proposed before of eating vegetables only at one meal and meat alone at another. A small quantity of watercresses is also a great resource.

The same observations apply with nearly equal force to tailors, but unfortunately they have not the refuge of the upright bench to fly to, and to a certain extent also to sewing-machine workers, in whose case I would suggest that a simple contrivance by which the legs of the instrument could be shortened or lengthened would enable the changes of posture necessary to health to be made.

However, it is indubitable that a great deal of the ill-health of all classes of artisans arises from the closeness of their workrooms, and a more philanthropic deed cannot be done for a deserving class than the bringing under the notice of the district health-officers instances of the violation of the law by masters.

Gardeners often are afflicted with water-brash, arising in a measure from the stooping posture deranging the viscera which receive the food, especially the lower end of the gullet. But I think that an additional cause is frequently the bad cooking of the vegetables they eat. Half-boiled potatoes and cabbage are as injurious as ill-prepared oatmeal is found among populations which are nourished on that diet.

The poisoning to which those who work in lead are exposed by their occupation may be almost always prevented by scrupulous cleanliness in taking food. There is abundant proof that the metal enters the system through the stomach, and there is but doubtful evidence of its entering by any other path. From dirty hands it gets into the bread, from dusty clothes it besprinkles the meat and drink, and thus acts as quickly and surely as if it had been brought in by the more usual way of the drinking-water. It is the most certain and noxious if by any peculiarity in the manufacture it is converted into a chloride salt, but the form which we generally meet with is the white carbonate, insoluble indeed in water, but unfortunately soluble in the fluids of the digestive canal, saturated as they are with carbonic acid. In the case of painters who employ white lead, it is quickly deprived of some of its noxiousness by mixture with oil, for in that condition it can only get into the food from the hands. But where the finely powdered or precipitated Kremnitz lead is employed, as for example in the manufacture of polished cards, the clothes become saturated with the dust and convey it to the victuals. Not only should the hands be washed, the hair brushed, and the outer garment shifted, before meals, but the men should not be allowed to bring their food within the poison-laden atmosphere of the workshops.

Plumbers are said to inhale lead in the fumes which arise in the process of casting, and "brass-founders' ague" also appears from the researches of Dr. Greenhow to be caused by the fumes of solder, consisting mainly of oxide of zinc, being drawn into the lungs. But it must be remembered that in both these handicrafts a great deal of dirt adheres to the skin and clothes and may thus pass into the food. I have never seen clean men affected.

Some handicrafts are noxious from the high temperature at which they are obliged to be carried on. In these cases, the frequent and free use of cold drinking-water is sanctioned by experience as the best preservative of health; the copious evaporation from the skin keeping down the heat of the blood. And the most cruel enemy to health is alcohol, which induces degeneration of the liver, heart, or kidneys, or all these at once, and prevents at any rate the due exercise of the lungs' functions, even if it does not directly disorganize that tissue. My own impression is that the emphysema and black deposit so often found in the lungs of artisans exposed to great heat is in no small measure due to alcohol and to the neglect of its antagonist water.

For example, the consolidation and subsequent breaking down of the lung peculiar to dry grinders is seldom if ever found in temperate men; a healthy mucous membrane has the power of rejecting the foreign particles of metal which adhere to it when congested and degenerated.

Colliers, who labor in the dark in a confined hot atmosphere deficient in oxygen, suffer from bloodlessness and indigestion. The bonesetters, the popular practitioners among this class, describe it as "a little bone broke" in the stomach, pummel the abdomen and make the patient give up work and drink for a season with successful result. Philanthropic coal-owners should arrange the shifts, so that a man may be put in turn on to night work and have his share of sunlight. And it is better for the men not to eat in the pits, but to make their principal meals when off work. The diet of colliers is generally too nitrogenous for a life of daily muscular labor.

Tea-tasters sometimes suffer from a special kind of nervous affection. The hand gets tremulous, there is sleeplessness, headache, anæmia, indigestion, with a flabby tongue covered with a smooth yellow coat. To avoid this, they should live well, and always take some food before exercising their office. Smelling the tea seems to be more injurious, and really less decisive, than sipping the infusion.

Evils consequent on other trades are not mentioned here, either because they are unconnected with diet, and not to be avoided by any special arrangement of diet, or else resolve themselves simply into temptations to intemperance.

CHAPTER VI.

ATHLETIC TRAINING.

ARE the dramatist and the novelist drawing from nature when they present us a picture of a well-born and well-bred athlete, stupid, immoral, selfish, case-hardened by his brute strength against the finer emotions of pity and honor, and blind to intellectual pleasures? If the original exists, he is happily very rare. He is certainly not conspicuous in the list of 294 rowers in University races collected by Dr. Morgan, which on the other hand is adorned with bishops, poets, public school-masters, leading barristers, devoted clergymen, elegant orators, scientific chemists, philanthropists, and other ornaments of the human race.[1] Eminent muscular ability evidently is not inconsistent with a superiority to the average in other respects, and the improvement of the body does not prevent the improvement of the mind.

A charge more serious, because more troublesome to answer against athletics, is that they lay a foundation for disease in after years, and thus shorten life. Likely enough the spectators know that the dropping down dead on the stage of an athlete, apparently in the height of healthy vigor, is a gross misrepresentation of nature. But yet the scene rankles in their memories, and they can with difficulty divest themselves of the feeling that the exuberant energy of a man in training wears out the vital forces, and is repaid by weakness which will cut short the days. We may know that the impression exists by the frequency with which the friends of patients assign athletics as the cause of all sorts of diseases, without any other reason than that the failure in health was first made manifest during some bodily exertion. Of course it is during bodily exertion that the discovery is made: no one finds out that his legs are weak till he tries to walk, or that his lungs or heart are injured till his wind fails him at a pinch. But that a man

---

[1] University Oars, by John E. Morgan, M.D.

previously in good health injures his constitution by training, so as to be more liable than ordinary persons to any peculiar class of disease or degeneration, is negatived by the laborious investigations of Dr. Morgan. He has followed up with personal inquiries the 294 "university oars" mentioned above, and he finds, as was to be expected, that since 1829, when his list begins, some have died, some have been killed, some have fallen into ill health, but 238 survive to describe themselves as hearty and strong. Of the deaths (39 in all) 11 were from fevers, 7 from consumption, 6 from accidents, 3 from heart disease, and lesser numbers from other special causes. Now it is heart disease which especially is attributed to athletic sports, and it is a surprise to find statistics showing that their patrons have suffered from it rather less than the rest of the population, and much less than the sailors whom we are so solicitous to keep in good health.[1] Deaths from fevers certainly cannot be considered as evidence of an injured constitution; indeed Dr. Graves of Dublin (a high authority in this matter) remarks, and the experience of most of us will bear him out, that when zymotic diseases attack strong men the risk is greater than is run by weaker frames. The end of 2 by drowning, and 3 by gunshot wounds, show the possession of energy and unselfish courage, seldom the characteristics of a broken invalid. The cases of the 17 who do not furnish a good account of their health are mostly somewhat vague. Among so many, several must have hereditary tendencies to disease; others say their medical attendants trace no connection between their complaints and previous muscular exertion, and in such a long period as forty years innumerable evil influences must have been in action; while in some families it seems traditional always to speak of their health as only moderate, and in others to look back upon the exuberances of their youth as follies. So that 17 is in fact a small number to be occasionally falling into the hands of the physician.

The best test of the value of anything is to reduce it to Arabic numerals, and pounds, shillings, and pence, as insurance offices act by our constitutions. Dr. Morgan has applied this test to the

---

[1] Mortality from heart disease in Registrar-Gen. Reports for 10 years 8 per cent.; in navy (1868), 13 per cent.; among university oars, $6\frac{1}{2}$ per cent.—*University Oars*, p. 28.

294 cases under consideration. According to Dr. Farr's Life Tables the expectation of life at 20, the average age of university oarsmen, is 40 years. But the survivors have still an expectation of life of 14 years before them, and this must be added on, while a calculated allowance must be made for those who have died, and an estimate also deducted for the 17 lives who reckon themselves damaged. The whole calculation is too long to be gone into here, but the result is decidedly favorable; for, taking the experience as it stands, the expectation of life of each individual comes out, not 40, but 42 years. So that any insurance office which had taken them all at ordinary rates would be making a handsome profit and exhibit a good prospective balance sheet.

The conclusion is inevitable that for young men in good health very severe athletic training strengthens the constitution and lengthens life.

It will of course strike every one that our example here is taken from a specially select class of humanity. True, the fame of the University would not be intrusted to one likely to break down and disappoint his colleagues. And herein lies a great advantage possessed by boat-racing above other athletic sports, namely, that it is to the interest of all concerned to exclude from the practice those who are likely to be injured by it. For that some are likely to be injured is never denied; and it probably would be wise if the crews, instead of acting solely on their own responsibility, were to insist on all who joined them having their fitness to undergo training tested by a medical man. Mr. Maclaren says he would not allow any one to pull in a college boat whose chest measured less than 36 inches, but it is evident that such an absolute rule must be fallacious, for the circumference of the chest must bear in a well-built man a proportion to the height. The better test is the vital capacity or aerial contents of the lungs, which Dr. Hutchinson's spirometer and tables enable us to measure so accurately.

Other forms of athletics have not the same safeguard. But still the good sense and good feeling of Englishmen is such, that a man very quickly finds out, or is told by his comrades, if anything renders the ambition of distinction in bodily exercise unsuitable for him. Where there is any suspicion of this being the case, parental authority may fairly be interposed, and the matter settled in a single medical examination.

It is not necessary that every one who trains should aim at being, or even wish to be, a distinguished athlete. There are modified forms or rather degrees of the same process, which cannot be trusted indeed to produce the extraordinary development of nerve-force needful for successful boat-racing and the like, but which nevertheless bring the body into a state of high health very conducive to comfort and usefulness.

The reading or other intellectual pursuits during training should be very moderate and (so to speak) mechanical. Hard head-work should not be carried on at the same time as hard training. It should be gradually given up at the beginning, and resumed gradually after the training has been gone through. But there is no reason why the systematic cultivation of the mind and body should not alternate to their joint advantage, and indeed it evidently has done so in the case of many of Dr. Morgan's heroes, whose names make a conspicuous appearance in the class lists of classical and mathematical honors.[1]

The usual time allotted to training is six weeks. The objects to be attained in this period may be described as:

1. The removal of superfluous fat and water.
2. The increase of contractile power in the muscles.
3. Increased endurance.
4. "Wind," that is, a power of breathing and circulating the blood steadily in spite of exertion.

The first object is aimed at by considerably adding to the daily amount of nitrogenous, and diminishing farinaceous and liquid food, and providing that it should be so consumed as to be fully digested. The second and third are secured by gradually increasing the demands made upon the muscles till they have learnt to exert at will all the powers of which they are capable, and for as long a period as the natural structure of the individual permits. Wind is improved by choosing as part of the training an exercise, such as running, which can be sustained only when the respiratory and circulating organs do their duty fairly.

---

[1] The 294 include, at Oxford 6 firsts and 11 seconds in classics, 1 first and 2 seconds in mathematics; at Cambridge 10 firsts and 5 seconds in classics, 8 wranglers, and 21 senior optimes in mathematics.

The muscles of the limbs become under a regimen of this kind more "corky" or elastic, and more prominent when "put up" in a state of contraction. They improve in quality and efficiency, but that they become larger is extremely problematical. Hypertrophied organs are well known to lose their shape and power; an enlarged heart, instead of circulating blood better, is an incumbrance; the muscles of the hollow viscera, when augmented in thickness, do not expel their contents freely; an hypertrophied finger, instead of being stronger than the rest, is weaker; and all these are extremely liable to degenerate prematurely and lose their vitality. So that if the muscles did by training grow bigger, as reckoned in a state of repose, it were a result not at all to be desired.

The skin becomes soft and smooth, and apparently more translucent, so that the red bloom of youth shines through it more brilliantly. The insensible perspiration is regular and even; while at the same time sweating is not so readily induced by bodily exertion, and it is never cold and sudden, even with mental excitement.

Superfluous fat is removed from all parts of the person, as is evinced by loss of weight. This requires to be carefully tested by the scales from time to time; for if the reduction be carried beyond a certain point, which varies in different men, a loss of power and of endurance is felt, and probably future evil results may arise. This point is technically called the "fighting weight," but the observation of it need not be confined to the pugilistic trade.

Training increases wonderfully the vital capacity of the chest, so that a much greater quantity of air can be blown in and out of the lungs and with greater force than previously. And this vital capacity endures longer than any other of the improvements, for I have found in examination for insurance several clients, formerly in training, but who had laid aside violent exercise for some years, still retaining that mark of vigor to a considerable extent above the average. It is evidence of the permanent elasticity of the pulmonary tissue, an efficient protection against asthma, emphysema, and other degenerations of the organ of breathing.

Indigestion, acidity of stomach, sleeplessness, weariness of life, nervous indecision, dyspeptic palpitations, and irregularity of the bowels disappear under training. But if they exist, the regimen should be entered upon with more than usual caution and under medical advice.

The following were the systems pursued at the Universities in 1866 as given by Mr. Maclaren in the Appendix to his "Training, in Theory and Practice," and I believe still carried on for boating-men:

THE OXFORD SYSTEM.

Summer Races.

A DAY'S TRAINING.

Rise about 7 A.M.		So as to be in chapel, but early rising not compulsory.
Exercise	A short walk or run	Not compulsory.
Breakfast, 8.30	Meat, beef, or mutton	Underdone.
	Bread, or toast dry	The crust only recommended.
	Tea	As little as possible recommended.
Exercise (forenoon)	None.	
Dinner, 2 P.M.	Meat, much the same as for breakfast.	
	Bread	Crust only recommended.
	Vegetables (none allowed)	A rule, however, not always adhered to.
	Beer, one pint.	
Exercise	About five o'clock, start for the river, and row twice over the course,[1] "the speed increasing with the strength of the crew."	
Supper, 8.30 or 9	Meat, cold. Bread; perhaps a jelly or watercresses.	
	Beer, one pint.	
Bed about 10.		

Summary.

Sleep....................About nine hours.
Exercise...............Walking and rowing about one hour.
Diet........................Very limited.

---

[1] The length of the course is nearly a mile and one-eighth.

ATHLETIC TRAINING.

Winter Races.
A Day's Training.

Rise about 7.30 A.M.		Early rising not compulsory.
Exercise		Not compulsory.
Breakfast, 9	As for summer races.	
Exercise (forenoon)	None.	
Luncheon about 1 P.M.	Bread or a sandwich. Beer, half a pint.	
Exercise	About two o'clock start for the river, and row twice over the course	Crews are taken over the long course to Nuneham, perhaps twice during their practice.
Dinner at five, in Hall	Meat as for summer races. Bread. Vegetables as for summer races. Pudding (rice), a jelly. Beer, half a pint.	
Bed, 10.30.		

N.B.—It is particularly impressed on men in training that as little liquid as possible is to be drunk, water being strictly forbidden.

Summary.

Sleep.........} As for summer races.
Exercise....}
Diet...............Nearly the same as for summer races; luncheon being about equivalent to supper.

THE CAMBRIDGE SYSTEM.
Summer Races.
A Day's Training.

Rise at 7 A.M.		
Exercise	Run 100 or 200 yards "as fast as possible"	The old system of running a mile or so before breakfast is fast going out, except in the case of men who want to get a good deal of flesh off.
Breakfast, 8.30	Meat, beef, or mutton. Toast dry. Tea, two cups, or towards the end of training a cup and a half only. Watercresses occasionally.	Underdone.
Exercise (forenoon)	None.	
Dinner about 2 P.M.	Meat, beef or mutton. Bread. Vegetables, potatoes, greens.... Beer, one pint. *Dessert.*—Oranges or biscuits, or figs; wine, two glasses.	Some colleges have baked apples, or jellies, or rice puddings.
Exercise	About 5.30 start for the river, and row to the starting-post and back.	
Supper about 8.30 or 9	Meat, cold. Bread. Vegetables—lettuce or watercresses. Beer, one pint.	
Bed at 10.		

Summary.

Sleep...Nine hours.
Exercise....................................About an hour and a quarter.[1]
Diet...Limited.

N.B.—On Sundays men generally take a long walk of five or six miles.

---

[1] The course is a trifle longer than at Oxford, and there is a pull of 1¼ mile to get to it.

Winter Races.

A Day's Training.

Rise about 7 A.M.
Exercise, ... As for summer races.
Breakfast, 8.30, " "
Exercise (forenoon), . . None.
Luncheon about 1 P.M., . A little cold meat.
    Bread.
    Beer, half a pint, or biscuit with a glass of sherry; perhaps the yolk of an egg in the sherry.
Exercise, ... About 2 o'clock start for the river and row over the course and back.
Dinner about 5 or 6, . . As for summer races.
Bed about 10.

Summary.

Sleep, . . }
Exercise, . . } Same as for summer races.
Diet, . . Nearly the same as for summer races, luncheon being about equal to supper.

There is nothing very terrible in the discipline here enforced, while some latitude is permitted to peculiarities and a wish for variety, and plenty of time is left for business and social intercourse. Other plans are objectionable from involving, without any resulting advantage that I can see, a complete *bouleversement* of the usual times and seasons adopted by the upper and middle classes in this country. For example, in Clasper's method dinner is to be at 12 o'clock, with nothing more than a very light tea afterwards and no supper. Then a country walk of four or five miles is to be taken before breakfast, and a couple of hours' row after, and another hard row between dinner and tea.[1] "Stonehenge" again requires the time between breakfast and dinner to be spent entirely in billiards, skittles, quoits, rowing, and running, in spite of another hour's row being prescribed at 6 P.M. He also requires the aspirant to athletic honors to sleep between ten and eleven hours.[2] Only professionals are likely to carry out such rules. The most doubtful point which a physiological critic

---

[1] Quoted by Mr. Maclaren from Rowing Almanac for 1863.
[2] Article "Bout-racing" in British Rural Sports, 1861.

would lay his finger upon is the exaggerated abstinence from fluids recommended in the Oxford scale. The use of water to the extent of the thirst felt by the individual promotes the vital renewal of the skin, kidneys, and digestive viscera, and cannot be injurious. But it should not be very cold, or swallowed in great quantities at once on a full stomach, or after extraordinary exertion, lest it should lower too much the bodily temperature. If the mouth be first rinsed out, and the draught imbibed calmly and deliberately, it quenches thirst much better than when rudely gulped, and is not likely to be taken in excess.

It is probably not necessary in the present day to enforce sufficient tubbing to keep the skin clean and fresh. It is in fact more necessary to deprecate excess in the use of cold water. If a bath is taken between exercise and a meal, the chill should at least be taken off, for there has been a considerable loss of temperature by perspiration, and more cannot be afforded. The use of a cold bath is to contract the cutaneous arteries, and by throwing the blood back suddenly on the heart and lungs to stimulate them to increased reaction, so that the living stream should flow vigorously to the extremities. If the skin is already pale and cool, as after exertion, it shows that they are already contracted and rather demand relaxation. The time for a cold bath is when the skin is full-colored, dry, and warm.

Nothing is said here of the training of jockeys and others whose object is to reduce their weight to its extreme minimum irrespectively of augmenting the strength, as that cannot be recommended on the score of attaining high health, nor is it likely to be voluntarily undertaken by healthy persons.

The university scheme may fairly be accepted as a typical regimen for fully developing a young man's corporeal powers to fulfil the demands of an extraordinary exertion. It is a standard which we may modify according to the circumstances for which the training is required.

Thus, for instance, in training for the moors or for the thorough enjoyment of partridge shooting, the reduction of fat should not be carried so far, as steady endurance for many days together is required, and a treasure of adipose tissue as a basis of molecular growth must be retained. Butter may be allowed, milk in the tea, and eggs as a change for the lean meat at breakfast. For men who have got into middle life running is needlessly troublesome, and quick walking may be substituted both for that and for the rapid rowing. Nevertheless the times should be observed strictly, and the amount of the walking may be raised gradually up to that wanted for the day's sport. It will be necessary, however, to allot to exercise a considerable longer time than is allotted in the college training scheme. For it must be remembered that quick rowing for an hour is a violent exertion and takes more out of a man, and practices the wind better than four hours' walking. The chest may be expanded by the employment of light dumb-bells or clubs (if heavy they strain the muscles). And the healthy action of the skin should be promoted by friction with rough towels and horse-hair gloves. The time of the training should not be so long as six weeks, for in point of fact it is carried on by the exercise of the sport, and if such an extended period as used for boat-racing is adopted, there is a risk of overdoing the discipline. A fortnight is quite enough for the purpose.

To those of our countrymen and women who have not the opportunity or inclination for spending their holiday in what is commonly called "sport," the fashion of mountaineering is a great boon. And even sportsmen, during the dead season when there is nothing to be killed, experience a compensation in finding something to go up. But a great deal of the advantage of the relaxation is often lost by not being already in training. At least the first week is wasted in getting into condition, and is a period of as much pain as pleasure. This may be obviated by a gradual adoption of the diet and discipline, modified as above, for a week or ten days before starting.

The pain in the back and sides which hunting men often experience at the commencement of the season, arises usually from imperfect expansion of the lungs in ordinary breathing. The muscles of the trunk are strained by the effort of expiration during exercise. The inconvenience may be prevented by a partial training. The diet should be drier than usual, and all sweets and pastry left off, the chest expanded by dumb-bells and running, and the habit acquired of keeping the lungs as full of air as possible. Women, being weaker-muscled than men, often feel this to the extent of giving up altogether the healthy amusement of riding. The simple adoption of modified training gets over the difficulty. The dumb-bells should be used in private before putting on the stays, and particular attention paid to the injunction of thoroughly inflating the lungs.

It used to be the custom before the commencement of a course of training to be bled, purged, and sweated. I do not think it of any service, and it induces constipation of the bowels, besides being weakening. Some take Turkish baths during athletic training, but they appear to derange the daily discipline, unless they are taken every day, which would be a decided excess.

Ladies who are going to try training for athletic purposes, will find some attention to costume expedient. If stays are worn (and there is no objection to them if well-fitted and not too tight) they should have no shoulder-straps. The drawers should not be tied below the knee. The best defences to the lower extremities in rough ground are stout Alpine shoes, and light leathern gaiters half-way up to the knee supporting the long socks without garters. A light woollen jersey should be worn next the skin. The skirt of the dress should be short and narrow, and the best materials are serge and homespun. Besides these the less drapery is worn the better.

Training is sometimes carried too far—the men describe themselves as "fallen to pieces." The most peculiar symptom is an occasional attack of sudden loss of power, after exertion. It is sometimes called "fainting," but there is no loss of sense, as in that state, and is quickly relieved by liquid food. It is pathologically an acute and temporary form of that consequence of overstrained muscle which constitutes "writer's," "turner's," and "blacksmith's" palsy. The obvious remedy is to leave off training.

The exercise and excitement combined of practicing for boat-racing will sometimes induce recurrent palpitations of the heart. A physician should immediately be consulted as to whether this arises from an organic cause; if it does not, rest and a dose or two of purgative medicine should be taken before a resumption of training; and it will be well to add a moderate quantity of port wine to the dietary. If the palpitations still return, there is no help for it but to give in, and acknowledge that nature has not cut out every one to the pattern of an athlete.

The unusual strain on the skin sometimes induces boils. The best preventive is to anoint the skin with a little sweet oil after the morning bath. If a spot gets tender and red, threatening a boil, touch it lightly every day with nitrate of silver, and give bark and chlorate of potash twice a day in the usual doses. (*Decoct. Cinch.*, fl. ℥j; *Pot. Chlor.*, gr. xv.)

A modification of training of considerable importance to notice is that which contemplates the reduction of superfluous plumpness, either for the sake of the appearance or the general comfort of the sufferer. There may be a question whether the health is benefited by it, unless the previous diet or habits had been irrational and improper.

Corpulence usually prevents exercise being taken to a sufficient extent for confidence to be placed in it as an efficient part of the treatment, and therefore the diet becomes a more essential feature. If an exhausting amount of bodily exertion be persisted in, the digestion of meat is interfered with, while at the same time the absorption of such fat as unavoidably exists in the food still goes on, so that the muscles and nerves lose strength while the adipose tissue grows. Besides this, if by violent means the weight is worked down, those violent means must be continuously sustained to keep it down; and if they are neglected in consequence of more absorbing occupations, the inconvenience rapidly increases to a greater degree than ever. Many uncomfortably stout persons are very active in mind and body, and really could not add to their muscular discipline without risk of injury.

The following may be taken as a modification of the training regimen suitable for the reduction of corpulence.

Day's regimen for a three weeks' course.

Rise at 7. Rub the body well with horse-hair gloves, have a cold bath, take a short turn in the open air. Breakfast (alone) at 8 or 8.30, on the lean of beef or mutton, cutting off the fat and skin, dry toast, or biscuit, or oat-cake, a tumbler of claret and water or tea without milk or sugar, or made in the Russian way with a slice of lemon. Luncheon at 1 on bread or biscuit, Dutch cheese, salad, watercresses, or roasted apples (without sugar or cream), hung beef, or anchovies, or red herring or olives, and such-like relishes. Drink, after eating, claret and water, or unsweetened lemonade, or plain water, in moderation. Dinner at any convenient hour. Take no soup, fish, or pastry, but plain meat, of any kind except pork, rejecting the fat and skin. Spinach, French beans, or any other green vegetable may be taken, but no potatoes, made dishes, or pastry. A jelly or a lemon-water ice or a roast apple must suffice for sweets and dessert. Claret and water at dinner, and one glass of sherry or Madeira afterwards.

Between each meal exercise, as a rule, in the open air, to the extent of inducing perspiration, must be taken. Running, when practicable, is the best form in which to take it.

The number of hours alotted to bed in the University schemes is too much for the purpose now proposed. Seven is quite enough, and if the person under training wants to retire before 12 o'clock, he ought to be astir before the time mentioned above. There are few things more weakening than remaining in bed, or even in a bedroom which has been closed during the night, when thoroughly woke up. During sleep little air is required; we all know the slow shallow breathing of a sleeper, by which the respiratory muscles are rested. Beasts get enough oxygen in their narrow dens, and man in his fusty garret. But once awake, both expand their lungs fully, instinctively demand fresh air, and suffer from the want of it more at that hour than at any other time during the day. If a Sybarite must indulge in the horizontal position, let him at all events open his window and take his tub before he does so.

If good Turkish baths are accessible, four or five may be used in place of exercise between meals at intervals during the reductive training. And thorough shampooing by an experienced hand should not be omitted.

The weight is to be accurately recorded at the commencement and every four days, so that its loss may not be too rapid or excessive. Six or seven pounds is usually as much as it is prudent to lose during the fortnight.

A more important sign of improvement is increased vital capacity of the lungs as measured by the spirometer.

After the fortnight's course the severe parts of the discipline may be gradually omitted, but it is strongly recommended to modify the general habits in accordance with the principle of taking as small a quantity as possible of fat and sugar and of the substances which form fat and sugar, and sustaining the respiratory function. Fat meat, rich milk, butter, malt liquors, pastry, starchy foods (such as potatoes, puddings), sweet vegetables (such as parsnips and beet-root), sweet wines (such as champagne) should be taken only most sparingly. An appetite should be acquired for lean meat, especially for beef, mutton, and venison, for game and poultry, for plain boiled fish, for poor new cheese, for green vegetables and salads, summer fruits, oranges, lemons and pomegranates, almonds (fried and sprinkled with salt and cayenne), roast apples, olives, lemonade, buttermilk, claret, and hock. Aerated bread, captain's biscuits, and dried toast, all in moderation, are the most appropriate form of flour.

Excessive stoutness amounts to a disease; it is a true hypertrophy of the adipose tissue, and it is not capable of removal by the means mentioned above, though in cases where it has been augmented by a previously inconsiderate diet, it may be considerably reduced. The subject will be resumed when the dietetics of disease are under consideration.

CHAPTER VII.

HINTS FOR HEALTHY TRAVELLERS.

A MODERATE course of training is a good preparation previous to travelling for business or pleasure, or for active military service; but it is well in these cases not to let the dietary become habitually too limited or careful. It is convenient to be able to eat without repugnance any food capable of supplying nutriment, even though dirty, ill-cooked, or of strange nature. There is often a choice only between that and going without.

When actually on a carriage or railway journey it is unwise to make large meals. They are sure to be swallowed in a hurried manner, and in a state of heat and excitement very unfavorable to digestion. The best way is to make no meal at all till the journey is over, but to carry a supply of cold provisions, bread, eggs, chickens, game, sandwiches, Cornish pasties, almonds, oranges, captain's biscuits, water, and sound red wine, or cold tea, sufficient to stay the appetites of the party, and let a small quantity be taken every two or three hours.

If this plan be adopted, not only is activity of mind and body preserved, but that heat and swelling of the legs which so often concludes a long day's journey is avoided. Attention to the matter is particularly necessary when the journey continues all night, and for several days in succession, since varicose veins and permanent thickening of the ankles have sometimes resulted from this exertion being combined with too long fasts and hurried repletion at protracted intervals.

The less stimulant a traveller consumes before he arrives at his sleeping-place the better. Then the habitual allowance is of advantage. If a good wine is made in the country he is passing through, he will probably prefer to fare the same as his hosts; if not, Bordeaux and Burgundy are the best vintages when procurable, and Marsala in Italy.

In France and Germany very good local beer is to be obtained, but landlords seem to object to its being publicly used as a beverage. We ought to insist on our rights as tourists on this point. In apple districts, cider is usually placed on the table gratis, and makes a good substitute for doubtful water.

The water is very apt to disagree with tourists, especially in volcanic, basaltic, mountainous, and marshy districts. A pocket filter is a great protection, and boiling the water makes all organic matters harmless, and gets rid of the greater part of the lime. But neither of these expedients removes the neutral and alkaline salts, which will sometimes act as a purgative.

In almost all country places out of England it is impossible to avoid the greasy dishes which are apparently preferred by all except our own countrymen. And a frequent consequence is rancid indigestion, with nauseous taste in the mouth, and flatulence or diarrhœa. A few drops of vinegar or lemon-juice, and a little cayenne pepper in the plate are the readiest correctives.

Another article of cuisine that offends the bowels, if not the palate, of Britons, is garlic. Not uncommonly in southern climes an egg with the shell on is the only procurable animal food without garlic in it. Flatulence and looseness are the frequent results. Bouilli, with its accompaniments of mustard sauce and watermelon, is the safest resource, and not an unpleasant one, after a little education.

By special favor potatoes can usually be obtained boiled with their jackets on (*en chemise*), but unless asked for are seldom produced.

Raw ham, which some persons seem to find a luxury, will be avoided by all sane travellers who have heard of the frequency with which it is infested with live measle-worm and trichina spiralis.

It is a great convenience to be able to eat olive oil, which is much wholesomer than doubtful butter in warm climates; and those who care for the future comfort of their sons and daughters will accustom them to the taste in youth, instead of encouraging a daintiness in this particular, as I have seen done. Repugnance to the flavor of goat's milk ought to be got over by those who ever intend to frequent lands where the pasture will not support cows. A preference also for boiled milk, or milk that has been boiled, is a safe fancy to indulge, where you are not acquainted with the yielder of the liquid, especially when typhoid fever is rife. Irish peasants scarcely ever drink it raw.

On the Continent the household bread is usually unwholesome and nasty, and captain's biscuits are never to be obtained. It is prudent to carry a store of them for use whenever the staff of life is especially abominable. This does not apply to Spain, where delicious white, firm, fine-grained bread can be procured in places where it is the only thing eatable by a dainty person.

A small tin of the usual tea employed at home is well worth the space it occupies.

"Liebig" is procurable in almost every civilized town, and a small store may be laid in when rough cookery is expected.

A knowledge of simple methods of preparing food is often a great comfort to a traveller. A friend of mine was once considerably nonplussed in Norway, after he had bargained for some lamb, by having the animal handed over to him bleating, with a request that he would return the skin in the evening. The task was accomplished under difficulties, but the details are unpleasant. This is an extreme contingency, which need not be provided against by all vacation barristers acquiring the art of butchering; but still it is worth while to learn in your own kitchen how to prepare an omelet, fry fish, eggs and ham, cut and grill a steak off a joint, boil and fry potatoes, scrabble eggs, mull wine (if it happens to be sour), boil coffee, make "Liebig" into good soup, etc. These accomplishments may be brought into play without wandering very far from home; and it is astonishing how popular they render those sometimes troublesome fellow-travellers, the ladies of the party.

Travellers, otherwise strong, are apt to get diarrhœa occasionally, partly from the unaccustomed diet, partly from the water, but very frequently also from the pestiferous state of the provisions for daily retirement in Continental inns. It is worth knowing that in many places, especially in France, the landlady has a small private establishment of her own, quite unobjectionable, of which she will lend the key to favored guests, especially Britons. In country places gentlemen will do well to worship Cloacina *sub Jove*. For this sort of mild dysentery will keep recurring again and again, easier induced by having occurred before; and not unfrequently it will leave traces of imperfect digestion in the bowels for weeks after returning home.

As a provision against accidental diarrhœa it is wise to be prepared with some chalk and opium powders (*Pulvis Cretæ aromaticus cum Opio, Pharm. Brit.*) made up in 20-grain packets, in thin gutta percha or oil silk, to keep them dry. In northerly latitudes half a packet, containing ¼ grain of opium, can be taken after each relaxation. But in warm countries a more efficient, at least a more permanently efficient remedy is to be found in lemon-juice. The patient should lie down flat, and keep sipping a mixture of half and half lemon-juice and water, or simply sucking lemons, till the symptoms have ceased, which will soon be the case. The nausea and narcotism induced by opium are thus avoided, and there is no danger in taking an excess of the fruit. It is a good thing to get accustomed to the acidity of the flavor, for there is nothing so wholesome and convenient as a drink.

Travellers in countries where the atmosphere is very dry, as in the vicinity of the Mediterranean, sometimes lose their appetite for breakfast from want of sleep. This inconvenience may be overcome by soaking a sheet or some towels in water and spreading them out on the floor of the bed-room, so as to diffuse moisture through the air breathed during sleep.

Long days' rides, especially in the heat, are liable to bring on an inert or semi-paralyzed condition of the stomach, so that if a full meal be taken immediately it remains undigested, and is frequently thrown up again. This may be prevented by a rest and a hot bath between getting out of the saddle and sitting down to table. If these cannot be had, it will be best to eat something very light, such as soup, eggs, bread, and in small quantities, and to make up the deficiency next day at breakfast and luncheon, which should be always the solidest meals in journeys of this sort.

Boils are sometimes very troublesome to equestrians. A small piece of nitrate of silver ought to be carried in the baggage, and on the first tenderness, redness, and hardness of the skin, the part should be damped and the caustic crossed twice over it. The object is not to make the cuticle rise in a blister, but to contract and render insensitive the cutis. This will usually cause the boil to die away.

Pedestrians will do well to make a good breakfast before starting, however early the hour may be. If tea or coffee are not relished on account of the time being so unusual, beef tea or soup will be found an excellent substitute. If prepared over night, they are easily warmed up in the morning.

Advantage should be seized of every day of rest to feed well, and fatten up as much as possible. This does not put the body out of training, but in fact keeps it in a condition fit for continuous exertion.

Beer, wine, and spirits should be avoided altogether during the day's work, but water, cold tea, or lemonade may be drunk according to thirst. An occasional pipe of tobacco seems to palliate better than anything else that dryness of mouth which constitutes false thirst. This false thirst naturally arises during exercise in a rarefied air, but in mountainous places it is often aggravated very much by eating snow or ice. Spring-water, though scarcely over the freezing-point, does not seem to have the same unpleasant effect.

Sea-voyages have a powerful curative effect on some invalids, but they do not generally bring healthy persons into very good condition. If it is calm, landsmen overeat themselves, take too little exercise, sleep badly, and get their bowels constipated. If it is rough, they suffer from sea-sickness and the increased badness of ventilation below. The remedies for these things, so far as they are remediable, are obvious.

Short sea-voyages do nobody any good, and a few people a great deal of harm. They are an inevitable evil for all islanders who wish to enlarge their ideas. Sea-sickness may, however, be considerably palliated by rational preparation for it. In the first place care should be taken to finish all preliminary arrangements as long before starting as you can, so that a day or two may be given to rest and a temperance somewhat more than usual. If the eyes or skin are dingy and yellow, take a purge of aloes or taraxacum. Go on board in good time, so as to secure a comfortable post. If it is evidently going to be rough, go below and lie down immediately. If you remain on deck, be very warmly clothed, and especially let no chill affect the abdomen or back. If the stomach feels empty, and still more, if any dry retching occurs, take bottled porter and biscuit spread with a little butter and cayenne pepper—which last article, by the way, amply repays the space it will occupy in a traveller's pocket throughout a journey, so useful is it on all occasions. Nutritious food should be taken when practicable, but loading the stomach with trash brings on sickness; though truly enough it facilitates the process of vomiting, and prevents the regurgitation of bile, which is always peculiarly painful after dry retching.

If the voyage is by night, and sufficiently long to make a night's rest of, say seven or eight hours at least, it is worth while to swallow a full dose of chloral on embarking, and to sleep through one's troubles. But if you have to wake up in two or three hours to disembark, you feel ill all the next day, if not longer.

Ice-bags, and all other charms for sea-sickness, have turned out mere trade puffs.

CHAPTER VIII.

EFFECTS OF CLIMATE.

The race of man exhibits great powers of resistance to external influences, and is able to occupy a length and breadth of the earth's surface such as is attained by no other animal or even plant. This arises not from any innate bodily strength, but from his being able to accommodate himself by the aid of reason to circumstances. Thus experience has led to the adoption of very different dietaries in different regions. An Esquimaux would find much difficulty in growing rice near his home, so he wisely dines on such meat as he can get or on whale-bubbler. A Bengalee could not obtain a supply of flesh food without immense labor, and finds rice grown easily, so he lives almost entirely on the latter. The curiosities of food afford examples of the boldness of man in not being deterred by their repulsiveness to his senses from converting assimilable substances to his use, enough to make the simple reader shudder; but I do not know that the philosopher gains much knowledge from such recitals. Man learns to swallow, *bon gré, mal gré,* whatever contains aliment, and the art of living lies in the learning so to eat it as that it shall serve his turn. Climate influences diet mainly by the supply it affords.

In most warm countries there is an abundance of starchy and sugary food, and but little animal. How shall this existent provision be made most available for the prolongation of life? Let us refer back to the principle on which were reckoned in the first part of this volume the requirements of the body for its daily work (p. 21), and draw the obvious inferences therefrom. The diet is, in hot countries, perforce one that entails the loading the digestive organs with a great excess of carbon in order that enough nitrogen may be obtained. In the first place, therefore, the carbon should not be in too rapidly digestible a form. Starch and vegetable fibre, as supplied by grain and green food, are better than oleaginous matter in warm climates; for while the former only overloads the intestinal canal, the latter overloads the blood and tissues with useless and deleterious products. Then, it is essential that no frequent calls should be made for unusual exertion: the muscles and nerves must not be worn out, for the materials of their repair are few. Moreover, the supply of food must be continuously copious and accessible; for starvation is badly borne by him who is hanging on to life "by the skin of his teeth." No sudden changes must be made in the dietary, even in the form of the vegetable food; for a new article is with difficulty digested by an unhabituated stomach, though it should be perhaps more ordinarily digestible than the usual nutriment. This is not the case with meat-eaters, who can bear change much easier from one kind of flesh-food to another.

The English reader's interest in his fellow-subjects will naturally suggest to him the important bearing which these considerations have on the duties of both government and individuals towards the inhabitants of our Indian possessions. Our first business is to keep them alive at whatever cost to ourselves; and the next to render them as little dependent as possible on the accidents of drought, flood, and other unfavorable contingencies of season, partly by storing grain, and (what is infinitely more important) water, the means of producing grain, from one season to another. Tanks, artificial lakes, irrigation works, and roads, stamped with the latest improvements of modern science, will preserve the memory of our rule when the bronze statues of our leaders are as unintelligible as the Memnon. Who cares for, or knows of, the martial exploits of the Pharaohs? Yet their successful efforts for regulating the food supply of Egypt preserve fresh, forever, our grateful remembrance of them. The heaven of nations is in the hearts of men.

Again, we must not expect to get work out of vegetable-feeders in return for our bounty. If required to exert themselves in any unusual way, when food is deficient, they simply die. The reason is evident: they have been living on their own tissues, and the small quantity of albuminous matter in grain is a long time in building them up again; so that for weeks or even months their muscles are in a state of atrophy. A broken watch must be repaired before you call upon it to go.

Also, any variations made in the nature of their food must be very gradual and well considered.

Still, there is no impossibility in the gradual introduction of changes, at least in the preparation of food. Some method might be popularized of augmenting the proportion of nitrogenous matter in the dietary by mechanically reducing the carbon, such as produces in Italy the highly nutritious macaroni. Starch is readily washed out of the grain, and is itself a valuable article of commerce for industrial purposes, as well as being capable of conversion into more digestible substances for use as food. Also the separation of starch and the storing it in a form less liable to decomposition and the ravages of insects than ordinary grain, would be a great source of safety to a graminivorous people.

I had a striking illustration of the different values of vegetable and animal food a few years ago in the case of a robust Hindoo gentleman, who habitually lived on rice and vetches, which he imported himself from Bombay, and had cooked by a servant of the same faith as himself, so that his meal should not be defiled by the touch or even the look of a Christian. The said servant went holiday-making to Greenwich, got drunk and into the lockup, so that his master had an involuntary fast of nearly two days. And then he was so weakened that the labor of opening his letters brought on hiccough, vomiting, and extreme depression, so that he could not take food when, at last, he obtained it. The mention of beef tea was an abomination to him; he said, his ancestors had not put in their mouths animal food for 6000 years, and he was not going to begin. But when the abominable substance was craftily introduced to the other extremity of the digestive canal, it seemed to flow directly into his veins, which filled with blood, and he was well. The absorbents had clearly not lost their natural habits by disuse for so many generations.

Where the circumstances of a country are such that plants suitable for food cannot be grown, while there is a sufficient supply of animals to nourish the population, the inhabitants are hardy, enduring extreme cold and heat, and capable of violent physical exertion. But steady daily labor wears them out, and is abhorrent to their feelings. We may instance, as under several varieties of temperature, the Esquimaux, the Indians and half-breds of the Pampas, the Tartar hordes, and the Arabs of the Nubian

Desert. These nations of meat-eating hunters and herdsmen are mightily strong and prolific, and have fulfilled to them the promise made to the sons of the wild Sheik Jonadab, of never lacking "a man to stand before the Lord forever." But that is only so long as they follow their ancestral traditions, and retain habits suited to none but sparsely inhabited lands. When the inevitable tide of civilization overtakes them, and they become cultivators and craftsmen, they fall under the natural laws of population and take their chances with the rest of mankind. Often, indeed, the day goes against them in the fight of innovation, especially if it has been sudden; they fade away childless under our very eyes, like that vast American tribe of which, it is said, the only remnants are a chief, a tomahawk, and six gallons of whisky. The only possible remedy for this terrible state of things is beef. Our progress is progressively poisoning off our weaker brethren; we are no more to blame for it than we are for crushing the harmless beetles and daisies that lie in our path. Still, we are bound in mercy to tide them over the struggle, to let them assimilate gradually with the more civilized world. Hunters should have facilities afforded them of becoming herdsmen, and, in course of generations, from herdsmen, farmers, and gardeners; for the immediate transition from a purely animal to a principally vegetable diet, though borne by the individual, is fatal to the race.

The action of climate on diet seems to be affected by the food produce which it enables the soil to bring forth. The fixed inhabitants grow accustomed to it, and, according to the law of the survival of the fittest for the peculiar circumstances, are prosperous and prolific as a race, and healthy as individuals, while following their ancestral habits. But that historic fact does not at all show either that the diet is the best abstractedly, still less that it is the best suited to foreigners. It seems absurd to argue that the inhabitant of the Polar Circle lives on fat animal food because it is so cold, and, at the same time, that the burning plains of the Pampas are a reason for thriving on flesh and water only; or that the climatic circumstances of North Norway and Southern Spain are the cause of the inhabitants living almost entirely on bread. The best diet in the abstract is a mixed diet, and mixed in the proportions selected by the experience of the most civilized nations. And it is also the best for the individual who is accustomed to it to adhere to, under whatever sky he may be wandering. The higher the health he enjoys, the more nearly he approaches to the true aim of being in training, the better he is able to resist the adverse circumstances he may be subjected to. Experience does not justify an agreement with those dieticians who desire us to alter our commissariat in accordance with the example of those among whom we dwell for a season, or in obedience to the thermometer, and M. Cyr is indubitably wrong when he blames Britons for "retaining their customary substantial regimen under other skies and in hot countries."[1]

In India and in Africa our soldiers suffer from fevers, ague, dysentery, and are liable to contract cholera and other epidemics. But the camps of our foes are usually still more severely ravaged at the same time; and it is observed that those suffer least who continue the habits of sensible men at home. Inflammations and degenerations of the liver also afflict our countrymen in the East, and a certain proportion of this evil is due to intemperance, as it is in England; but the great majority of the cases are traceable to the consequences of malarious fever. It does not appear that those who make a rational use of alcohol, as they would have done at home, suffer more than the abstinent. The principal thing to be remembered is that as the outgoings of water by skin and lungs are very great, the ingoings must be great also, and therefore that the fermented drinks must be taken in a dilute form, otherwise, thirst will cause an excess of stimulant to be consumed. This applies equally to the warm summers in dry temperate climates, such as Italy, as well as to tropical regions.

The object of attention to diet in unaccustomed climates should be to accommodate to the demands of the system the food which can be obtained, and to which we are obliged by necessity to restrict ourselves. If starchy food is to be got, solely or mainly, a great deal of it must be eaten, and the digestion of this unusual mass is facilitated by being taken alone and not mixed with meat; and the meat, when it comes to hand, should form a separate meal. Thus the full force of salivary digestion is brought into play. Involuntary vegetarians are apt to starve themselves from want of inclination towards the flavorless viands. They should be warned

---

[1] Traité de l'Alimentation, p. 221.

of the danger of this. If nothing but animal food is within reach, again, still more is it imperative to eat largely, if the body is to be preserved in its integrity, as has been argued in a previous chapter (p. 21). Sir John Ross found the Esquimaux devouring about twenty pounds a day of meat and blubber. And his experience among his own men leads him to urge the desirability of acquiring, previously to a contemplated winter residence in Polar regions, a taste for Greenland food, the large consumption of it being the true secret of life in those frozen countries. "The quantity of food," he says, "should be increased, be that as inconvenient as it may."[1] Again, Sir Francis Head, in his famous journey across the burning plains of the Pampas, where beef and water were the only victuals to be had, got himself into magnificent condition, not by dint of the limited slices of civilized society, but by eating flesh, morning, noon, and night, liberally.[2] Under both circumstances the addition of a small quantity of vegetable food would have rendered needless the excess of nitrogenous aliment.

When the Englishman is in foreign countries it is more necessary than at home to pay that attention to diet which will insure the highest attainable health and condition. For to his constitution, at any rate, if not absolutely, every place is less healthy than England. Plagues of all sorts, terrestrial and celestial, beset his path, and he must walk warily if he would return sound. Perhaps, at home he may have lived carelessly, and been lucky enough not to suffer, but he cannot hope for the same good fortune under less favorable auspices. This caution is not required by the sensible readers of these pages, but it may be useful in its application to their less wise dependents and clients, who, in countries where one is always thirsty and there is abundance of drink, are as apt to yield to temptation as in England. The punishment of stupidity is surer and heavier than they are led to anticipate by former experience.

Exercise and clothing should be accommodated to the food. We should not in these particulars copy the manners of natives any more than we do their dietary. Active muscularity and field

---

[1] Ross, Second Voyage for the Discovery of the Northwest Passage, p. 413.
[2] Head's Journey across the Pampas (1828).

sports render the body less likely to suffer from the solar and malarious influences to which they, to a certain extent, expose those who pursue them; and the simple precautions of keeping the skin dry and warm after exertion, and of taking small preventive doses of quinine, will make these healthy pleasures nearly as safe in the tropics as in Europe, times and places of extraordinary risk being avoided.

In the selection of fit persons to undergo, with safety to themselves and others, exposure to extremes of either cold or heat, the surest guide is their power of gaining weight and condition under a course of training. These are not always persons of the biggest muscles and bones; indeed, a moderately sized frame is the toughest as a rule. Sheer pluck will sometimes enable a most unfit subject to pass undetected through tests of endurance; and doubtless such a temper is valuable in a colleague; but it will not supply the place of hardihood. The surest proof of hardihood is improvement under training.

As the women desiring to undergo bodily hardship are more exceptional than men, so is it all the more necessary to test them thoroughly. Their desire is almost always the self-sacrifice of love; but they do not wish to burden others with bitter memories or to injure the object they profess to aid, as happens if they break down. The world has less direct claim on their assistance, and therefore they should not offer it without being sure that it is really worth having. This specially applies to the wives of missionaries, travellers in new countries, emissaries to barbarous nations, and the like. They are of incalculable service while sound, but a serious impediment when sick. Their enduring courage may be taken for granted, as it is proved by their volunteering. But unless they grow in strength and weight, or at least preserve their weight, under a course of training, their place is home.

The standard chapter in dietetic treatises on the due influence of the seasons on the selection of food in temperate climates does not exhibit any practical contributions of science. We hardly require to be told to indulge more in weak potations in July than in December, or to eat a better dinner when our appetite is braced up by a frost. Some of their refinements remind one of the dandies in Imperial Rome who wore heavy finger-rings in winter and light in summer, and are beneath the notice of a healthy man.

Some are positively repugnant to experience, as, for example, the recommendation to keep out the cold by eating sugary and starchy dishes, on the ground that they are producers of heat by combustion. The loss of heat by evaporation makes it quite as necessary to sustain the temperature of the body in summer as in winter, and the same amount of force has to be elicited; so that, as a matter of fact, meals of bread and pastry and sweet fruits are more seasonable in warm weather than in cold, for they cause less feverishness and excitement than meat does. The succession in their due season of marketable articles affords a sufficient guide to their selection. As an almost universal rule they are wholesomest when cheapest, if the simple directions given already for securing their soundness and freedom from adulteration be adhered to.

Sir James Clark[1] remarks that "change of air is not more valuable as a remedy in the cure of disease and its consequences, than as a preventive of disease, more especially in childhood and youth." It is upon the appetite that its effect is first marked, and no doubt this is most prominent when the change is from an impure to a purer air. Yet I have known the mere change alone to have a beneficial influence, as, for instance, a removal for a time from the seaside or the fresh breezes of the Worcestershire hills to London. Observation does not incline me to have faith in the doctrine of acclimatization. It seems to me that a long residence in a climate, instead of rendering it more salubrious to the resider, makes it less suitable in close proportion to its length. I cannot at all join Claudian in his praise of the old man of Verona, who attained the age of ninety without ever going out of the suburbs.[2] He used his natural toughness to set a very bad example to his neighbors; and if many followed it, I am sure some must have suffered in mind and body.

In choosing a place of education for children, it is desirable that the climate should be decidedly different from that enjoyed at home during the holidays. Denizens of the stagnating, oft-breathed atmosphere of a metropolis will do well to select a country school; dwellers on the high ground of central England will find what suits them best on the coast; while both the seasiders

---

[1] Cyclopædia of Practical Medicine, i, 40.
[2] De sene Veronensi epigramma.

and country folks may venture, without risk of deterioration, to secure for their growing families the many advantages of instruction in a town.

Clergymen whose health is below par and even verging on disease, will often gain wonderfully by a mutual exchange of duty, *provided the climates of their several spheres are different.* Satirists say that parsons' livings always disagree with them; and there is a strong spice of fact in the statement; it is not fancy, but a real stagnation, from monotony in their aerial and other surroundings. The remedy is easy and cheap, but the physiological conditions of it should be clearly understood. It would be a profitable subject for bishops and archdeacons to dilate upon in their charges, as it is quite as important to the public that clergymen should be kept in repair as that churches should be so attended to.

CHAPTER IX.

STARVATION, POVERTY, AND FASTING.

Starvation.

THERE has always been a certain amount of importance attached to the diet and regimen of the sick, but not till the present generation do we find any notice taken by men of science of the consequences to the healthy, of its insufficiency or imperfection. The remarkable researches of Chossat on Inanition[1] form the first important work published on this point. The results deduced by this physiologist from his experiments have only been confirmed and expanded by later observers.

The first and most important principle established by Chossat is that absolute deprivation of food and deficiency of food are physiologically identical in their action on the animal life. One acts quicker than the other, but the difference is merely one of duration and degree. Both are equally fatal in the end, if not interfered with; and the end in both is regulated by the same law. Death arrives when the body has lost $\frac{6}{10}$ of its weight, whether that happens after days, or months, or years.

The loss of temperature is a feature common to and identical in both. Starvation or abstinence proved almost always fatal, in Chossat's observations, whenever the animal warmth fell to about 76° (Fahr.) in a red-blooded creature. The importance of the loss of temperature was shown by the fact that a renewal of consciousness and nerve-power could be effected, even from the torpor preceding death, by the application of external warmth. This fact affords a most valuable hint for the management not only of absolutely starved but of poorly fed individuals.

The sensations of hunger need not be described. After a time these are appeased, and are succeeded by a working and a grumbling and a dull aching in the small intestine. Then the secre-

---

[1] Chossat, sur l'Inanition, Paris, 1843.

tions of the abdominal canal diminish and finally nearly cease. Digestion becomes more and more difficult, the longer the abstinence. In fact, an insufficient diet is not only hurtful immediately, but it brings on an additional danger which acts in the same direction, namely—an impediment to absorption. The appetite falls off, and it is only from habit that the sufferer is induced to seek the food for want of which he is perishing.

There is absolute constipation, as may be daily noticed in hysterical subjects and others who do not take nutriment. And the forcible relief of this constipation by drugs makes matters worse, as we may also observe in the same hysterical subjects under foolish maternal discipline.

The breathing goes on gradually getting slower and less deep. In a case of starvation for twenty-four hours, Dr. Smith found the exhalation of carbonic acid fall from 34 ounces per diem to 22 (Philos. Trans., 1859). Fasting ecstatics hardly seem to breathe at all; and if you watch at rest a London needlewoman, or pauper before she goes into "the house," the motion of the ribs cannot be seen, so little air does she draw in. Contrasted with this is the gasping rapidity with which the respiratory muscles act when forced exertion sets them in motion.

The alterations in the blood consequent on insufficient food depend on whether there has been a deprivation of water or not at the same time with solids. If there has, strange to say, the proportions of the ingredients of the circulating fluid are not affected, but the whole quantity is diminished. In animals starved to death, three-quarters of the blood was gone. But where water is abundant, and the starvation gradual and not immediately fatal, it is easy to see by the coloring of the lips that the change consists in the dilution of the nutrient stream with aqueous fluid. In a ghastly picture exhibited at the last season of the Royal Academy (1874), "The Door of a Casual Ward," it was singular to notice how the secret of the weird and true effect of color lay in the omission of vermilion from the flesh tints. This abstraction was natural, for the reason that it was an imitation of the inner work of nature, not of the mere outside.

The nervous system shows how it suffers from inanition by giddiness, fainting, hallucination, and delirium. During sleep the dreams are most characteristic, presenting wondrous scenes of festivity and sensual enjoyment. Lately Mr. Parrot[1] has believed he has traced to inanition certain tissue changes in the nervous system, as well as in the other viscera, which would appear to be the material expression of the functional derangements first named.

In an account by Mr. Brett in the "Medico-Chirurgical Review" (1841) of the denizens of the Indian prisons at Moorshedabad, Cawnpore, and Shahjehanpore, one of the most notable results of insufficient food was a peculiar glassy appearance in the eye followed by inflammation and ulcer of the cornea, a copious secretion of tears and from the Meibomian glands, and finally blindness from destruction of the eyeballs, and death by emaciation. Strange to say, these dreadful symptoms in such a sensitive part were accompanied by no pain. The same phenomena were observed long ago by Magendie in dogs starved to death by deprivation of all nutriment, and they occur also when gelatin only is supplied.

Among the poor, especially among children badly fed, very similar affections of the eyes are frequent. They rarely, however, go on to the lengths mentioned above, for the sufferers come under medical care, are sent into hospitals or parish infirmaries, and are reinstated with nourishing diet. A recognition of the true origin of many ophthalmic epidemics in reformatories, pauper nurseries, and such-like collections of infantile weakness is very necessary to be impressed upon the managers of those institutions. Not being medical men, they are prone to apply, universally, principles of regimen suited only to their own over-fed nurseries, and to cure all inflammations by restriction, till they discover their mistake by sad experience.

If the lack of nitrogenous food is too prolonged, since it is the blood which is subjected to the greatest drain, and since in fat persons the blood errs by defect rather than by excess, inanition can exist while the body still retains its fat. In the case of the Welsh fasting-girl[2] the body was found after death plump and

---

[1] Compt. rend., Acad. des Sciences, 1868, t. li, p. 412. Dr. Brown-Séquard calls this Stéatose interstitielle diffuse de l'encéphale. By long-lasting insufficiency of food the brain is more or less converted into a suety substance.

[2] This was a case of notorious imposition, which the daily papers in 1869 detailed very fully. The parents made a show of her, docking her out like

with a considerable quantity of adipose tissue upon it, though the jury was quite right in finding that she was indubitably starved to death. This is a fact of great importance in practice: one must be very circumspect in starving or bleeding a stout patient; doubtless the store of adipose tissue may serve partly to nourish him, but it will not keep him alive. In our treatment of the poor we must apply the same reasoning, and not suppose that because a client, especially in old age, is fat, that he is therefore in good case and capable of bearing restricted diet. Jailers have sometimes attempted, really more from mistaken kindness than harshness, to "tone down" an obese prisoner with hard fare; and the consequence usually is that the country is put to the expense of sending him to the infirmary.

When entirely deprived of nutriment we are capable of supporting life for little more than a week. The Welsh fasting-girl lived for eight days from the commencement of the time she was carefully watched. But so many circumstances influence the amount of resistance which the body can exhibit, and make it vary so much, that no practical importance attaches to the theoretical limit of possible existence.

One thing which remarkably prolongs the duration of life is a supply of water. Dogs furnished with as much as they wanted to drink were found by M. Chossat to live three times as long as those who were deprived of liquids and solids at the same time. Miners who have got shut up in damp headings have experienced much relief by getting moisture from the walls of their prison. Even wetting the skin with sea-water has been found useful by shipwrecked sailors. The fact is water is a food, necessary to the building up of the body, and I think a great mistake is made when we caution the underfed against a free use of it. And,

---

a bride on a bed, and asserting that she had eaten no food for two years. Some reckless enthusiasts for strict truth set four trustworthy nurses to watch her; the Celtic obstinacy of the parents was roused, and in defence of their imposture they allowed death to take place in the usual time which it does after total deprivation of food. They were found guilty of manslaughter with perfect justice, for the law rightly supposes everybody to know that a human being without food must necessarily die. There was no need to prove the fact by a cruel experiment, and to visit a poor crazy swindler with the punishment of death. Thus to take the moral government of the universe into private hands is quite unjustifiable.

moreover, water prevents that concentration of the circulating fluid which impedes absorption, and which is a serious hastener of fatal results in those who have not enough to eat.

Though, as was before noticed, a certain amount of fat on the body does not prevent the blood from being impoverished, and by its impoverishment leading to injurious results, yet fat is a considerable protection against starvation. The old tale of the pig which was buried by the fall of a cliff at Dover, and was dug out alive after 160 days, as recorded by Mr. Mantell, the naturalist, sixty years since, in the "Transactions of the Linnean Society" (vol. xi), is a standing case to cite. The loss of substance, amounting to three-quarters of the entire animal, is probably an exaggeration, as the weight previous to the imprisonment is estimated by guess; but still the duration of the starvation, palliated only by the moisture which oozed through the sides of the sty, is very remarkable, and may certainly be taken as evidence of the protection afforded by the adipose tissue.

Life can be supported by a very minute quantity of food as long as complete inactivity of body is maintained. But under such circumstances the slightest exertion will bring on a fatal result. I attended for a long time a surgeon whose power of swallowing was completely lost by cancer of the œsophagus. He was cheerful and happy as long as he lay in bed, but at last during my absence from London, he thought a trip to Greenwich in a steamer would be an agreeable change, and died immediately after the exertion.

An even elevated temperature and the exclusion of the sun's rays, as in the "dim religious light" of a closed bed-room, will enable life to go on slowly for a marvellous period; as in the case of ecstatics, fasters, hysterical and insane persons. But if these are suddenly routed out, mentally excited, and forced to live like other people without due preparation, they die in spite of the soundness of all their organs. It is the same with convalescents from acute fevers, if they are injudiciously roused to mental or bodily activity.

Sex seems to have no appreciable influence on resistance to deficient diet. But it is not so with age. It is an observation of the rough old times when famine was oftener seen (and therefore we may trust Hippocrates)[1] that the younger a human being is, the easier is it starved, till we come to extreme old age, when the powers of life are considered by some physiologists, Celsus among the number, to give way quicker under famine than those of middle-aged men. However, there is no difference of opinion as to its effect upon children. Very large is the number of victims to this law of nature, and Malthusian optimists may admire its equity in balancing population and food. Nevertheless, so long as wide tracts of the earth lie uncultivated, we may not imprudently suspend its operation as much as we can. Dispensary and parochial practitioners are sadly familiar with a mass of infantile sickness and death which they classify on paper to a certain extent, under heads prescribed by the Nomenclature of Disease, but which appeals to their hearts by its real terrible name "Starvation." Europe is deeply indebted to M. Parrot for the bold outlines with which (in the "Archives de Physiologie" for 1868) he sketches the condition in which an innumerable army of speechless martyrs are found in the great centres of civilization.

"In these little creatures," he says, "we see the functions growing weaker and weaker with extreme rapidity, though in a gradual manner. The temperature, often lower in the interior of the body than in the axillæ, falls below 92° (Fahr.) and is never above 95°. (We all know what 'blood-heat' is.) Usually not more than 90 beats in a minute can be counted in the pulse; once the number was over 100, in another case it was below 64. The respiratory movements were less frequent than in the normal state and often very weak. The cries, which in some were at first intense and prolonged, ceased little by little. The secretions, always scanty, ended by disappearing; more than once the napkins put on in the morning were taken off in the evening unsoiled and dry. The skin, rigid, dry and cold, often oozed out a serous fluid, especially in the dependent parts of the body. Motionless in their cradles, icy cold, with the face livid and drawn, as if they were mummied, these still living babies looked like corpses. The beating of the heart could not be heard, and were it not for an occasional

---

[1] Aphorism xiii. Hippocrates does not define who are the γέροντες, who he says bear starvation best; but as an Ionian he would probably include all over 45.

movement of the breath at long intervals, it would have been thought that a body some time dead lay under the eyes. In very truth, death had already taken possession, slowly indeed, and as it were molecule by molecule, but with a sure and fatal grasp. . . The death of these children is due to starvation; and the diseased appearances revealed by a post-mortem examination, should be looked at not as the cause of the malady, but as its inevitable consequences."

A common symptom, in children, of a diet deficient in nutriment is diarrhœa. It assumes, when severe, the dysenteric type, streaks of blood appearing in the light green, mucous evacuations. And so it gets entered in the register as "dysentery" or "inflammation of the bowels" or "enteritis." But it does not appear to be contagious or epidemic, like the dysentery of camps; and in hard times may be observed confined to the children of a district, to the children in arms, even, according to the mode in which the pressure of scarcity falls on the population.

Yet the sufferers are probably neither deliberately, nor completely, perhaps not even knowingly, deprived of nutriment.

> Evil is wrought
> By want of thought,
> As much as by want of heart.

Ask what they are fed upon, and compare it with what rational experience prescribes, controlled by the universally spread teachings of physiology, and you will see immediately that the dietary errs in not containing the essentials of existence. The food is not, strictly speaking, unwholesome, but it errs by defect. It is not food at all. Reference to Chapters I, and II,[1] makes this clear. It is there pointed out what children *should* eat, and everything else is what they *should not*.

Not only youth, but health and vigor render the body less tolerant of abstinence. Invalids bear it better than strong people, with a few special exceptions. Such of us as have passed middle life cannot but remember, during the old reign of depletion and restriction, when for fear of victualling the disease they famished the patient, instances of the extreme toughness of the human race.

---

[1] Pp. 125 and 134.

Among the sick, not the sound, these instances are recollected. It is an act of mercy, when, in sieges, in shipwrecks, and such-like disasters, meat and drink are scarce, to supply those on the invalid list first and most bountifully—no doubt it is an act of mercy, and to the honor of mankind will probably always be done —but the physiologist must pronounce it highly imprudent and far from being conducive to final success; for the vigorous and active, whose blood is circulated rapidly, and whose muscles are in constant movement, really suffer most, not only in feeling, but in their future health and strength.

In order to preserve life as long as possible where sufficient food for subsistence cannot be obtained, our aims, then, should be to secure water, warmth, and complete inaction of the muscles. The direst famine does not necessarily exclude these preservatives, and in the more insidious scarcities whose effects are noticed chiefly in children, we can very often readily get as much of them as we want.

In the apportionment of a spare supply, the youngest should have the nearest approach to a full ration, the active men and women the next, and the invalids and aged the scantiest allowance. So is distributive justice best satisfied.

It is a strange thing, and one which at first sounds paradoxical, that the supply of the stomach even from the substance of the individual body itself should tend to prolong life. A case of starvation for twenty-two days in an open boat was recorded in the periodical prints last spring (April 30 and May 1, 1874) in which the poor victims fought in their delirium, and one was severely wounded.[1] As the blood gushed out, he lapped it up; and instead of suffering the fatal weakness which might have been expected from the hemorrhage, he seems to have done well. I would not build much on the rough memories preserved during such awful sufferings, were it not for the support afforded by some experiments by a French physiologist, M. Anselmier, to whom the idea occurred of trying to preserve the lives of some dogs by what he calls "artificial autophagy." He fed them on the blood taken from their own veins daily, and he found that the fatal cooling

---

[1] Three men and two boys were out for 32 days with only 10 days' provisions, exclusive of old boots and jelly-fish.

incident to starvation was thus postponed and the existence consequently prolonged. Life lasted till the emaciation had proceeded to $\frac{6}{10}$ of the weight, instead of $\frac{4}{10}$, as in Chossat's experiments, and was extended to the fourteenth instead of the tenth day, which was its limit in those dogs who were not bled.[1]

It is not likely that anybody will feel himself called upon to repeat M. Anselmier's dreadful experience, but no one of us who run to and fro in the earth is secure from the possibility of shipwreck, or being buried alive, say in a mine or railway cutting, and thus involuntarily contributing a self-sacrifice to that knowledge which may save the lives of others. For one famished crew that is picked up, there are found dozens of empty wrecks, of which it is never known how the once living freight fared; and for each empty wreck there are dozens which have left no trace. The prolongation of life without provisions is by no means a mere speculative discussion. Were I in such a strait as above referred to, my reason would counsel me, and I hope I should have the courage to wound my veins and suck the blood.

After starvation, either complete or partial, a sudden return to full diet is not to be attempted. Small quantities at a time of the most digestible food must be given. And the temperature should be artificially sustained till such time as the system is able to generate its own heat, especially in the interior of the body. Tablespoonfuls of hot beef tea and of milk constantly administered ("tea-cup diet"), form the most appropriate nutriment. And it will be much assisted by admixture with a small quantity of pepsin; but not too much, or there is risk of diarrhœa.

Poverty.

The valuable calculations of Dr. Playfair, "On the food of man in relation to his useful work,"[2] enable us to arrive at a very practical estimate of what amount of solid victual is required by an

---

[1] Archives Gén. de Médecine, 1860, vol. i, p. 109. M. Anselmier conjectures that the formation of heat was due to the keeping up of a certain degree of activity in the gastro-intestinal absorption, and the consequent chemical action in the interior of the body, where it would not be lost by radiation.

[2] Lecture delivered at the Royal Institution, London, April 28th, 1865, published by Edmonston and Douglas, Edinburgh.

adult living by bodily labor, to preserve his health under various circumstances. The circumstances which chiefly influence the required amount can be classified as follows:

1. Bare existence;
2. Moderate exercise;
3. Active work;
4. Hard work;

1. The diet of bare existence—"subsistence diet"—is calculated from the mean of sundry prison diets, of the convalescent's diet at hospitals, that of London needlewomen, and of that supplied during the Lancashire cotton-famine, as reported by Mr. Simon. The result is that in a condition of low health without activity, $2\frac{1}{4}$ ounces of nitrogenous matter (calculated dry), 1 ounce of fat, 12 ounces of starch, and $\frac{1}{4}$ of an ounce of mineral matters per diem are necessary. The contents in carbon are 7.44 ounces. This, being interpreted, means that a man will die gradually of starvation, unless his provision for a week contains three pounds of meat with a pound of fat on it, or the same quantity of butter or lard, two quartern loaves of bread, and about an ounce of salt and other condiments. If he cannot get the meat, he must supply its place with, at least, two extra quartern loaves, or about a stone and a half of potatoes, or between 5 and 6 pounds of oatmeal—unless, indeed, he is so fortunately situated as to be able to get skimmed milk, of which five pints a week will fairly replace the meat. Let it be understood that all these articles must be good of their sort, and contain no indigestible matter or adulterant. And there is no economy in substituting for real nutriment things which merely stay the hunger by occupying the stomach for a longer period. The completeness of the digestion is thus interfered with, and a morbid derangement of the function induced, which causes part of the food to be wasted.

A person brought to bare existence diet can undergo no toil, mental or bodily, under the penalty of breaking down.

2. By moderate exercise is meant the equivalent of some—say from 5 to 7 miles—walking daily. Dr. Playfair takes as fairly representing the appropriate food of this class the dietaries of English, French, Prussian, and Austrian soldiers in a time of peace.

The English soldier on home service, according to Dr. Parkes, receives from government five pounds and a quarter of meat and seven pounds of bread weekly, and buys additional bread, vegetables, milk, and groceries. Now, such a diet as this is amply sufficient for anybody under ordinary circumstances of regular light occupation. But should extra demands be made upon mind or body, weight is lost, and, doubtless, if the demands continue to be made, the health would seriously suffer. Mr. Buckland, of the Guards, remarks (Soc. of Arts Journal, 1863, quoted by Dr. Playfair), that though the sergeants fatten upon the rations, the quantity is not sufficient for recruits during their drills.

3. Active laborers are reckoned those who get through such an amount of work daily, exclusive of Sundays, as may be represented by a walk of twenty miles. Of this class are soldiers during a campaign, letter-carriers, engineers employed in field work or as artisans. These habitually consume on the average about a fifth more nitrogenous food and twice as much fat as the last class, while the quantity of hydrocarbons is not augmented, except in the Royal Engineers.

4. Hard work means that which is got through by English navvies, hard-worked weavers, full-fed tailors and blacksmiths. It is difficult to get exact information, but it would appear from Dr. Playfair's estimates that the addition to the diet is entirely in nitrogenous constituents. The higher their wages the more meat the men eat.

This neglect of vegetables by the two last classes is in a physiological point of view imprudent, and possibly may be a contributing cause of that inordinate desire for alcohol which impoverishes and degrades them. The discovery of the production of force from the assimilations of starch leads to a knowledge, opposed indeed to old prejudices but supported by experience, that the raising of the energies to their full height of usefulness may be effected by vegetable food proportioned to the increase of requirement, quite as well as by the more stimulating and more expensive animal nutriment.

Deficient diet, like all morbid conditions both corporeal and mental, is a vitiating and degenerating influence. Famine is naturally the mother of crimes and vices, not only of such sort as will satiate the gnawing desire for food, but of general violence and lawlessness, ill-temper, avarice, lust, and cruelty.

The love of purposeless destruction exhibited by the Parisian communists in our own day may be fairly credited to deficient food. No well-fed people could have wrecked the Vendôme column or burnt the town-hall and Tuileries, of which they were so proud. They were like hungry children smashing their dolls. And Thucydides, Boccaccio, and Defoe are all agreed as to the hideous wickedness exhibited at Athens, Florence, and London, during their famine-fevers. The exceptional instances are those where individuals or nations have conquered by courage and self-restraint their natural selfishness, and have made the interests of others paramount to their own. Am I blinded by love of my country, or may I justly quote the history of the Lancashire cotton-famine as a case in point?

In all physiology there is no more convincing proof of the benevolent government of the universe and of its perfecting influence upon our race, than the fact that directly a man begins to care for others in preference to himself alone, his care ceases to wear and exhaust him. It rather seems to be a sustaining force. Observe a lunatic, induced to work hard for some worthy object, he grows fat; let him sit still and brood over his wrongs, and he dies emaciated. Let a hypochondriac's wife or child fall ill, he is cheerful and well directly; but he relapses as they become convalescent, and grows as thin and miserable as ever when he turns his attention to his own health. This is the reason why in sieges and famine medical men have often remained sleek and plump, while their neighbors pined, and perhaps also why military officers bear short rations better than the men. Like all divine truths, "to love your neighbor as yourself" is found to be taught by material nature as well as by revelation.[1]

Fasting.

Fasting is the voluntary restriction of the diet for the express purpose of developing the higher features of the mind. It is "a

---

[1] Professor Maurice well remarks in the preface to his Moral Philosophy that the difference between revelation and discovery is shown by the words themselves to be very slight; one is removing a veil, the other is removing a cover.

means of grace," and approved as such by religious men of most opposite creeds and diverse nationalities. I am not going to question their experience, and I should bow to it even if it differed from my own, which is not the case. Like all the material machinery by which the bodily man influences the spiritual man, this "means to an end" not only admits of, but requires frequently to be brought into harmony with, the progress of knowledge concerning physical life. And therefore I think that in a manual of this sort mention of it should not be omitted.

In the first place, to be useful, fasting must be wholly *voluntary*. If forcibly imposed even by imperious custom or enjoined as an end in itself, its principal effect is to sour the temper and narrow the intellectual apprehension for the time being. In fact its physiological action is a minor degree of starvation.

To secure its being wholly voluntary, it should be private, as advised by the highest authority of all.

At the same time that the act is voluntary, the manner of it is best prescribed by another. If the peculiar form and degree of abstinence are self-imposed, they are apt to be excessive, and to do harm without any corresponding advantage.

One of the highest mental developments to be expected from fasting is the power of self-control by voluntary effort. If a man finds himself during a fast weakened in his ability to concentrate his thoughts and to keep up his attention, it is doing him no good. If the instinctive appetites, implanted by nature, fail after it, injury to the health and individual degeneration follow. For these reasons it is best that the matter on which abstinence is exercised should be rather a luxury than a necessary. It is quite safe for a healthy man to leave off tobacco, wine, or beer, spices and sauces, hot meat, or even meat altogether, or vegetables altogether (if vegetables are a luxury to him), or pastry, or sugar, or butter for a day or two at a time; especially if he withdraws himself from the bustle of the daily occupations. But it is not wise to do this too frequently, or to let the low diet become habitual. Like all acts of free will, it is most powerful in its effects on the mind when rare, and when it presents the strongest contrast to the usual life. If once they become habits, the spiritual influence of acts ceases.

Those who never feast will find a difficulty in advantageously fasting.

CHAPTER X.

THE DECLINE OF LIFE.

EVERYBODY who has passed the age of fifty (or thereabouts) with a fairly unimpaired constitution will act wisely in diminishing his daily allowance of solid food. At the "grand climacteric" (as this turn of life is pompously called) the movements of nutrition are retarded, and the constructive and evacuating actions of the system being diminished, there is less call for materials of repair. It becomes a moral duty to avoid all articles of diet which personal experience has shown to be difficult of solution, to make smaller meals, and, if need be, more frequent meals, so that the stomach may be never overloaded or too long idle. The saving up an appetite for the enjoyment of an abundant repast may be conceded as a harmless folly in our juniors, but it is a shame to a gray head. If custom has made a man a large eater, he should endeavor to "spoil his dinner" by a late luncheon, and to prevent his appetite being too keen at midday by breakfasting not over early.

Very aged people, however, and those who have lost their teeth run some risk of not being sufficiently nourished, from swallowing their food rapidly. They are hurried over their meals through the thoughtlessness of those around them, and since they chew slowly and secrete saliva slowly, the food remains undigested. Their juniors should remember this, and govern themselves accordingly. A kindly British matron, who spent more hours at table than was good for her, told me that if she did not do so, she "should be a widow in a week," and that she habitually ate too much to keep her aged husband in countenance. Not a word could be said against such pious gluttony.

The dishes of meat should be as soft and tender as possible, and the firmer kinds should be finely cut with a mincing knife. But vegetables should not be over-softened in cooking: there should be sufficient resistance in them to make chewing imperative, so as to excite the secretion of the fluids of the mouth, which are required for their solution. Soups and broths are nutritious, but should not contain solid vegetables, except just enough to flavor them. Puddings and pastry are not of much use, and overload the tired stomach.

I do not know the authority for the old proverb *vinum lac senum*, but it seems a very dangerous one, as it may lead our ancient friends to think they may treat it like mother's milk, and measure its benefits by the quantity imbibed. The saying, however does partially embody a truth, namely—that in the decline of life the advantages derived from fermented liquors are more advantageous, and the injuries it inflicts less injurious than in youth. The effect of alcohol is to check the activity of destructive assimilation, the rapidity of that moulting of the body's substance by normal secretion which in healthy youth cannot be excessive, but which in old age exhausts the frame. Alcohol calmly arrests the energies of the nervous system which would fret the tissues to decay, and would seriously weaken them, were not the wear and tear to be continuously replaced by new material. Now, with years, the replacement by nutrition is much diminished, and we, nevertheless, are apt to persist in using our brains as before. We shrink, rightly enough, from being shelved just when the rewards of our exertions are becoming due; and we do not care to rival the centuries of the olive or the yew, unless we can, like them, "renew our age" and bear fruit unto the end. Here, then, alcohol steps in as a help in need, and it is strictly in accordance with the teachings of physiology to increase, as years increase upon us, the moderate quantity we had been taking previously. I do not write for habitual revellers and muzzy dram-drinkers (the sooner they become teetotallers the longer they will live), but for the temperate users of natural good things, and I am sure that they may reasonably obey the instinctive desire to take more and stronger wine as they grow in years.

The physiologist Moleschott, moreover, states that "a glass or two of good old wine augments the amount of gastric juice, the liquid which performs mainly the digestion of albuminous aliments."[1] So that here we find another reason for indulging the instinct.

---

[1] Kreislauf des Lebens, letter 6.

Elderly people are able to do with less sleep than younkers, and need not be alarmed at a certain shortening of their night's rest, which is natural. But sometimes the shortening goes too far, even in health; they cannot get to sleep for a long time after going to bed, and are worn out with restlessness and rolling about; not to mention that they disturb others also in many instances. This inconvenience may often be obviated by having an egg, a sandwich, a few biscuits or other light repast the last thing, accompanied by a glass of bitter ale, or sound wine and water. Sweet, strong wines are those usually recommended for this purpose: Hufeland recommends Malaga; and Burgundy or Port, warmed, spiced, diluted, and sweetened, is not a bad drink; but probably the best for each individual is that which association or whim makes most agreeable to the palate. Some prefer gruel or arrowroot to be the vehicle of the alcohol, and it certainly dilates and distributes the virtues of the draught efficaciously, and also warms up the stomach comfortably; but it has a coddling invalidish look which should be avoided.

Dr. Welsted considers that it contributes to length of days to associate as much as possible with young people, and to adopt such habits and manners as may attract rather than repel them, to which last there is a temptation in old age. And, of the young people, he holds, that the best companions are those whose spirits are high and joyous, and whom we can induce to rally round and infect us with their life. "For," as he says, "that solitude which is associated with fear and sorrow breaks up the strength of both mind and body."

It is almost needless to say that ease of mind, contentment with the present, and calm confidence in a future happy renewal of the worn-out body and soul, are specifics suitable for all cases. It is not hard work that kills the active, nor idleness that kills the man of leisure, be he old or young, but worry. "Be careful for nothing, the Lord is at hand," is a motto which will prolong the lives of all.

CHAPTER XI.

ALCOHOL.

§ 1. Physical Effects.

Many a man has asked himself, like Horace, "Quo me, Bacche, rapis Tui plenum?" To what does the natural thirst for alcohol lead? He sees around him disease, death, and misery directly traceable to the abuse; but he sees also disease, death, and misery, arising from the abuse of other instinctive desires which it were profane to call evil. Is there a use? What happens to a temperate consumer, likely to shorten or prolong his days? Or does he merely gain a pleasure without any consequent good or evil?

It was with a view of getting a basis for the satisfaction of my mind on these points that about fifteen years ago I engaged a laboratory assistant, and carried on the following experiments:

Experiment I.—A. M. aged 38, weight 254 lbs.—taken at noon every day—Habits of life extremely regular. He walks half an hour before breakfast daily, breakfasts at eight on two cups of coffee, bread and butter, and a slice of cold meat: dines at one on beef and mutton in regular quantity, potatoes and pudding; has tea at five, two cups with bread and butter; sups at nine on bread and butter, or cheese, with half a pint of ale. He sleeps six and a half to seven hours. His bowels are open once daily. An idea of the normal amount of metamorphosis in his body is afforded by the following table:

	Quantity in cubic centimetres.	Specific gravity.	Urea in grammes.	Chloride of sodium in grammes.	Sulphuric acid in grammes.	Phosphoric acid in grammes.	Uric acid in grammes.
Amount of urine and its several parts made in 23 days, in perfect health and on usual diet,	24.970	1.022	728.437	174.625	51.307	44.719	....
Do. in 15 days,[1]	....	....	....	....	....	....	2.813
Mean daily amount,	1.085	1.022	31.671	7.592	2.230	1.944	.187

[1] On eight days the uric acid was not weighed.

The effect of the addition of a certain quantity of alcohol to the daily meal is shown by the next table:

Date.	Quantity in cubic centimetres.	Specific gravity.	Urea in grammes.	Chloride of sodium in grammes.	Sulphuric acid in grammes.	Phosphoric acid in grammes.	Uric acid in grammes.	Daily quantity of best French brandy, added to meals in ozs. by measure.
Sep. 13,	1,020	1.024	30.708	7.140	2.017	1.469	...	4½
" 14,	1,570	1.022	39.746	10.990	2.579	.848	...	3
" 19,	1,050	1.026	38.795	8.400	2.456	1.890	...	⎡ 6, viz., 1½ at
" 20,	1,200	1.025	42.695	9.600	2.622	1.944	...	⎢ breakfast,
" 21,	1,110	1.028	37.974	6.937	2.212	1.798	...	⎢ dinner, tea,
" 22,	770	1.026	30.030	6.160	2.065	1.386	...	⎣ and supper.

On September 23, the appetite for food was observed to be somewhat less than usual, and the experiment accordingly ceased; for any change of usual weight, health, feeling, or habits, of course would vitiate the result of an investigation conducted in this form.

Here are shown the effects of mixing with the daily meals such an extra quantity of alcoholic liquid as is very usual with moderate consumers. Such a quantity seems to put them at ease with themselves and with the world around, without causing any immediate injury to the general health, any untoward exhilaration, or any subsequent depression.

We may observe:

1. *The aqueous part of the urine daily excreted is not increased beyond the extent of the extra fluid injected.*
2. *The quantity of urea is increased after the first twenty-four hours.*
3. *The chloride of sodium and the sulphates are slightly increased.*
4. *The phosphates are diminished.*
5. *The augmentation is temporary, and after a time is succeeded by a reduction to the normal measure, which reduction is coincident with a loss of appetite.*

Alcohol, then, is not a diuretic in the commonly received acceptation of the term, and the repute which it has got of belonging to that class of medicines, must be due to some other ingredient of the compound forms in which it is usually swallowed, or to its relieving (in certain cases of diminished excretion) impediments to the due action of the kidneys.

At the same time there is represented by the increased formation of urea, a more active destructive assimilation, and (inasmuch as the weight of the whole body is not lost) a more active reconstruction of the nitrogenous elements of the tissues. Old flesh is removed, and meat food is appropriated as new flesh, somewhat quicker than when no alcohol is taken.

It is a matter of common observation that the appetite of an average healthy man has a keener edge put to it by a moderate quantity of beer or wine with meals. So that the abovementioned enhancement of the interstitial growth may be fairly credited to a temporary rise of the digestive powers of the stomach.

The decrease in the excretion of phosphorus is small indeed but indubitable. The chief source of that ingredient of the urine may be reasonably supposed to be the substance of the brain and nerves, and it can hardly be thought a mere coincidence when a reagent whose effects are most peculiarly manifest on the functions of those organs, diminishes what we believe to be the result of their chemical changes. Every one recognizes in alcohol a power of blunting sorrow and pain, of checking the sensation of weariness, mental or bodily, of rendering carnal love coarser, less keen, and less discriminating, taking the points off the stings and buffets, discomforts and nastiness of daily life, but also of corrupting the delicate appreciation of its higher delights, in short, of diminishing the sensibility to impressions in mind and body, of lowering the receptive functions of the nervous system.

In a series of experiments conducted for another object, Dr. Edward Smith has recorded minutely the sensations experienced by a healthy man on taking moderate quantities of brandy.[1] They consist essentially of lessened consciousness, lessened sensibility to light, to sound, to touch. The higher the sensibility of the part under ordinary circumstances, the more obvious its anæsthesia under the influence of brandy. For instance, there was in the upper lip and cheeks a feeling of stiffness and puffishness, which is one of the first symptoms of lowered sensation, and is familiar to all in cases of partial paralysis. The dartos also and

---

[1] Transactions of Royal Society, 1859, p. 732.

other muscles connected with the reproductive system were relaxed; as was also the sphincter of the bladder, accounting for the increased micturition during indulgence. The pulse also was quickened, a phenomenon which always attends temporary debility of the heart's action. Indeed, in all motion of a purely involuntary character, quickness always indicates diminished muscular force. (The same result follows tobacco smoking.) The sensation of swelling in the upper lip is the premonitory stage of the muddy flush by which the artist and the actor mark the face of the "fuddled" man (*ebriolus*). When he is completely intoxicated (*ebrius*), it is oftener pale or livid. The venous congestion and arrested circulation is then transferred to the capillaries of the cerebellum (Flourens),[1] and stomach (Brodie),[2] and a secondary series of phenomena follow, dependent upon the local affection of these organs. There is a staggering gait, a want of co-ordination in the movements, and often vomiting.

Life and warmth are so closely connected together in scientific as well as in popular notions, that perhaps the most striking evidence of diminished vitality is the lessened capability to generate heat. And we have this evidence in the case of alcohol. MM. Dumeril and Demarquay published in 1848 their observation, that intoxicated dogs exhibited a great loss of temperature; and Dr. Boecker[3] and Dr. Hammond find in their own persons the same result from even moderate doses of spirits. This accords with and explains the experience of Dr. Rae, that alcoholic drinks give no satisfaction to Arctic voyagers, and of Dr. Hayes (Surgeon and Commander in U. S. second Grinnell Expedition), that they actually lessen the power of resisting cold.[4] The "warming of the stomach" which dram-drinkers speak of with such gusto is, in fact, a fallacious sensation arising out of insensibility to external influences.

When the poisoning is still more profound, the congestion of the nervous centres produces complete apoplexy, and that of the stomach, gastritis. The traces of these morbid conditions are found

---

[1] Recherches sur les fonctions du système nerveux, Paris, 1824.
[2] Philosophical Transactions, 1811.
[3] Beiträge zur Heilkunde, i, 250.
[4] American Journal of the Medical Sciences, 1859, p. 117.

after death, but the loss of life seems due to the paralysis of the respiratory muscles and the lowered temperature of the body. I am unwilling to enlarge further upon dead drunkenness and fatal alcoholic poisoning, as I have no personal observations on the subject to record, and their commonly reported characteristics are repeated here principally to show their essential agreement in physiological nature with the healthful and beneficial, or, at all events, not pernicious action of the substance. The effects in a healthy man would seem to be the diminution of the energizing wear of the nervous system, especially that employed in emotion and sensation. Thence there ensues a raising of the digestive powers and appetite, should they have been anywise unconsciously blunted by the psychical movements abovementioned. Just as often, then, as the zest for food is so far lowered, that it is found to be raised to the normal standard by a little wine or beer with a meal, the moderate drinker is as much really better, as he feels the better, for his liquor. In cases, however, where the food is as keenly enjoyed without the aid of any of the products of fermentation, their consumption is certainly useless, and possibly injurious. So long as alcohol, in the indirect mode mentioned, augments vital metamorphosis, it ministers to the force of the body. But it is not a source of force, and its direct action is an arrest of vitality. This should be clearly understood by all those who try to have rational rules to guide them in proffering advice as to the dietetic use of a "stimulant." That word means a "spur," and inasmuch as a spur is employed either before or during the exertion it is supposed to bear upon, liquor is often resorted to as a means of invigorating a temporary effort. If it has any effect at all on a healthy man, it can but weaken nerve power, while at the same time it lowers the bodily temperature which contributes much to the capacity for muscular exertion. Instinct or experience has taught this to men whom we are pleased to call barbarous. The Indian porters in South America (it is stated by Mr. Salvin, the ornithologist), when they prepare for a stiff journey under one of their heavy loads, carefully eschew all strong drink, and swallow large quantities of water as hot as the stomach will bear. And Aristotle states that the Carthaginians (the only African race ever fit to fight with Europeans) used when out on military service to abstain from wine.[1] The unhappy adoption of the word "stimulant" has demoralized the notions of civilized communities on this head, and led them to reckon on priming themselves up to unwonted strength with anæsthetics. Let them rather say with Byron's Sardanapalus, "The goblet I reserve for hours of love, But war on water," and victory will be much surer. There is, however, another aspect to the question. Man must be viewed not only as a possible victor, but as liable to the reverse fortune, on the battle-field or in the struggles of life. Both sides cannot win. During the Crimean war the Russian surgeons were reviled for serving out spirits to their troops under fire; but were they not wise to calculate the after consequences of an engagement, and to prepare the bodies of their clients for successfully bearing blows and wounds? "They have stricken me and I knew it not," says Solomon's drunkard; and our accident ward at Christmas time will seldom fail to show proof of the power to resist severe injuries conferred by unwonted indulgence in the joys of Bacchus. A healthy man who gets the worst of it in any way, whose intellectual or muscular energy goes down under the pressure of the work demanded, gets the worst of it in a less degree by the aid of strong drink. Give it him when ready to perish from the drain on his nerve tissues, and his life is saved. The laborer whose limbs are stiff with his day's toil, and the brain-worker who still more acutely feels the wear and tear of bread winning, are not wasting the money they earn, which they spend on a fair ration of beer or wine at their evening meal. But if they take spirits of a morning (it is usually spirits which are then taken), never let them hope for the success in the undertakings for which they seemed designed. Both body and mind will be incapacitated, the life shortened, and all the keenest joys taken out of what remains. One of the most telling questions that can be asked of a life proposed for insurance is, "Do you ever take spirituous liquors in the forenoon?" If the answer is in

---

[1] Aristotle, (Econom., i, 5. I have not seen this statement alluded to by historians, though it seems to account for the enormous quantity of vinegar which Livy found no difficulty in believing Hannibal had in his commissariat and used for decomposing Alpine rocks (Liv. xxi, 37). It was made tasty with pears (Pallad. in Februar., xxv, 11), and employed to flavor the soldiers' drinking-water.

the affirmative, an immediate rejection is the only safe course for the office. As to small quantities of beer or wine that are consumed between breakfast and the midday meal, the evidence against it is not quite so decisive, but, at all events, it renders a man less fit for his daily work than he would otherwise be, and is a dangerous downward movement towards the abyss of dram-drinking. It will generally be found to have been the first fatal step in the cases of women of the upper classes who have adopted the practice. And, to the shame of our profession, it must be confessed that many of us have erred most unhappily by recommending or sanctioning the habit in weakly or self-indulgent women. The weaker they are in body or mind, the more helpless they are, and the more hysterical, the more reason there is for withholding the temptation. We must be proof against tears and sighs, blandishments, entreaties, and reproaches, or we are not fit to bear the rod of Æsculapius.

See what happens if, instead of being drunk with the meals, alcohol is taken in small divided doses:

Experiment II.—The same man described in Experiment I, at another time consumed daily, between 9 A.M. and 9 P.M., six ounces of brandy in doses of half an ounce every hour. No effect was produced on the general health and feelings, and the usual employments were followed except on the last day of the experiment, when M—— was bustling about in a great state of excitement, packing up to leave for a country holiday.

The analysis of the urine was as follows:

Date.	Quantity in cubic centimetres.	Specific gravity.	Urea in grammes.	Chloride of sodium in grammes.	Sulphuric acid in grammes.	Phosphoric acid in grammes.	Uric acid in grammes.
August 18, . . .	1,520	1.013	30.465	5.320	2.210	1.299	.008
" 20, . . .	910	1.025	33.077	6.370	2.375	1.474	.259
" 21, . . .	1,070	1.022	32.945	6.687	2.246	1.637	.193
" 22, . . .	1,000	1.021	23.735	6.750	1.897	1.440	.135
" 23, . . .	1,310	1.015	25.097	7.205	1.649	1.061	.196
" 24, . . .	1,530	1.021	41.867	9.945	3.064	2.203	.300

One day was an interval in the experiment, and only the usual amount of daily diet, without extra alcohol, was taken, when the numbers stand as follows:

Date.	Quantity in cubic centimetres.	Specific gravity.	Urea in grammes.	Chloride of sodium in grammes.	Sulphuric acid in grammes.	Phosphoric acid in grammes.	Uric acid in grammes.
August 19, . . .	920	1.026	35.88	5.750	2.374	1.004	.281

It is very clear from these observations that alcohol taken in the dram-drinking fashion—namely, in small divided doses—by no means increases metamorphosis. It rather tends to diminish it, so that during the first five days quoted the mean quantity of urea excreted is 29.063 instead of 31.671 grammes. And this diminution is not sudden or immediate, but is more and more for a certain period, till the retention reaches a point at which a critical evacuation takes place in healthy persons.

This evacuation may take place in consequence of the alcohol being left off, as may be observed on the day of interval. Or again it may result from increased exertions and unwonted expenditure of nervous energy, as may be noticed in the last day of the experiment. In neither case does the amount, even of the urea, replace the amount arrested during the days of arrest; and the phosphates are still less unable to make up their lost figures in the ledger.

It cannot be concealed then, that even without at all infringing on high health, alcohol in small divided doses and between meals, dram-drinker's fashion, deranges the metamorphosis of the tissues, and in the direction of arrest. And persistence in the habit must lead in the end to permanently diminished organization, degeneration, atrophy. Just as a disused limb wastes away, so must the unrenewed tissues die off gradually, till they become unequal to the support of a healthy man's existence.

The first action of alcohol, then, is on the stomach, enabling more food to be digested, and enhancing the origination of force. But if advantage be not taken of this first action, its secondary effect is a diminution of the vital functions in general and of digestion among their number.

It might be suggested that the subject of the experiments was so little used to stimulants that an abnormal effect might be produced upon him by them, but that is rendered unlikely by the succeeding observation on a person whose habits may be described as closely verging on "full living:"

Experiment III.—C——, aged 43, healthy, though not muscularly robust, of regular life and habits, took daily during the days named in the table a quantity of food proportioned to appetite, viz., about a pound and a half of meat, half a pound of bread, a pint and a half of tea, with milk, sugar, butter, sauces, etc., *q. s.*, half a pint of water, and from five to seven glasses of port or sherry,[1] care being always taken not to annoy the temper, and so nullify the experiments, by overstrictness.

	Quantity in cubic centimetres.	Specific gravity.	Urea in grammes.	Chloride of sodium in grammes.	Sulphuric acid in grammes.	Phosphoric acid in grammes.	Uric acid in grammes.
Amount of urine and of its several parts, made in fifteen days, in perfect health and on usual diet, .	18,800	1.022	493.852	137.655	....	....	....
Do. in fourteen days,[2]	....	....	......	......	26.487	27.683	3.839
Mean daily amount, .	1,253	1.022	32.923	9.177	1.891	1.977	.274

The effect of taking in addition, between meals, a moderate quantity of alcohol in divided doses, is shown in the following table:

Date.	Quantity in cubic centimetres.	Specific gravity.	Urea in grammes.	Chloride of sodium in grammes.	Sulphuric acid in grammes.	Phosphoric acid in grammes.	Uric acid in grammes.	Daily quantity of best French brandy taken between meals.
Nov. 16,	1,180	1.021	30.090	11.210	1.954	1.770	trace.	3½ fluid ounces.
" 19,	1,800	1.013	28.854	9.900	1.906	1.800	.258	8 " "
" 22,	1,150	1.025	32.775	12.075	....	....	...	7½ " "
" 23,	980	1.025	27.930	9.310	....	....	...	7½ " "
Dec. 3,	1,060	1.023	28.620	9.540	1.785	1.696	.339	3 " "
" 5,	1,320	1.019	30.875	9.900	1.865	1.980	.330	8 " "
" 6,	1,110	1.021	30.025	9.435	1.713	1.665	.299	8 " "
" 17,	1,180	1.020	30.208	9.440	1.586	1.652	.343	4 " "

[1] Which may be reckoned to contain from 33 to 35 per cent. of proof spirit.
[2] Of one day the record was imperfect, the sulphuric acid, phosphoric acid, and uric acid not having been weighed.

It is very clear from these figures that vital metamorphosis, as evidenced by the amount of the principal solids of the urine, is diminished by thus taking more alcohol than the healthy instinct prompts, even in a person used to the full quantity that temperance permits. Not only are the whole mean amounts low, but on no day do they come up to the average. The only exceptions are the chloride of sodium, which is slightly increased, by what agency I cannot tell, and the uric acid, whose augmentation is, probably with justice, considered an indication of an approaching abnormal state. It may be remarked that a greater quantity of brandy than that recorded above spoiled the appetite and prevented the usual diet being taken with pleasure. I considered that this would nullify the experiments as representing the effects of alcohol in health, for it placed the body in an abnormal state, and I therefore discontinued them for a time.

The only marked effect of the quantity taken on the full days, was a certain degree of insensibility of the facial skin to changes of temperature; the fire did not scorch, and the east wind did not chill. So that probably some temporary power of resistance to external influences was gained, to be followed by reactionary sensitiveness.

Let us observe now the effects of spirituous liquors taken in great excess, namely, in quantity sufficient to destroy the natural sensitiveness to mental emotion:

Experiment IV.—Letitia ——, a healthy prostitute aged 23, acquired the habit, during a year of her being on the town, of frequent tippling to drown care. Standing by her bed at home on August 14, she suddenly fell on it, not losing her senses, but having complete paralysis of the right leg and arm, without a fit. On her admission to a hospital two days afterwards, the power had returned in a great measure to the limbs, but the right lingual and facial muscles were still quite paralytic. As far as one could judge by external phenomena, all the viscera except the brain were in a healthy state, and she showed no signs of hysterics. She stayed in the hospital till September 6, when she was offered a place as servant, and a slight impediment to speech remaining I considered not sufficient reason for her passing over so good a chance of bettering her social state. During the time she was under observation no drugs were prescribed for her; she lay on her bed the greater part of the day, and sauntered about the ward and garden the rest; she was kept on broth diet,[1] and on and after August 30, three ounces of brandy were allowed daily. The amount of urine and of its chief constituents excreted by her on all the days when circumstances allowed it to be all collected, is shown in this table:

Date.	Quantity in cubic centimetres.	Specific gravity.	Urea in grammes.	Chloride of sodium in grammes.	Sulphuric acid in grammes.	Phosphoric acid in grammes.
August 17,	252	1.018	5.915	1 389	?	.409
" 19,	880	1.006	12.729	4.400	?	.396
" 20,	240	1.014	4.529	2.040	?	none.
" 21,	270	1.007	3.429	.337	.206	.061
" 22,	360	1.011	5.280	2.340	.212	a trace.
" 23,	1,000	1.007	15.353	3.500	.871	.540
" 24,	1,280	1.007	14.504	4.480	.715	a trace.
" 27,	570	1.008	9.405	1.425	.436	a trace.
" 28,	1,030	1.007	10.979	3.862	.596	.404
" 29,	730	1.010	9.252	3.285	.423	a trace.
" 30,	1,320	1.008	13 645	5.120	1.039	?
September 1,	....	....	....	....	....	....
" 2,	1,650	1.008	22.027	1.358	7.425[2]	?
" 3,	....	....	....	....	....	....
" 4,	....	....	....	....	....	....
" 5,	900	1.010	13 231	2.925	.790	.162

The metamorphic organs of this girl were clearly in a most perilous condition. So little phosphorus was excreted that the nervous system can have been hardly renewed at all. Sometimes not more than a tenth, generally about a third, and occasionally half, of the due proportion of urea appeared in the urine. When as much as half of the proper quantity of urea was formed, there was an extraordinary gush of the aqueous constituent accompanying it. But at other times the total bulk of the renal discharge was very deficient. The habits of the urinary department in the patient's economy closely approach to those which we find in that peculiar metamorphic (or functional) disease of the nervous sys-

---

[1] Tea, 2 pints, with 3 ozs. of milk, and sugar *q.s.* Bread, 12 ozs ; butter, ¾ oz.; broth, 1 pint with 4 ozs. of boiled meat; gruel, 1 pint.

[2] There was a suspicion that on this day the patient had surreptitiously taken a dose of Epsom salts. On the next two days the catamenia were present for a short time as in anæmic women.

tem, the various symptoms of which we roughly class as "hysteria." If her family had been cursed with the *damnosa hereditas* of that disease, she would most probably now have developed it. Sydenham shrewdly observes that if there is any difficulty in coming to the diagnosis between hysteria and other possible causes of ailment in a particular case, two questions answered affirmatively are enough to decide in favor of the former, viz.: "Do weariness and worry especially excite the symptoms?" and "Does the patient make sometimes an unusual quantity of pale urine?"[1] —an instance thus of how

> Old experience doth attain
> To something of prophetic vein;

for those are the very hooks on which the most advanced science of the present day would hang all the pathology of hysteria. For it holds it to be—1. A disease of the nervous system; 2. A disease of an exhausted nervous system; and, 3. A disease in which an exhausted nervous system fails to complete the evacuation of the débris of metamorphosis—the very points made in Sydenham's aid to diagnosis.

In the following instance alcohol, acting on a predisposed person, produced marked hysteria, and was identified as the guilty agent by the disease ceasing on its being left off: A clergyman's wife, of a refined sensitive nature, became subject to attacks of nervous sobbing and depression of spirits, especially during changeable weather. She also frequently vomited her food immediately it was swallowed, and at last entirely lost the tone of her voice, being unable to speak above a whisper when under the excitement of a consultation with her physician; but yet she had a brazen-faced repellent kind of manner which sat very oddly on a really modest Christian lady. Long and ingeniously did she elude the discovery that she was given to secret dram-drinking, but was at last persuaded to place herself under the care of a kind Quaker couple, who saw that she had no intoxicating liquor for two months. All the aphonia, vomiting, and other hysterical symptoms went away during this time, and it is believed have not returned. Such is the most usual effect produced on a woman, previously healthy, by frequent small doses of alcohol. She becomes

---

[1] Sydenham, Dissert. Epistol. de Affectione Hystericâ, sec. 78.

hysterical; but rarely does she get delirium tremens from it alone. That form of nervous tornado attacks the female sex only in old age, when the feminine characteristics are annulled, or in consequence of intense mental excitement. But when it does attack them, they are very apt to sink under it. (As *e. g.*, an upper housemaid at a large hotel, falsely accused of theft, died under my care at St. Mary's in two days of furious delirium tremens. Her boxes were filled with empty brandy bottles, though she had never shown signs of drinking.) In man delirium tremens is much more readily induced, and hysteria only exceptionally. It is not necessary to do more than allude to the well-known deficiency of both the watery constituent and the phosphatic salts of the renal secretion in these cases during the violence of the disorder, as that point has been ably illustrated by Dr. Bence Jones in his paper on the subject.[1] But it is worth while to draw attention to the fact, that exactly contemporaneous with a restoration of health is the reappearance in the urine of the phosphates in his third case. It is possible that the delirium may be a clearing storm, the consequence of which is a return to the normal state of phosphatic metamorphosis sooner than happens in cases where the delirium does not occur. For it may be observed in Experiment III in this chapter, where there was no critical paroxysm, that the phosphates remain almost wholly absent from the analysis, in spite of apparently recovered health. And this accords with the familiar fact that hysteria is a much more obstinate disease than paroxysmal alcoholism.

Allusion has been made to the use of alcohol as a protection, in cases of traumatic injury and mental exhaustion, against the pernicious over-activity of the nervous system. And it is possible, too, that it may oppose the entrance of morbid and other poisons by retarding absorption, as is indicated in its popular reputation as a preservative against malaria and noxious fumes. But if it has failed in its mission, if the wound, or the worry, or the virus have been too strong for their force to be broken by such a buffer, then there seems some doubt whether the patient is, or is not, worse off for having taken it. Just as after an operation under anæsthetics there may in some cases, be a similar doubt. But an

---

[1] Med. Chir. Transactions, vol. xxx, p. 21.

habitual tippler, as distinguished from an occasional exceeder, is certainly in more danger from disease than a temperate man under the same circumstances. The venerable Dr. Christison, in a letter dated 1870, writes: "How can we ever hope to express numerically the influence of drunkenness (habitual) in aggravating the mortality from fevers, cholera, dysentery, and other zymotics? . . . Let me conclude with one illustrative fact. I have had a fearful amount of experience of continued fever in our infirmary during many an epidemic, and in all my experience I have only once known an intemperate man of forty or upwards recover. He was the *exceptio quæ probat regulam*."[1]

The action of alcoholic drinks upon the internal viscera seems to be an immediate, and if continued, an habitual paralysis of the vaso-motor nerves; as may be seen by the naked eye in the flushed face of the drunkard. Venous congestion follows, and the glandular structure becomes loaded with black devitalized blood. The presence also of the alcohol in the blood further diminishes the rapidity of nutrition and disintegration, and consequently diminishes the dependent functions of elimination and calorification, and as long as it remains in the body.[2]

Secretion is retarded, and the organic material which ought to form it, grows into a semi-organic morbid tissue, bloodless, insensitive, useless, which we call *cirrhosis*. This gradually dries up and shrinks, like a scar, and thus puckers up the gland it is in, forming "nutmeg" liver, "granular" kidney, "fibrosis" and "chronic condensation" of lung, thickening of the mucous membranes in larynx and elsewhere, and arterial degeneration.

In the case of the liver no pathologists ever seem to have expressed a doubt as to the influence of alcohol upon it; but in regard to the kidneys there has been a controversy as to whether the prevalence of Bright's disease among the working classes should not rather be attributed to their exposure to weather; and Dr. Dickinson, of St. George's Hospital, has contributed somewhat to relieve alcohol of exclusive responsibility. He compares

---

[1] Second Annual Report of the State Board of Health of Massachusetts, Jan., 1871, p. 339.

[2] Dr. N. S. Davis, in the Chicago Medical Examiner, Sept., 1867, brings evidence from sphygmographic observations as to the depressing influence of alcohol upon the circulation.

numerically the frequency of renal degeneration in the bodies of a class especially open to the temptation of excess—consisting of pot-boys, publicans, spirit-dealers, cellar-men, waiters, etc.—with what was found by post-mortem examination in an equal number of other hospital patients of various occupations. The kidneys were reported free from disease in almost exactly the same number of the two classes.[1] Now the principal flaw in this argument is that many of those whose trade is among spirits are nevertheless very temperate persons (my own wine merchant takes only a glass of currant wine with his dinner); and "barmaids" and "waiters," included in the black list, are only exceptionally intemperate. And we must add to this that several of the employments of the other class are really more likely to lead to tippling than the trade in alcohol is. (Who, for example, ever knew a "traveller" who was not obliged to liquor with his customers, as part of his daily work?) Dr. Dickinson rightly points out that the non-alcoholic class are much more exposed to the weather than the others, and this noxious influence, balancing to an unknown degree the influence of alcohol, vitiates the value of the statistics. The only fairly reckoned non-alcoholic persons are total abstainers, who now are to be found in considerable numbers, and might be compared justly with the rest of the population. While I think Dr. Dickinson to be right in his general conclusions, as also in attributing greater importance than some others do to climatic influences in renal degeneration, I do not consider that his statistics bear him out. They are too raw. At present, it seems to me that the physiological knowledge which we possess of the influence of alcohol in health is a more accurate guide to its morbid consequences than any statistics yet put before the world. It is impossible to trust a patient's statement as to drinking habits, and there is no other attainable datum for classifying him as temperate or the reverse.

The "chronic condensation" of the lung induced by alcohol seems to be a form of atrophy not dissimilar to that brought on by the presence of fluid in the pleura, emphysema or frequent bronchitis. It arises from the imperfect functioning of the lung

---

[1] Med. Chir. Transactions, vol. lvi, p. 51 (1873).

tissue, which, like all imperfectly functioning tissues, becomes less vitalized, less fully renewed by interstitial metamorphosis.

This imperfect functioning is a necessary consequence of the limitation of the aerating area of the lungs by the sanguineous congestion which has been already alluded to. The extent to which the habitual congestion of the tissue and limitation of area is carried, even without apparent injury to health, is proved by the diminution of the "vital capacity" for air as shown by the spirometer; no drinker of drams between meals can ever come nearer than 20 or 30 cubic inches to the amount of air the lungs ought to expire in health. Yet so little does he notice this shortening of wind, that it is no uncommon thing at insurance offices (when they use the spirometer) to have proposers make their appearance stating that they have never been seriously ill, and never felt better than at that moment, who are caught out as dram-drinkers by the failure at the spirometer. The partial condensation of the lung from this cause is not so likely to lead to fibrosis as to inelasticity and fatty degeneration of the tissue. The surface of the organ when examined post-mortem takes the mark of the finger pressed upon it, is saturated with moisture, hardly crepitant at all and hardly able to swim in water. But it does not, as a rule, exhibit the whitish gristly stripes, which squeak under the scalpel, and consist of a dense fibrinous substance, nor the contracting scar-like depressions which characterize chronic pneumonia. Nor does it suppurate, I believe.

The chronic condensation of the pulmonary tissue is partly the cause of the wheezy broken speech of the intemperate; but this also depends very often on an irregularly thickened condition of the laryngeal and tracheal mucous membrane, following an habitual sluggish circulation through its capillaries.

The last-named injury to health from excess in alcohol, arterial degeneration, is perhaps the most doubtful. It is so common in all classes that it is very difficult to arrive at direct proof of the tendency to it being increased by this cause. The principal fact in evidence is that it is more common among men than among women, and so is also drinking. But when I review the physiological instead of the statistical aspect of the argument, I can hardly hesitate to assign a serious influence in the induction of the lesion to alcohol. Alcohol, which lowers all the vital actions we know of, is just the agency one would think most pernicious, in replacing elastic and fibrous tissue by that of an inferior grade, friable and unorganized. Under arterial degeneration there ranks itself a long list of evils, of which those best known by name to the public are apoplexy and dropsy, justly reputed to be due to drink. To the medical mind they naturally suggest themselves, and it is needless to enumerate them here.

When we consider how completely the whole body is saturated by alcohol taken in any marked excess, we cannot be surprised to hear that the evil effects of it are not rarely transmitted to future generations. In a recent work,[1] Dr. Auguste Voisin has published the results of an investigation into the consequences to the children of the father's addiction to intoxication, and very terrible they are. In seventeen cases of the spark of life being struck during the drunkenness of the father, infantile convulsions caused the child's death in eleven. Less fortunate were those who survived; three were idiots, two were epileptic, and one the subject of chronic myelitis. Of course this list is not a full statistical representation of the consequences of the parents' excess, for those who escape do not become subject to medical examination, and there is no means of knowing how many they may be. But it is a very terrible list for all that.

Chronic alcoholism in the parents is still more cursed than the action of the acute condition above alluded to; for it does not cut off the victims by a sudden blow, but allows them to live on in indefinite wretchedness. Of eighteen cases traced by Dr. Voisin, eight were idiots and ten were epileptic.

Having attributed so many ill effects to alcohol, it is but fair to inquire whether something may not be found as a set-off beyond the mere dietetic use mentioned at the beginning of the essay. If it causes disease, may it not also prevent? There is a prevalent notion that it excludes the formation of tubercle. The fact given in proof of the notion is that publicans do not often die of phthisis, and Dr. Walshe indorses the truth of it.[2] My own limited experience certainly confirms that observation, but yet it goes rather to show that their not dying of phthisis depends, not on the ex-

---

[1] L'Alcoolisme et la Séquestration des Aliénés, p. 3
[2] Diseases of the Lungs, 4th ed., p. 453.

clusion of tubercle, but on the postponement of the second stage of the malady. The tubercles do not so soon break down into suppuration. A publican who was a patient of mine for several years, off and on, died of hereditary consumption at a much later period than the rest of his family, and indeed after he had ceased to be a publican; but the first development took place while he was actively engaged in aiding the spirit-trade by precept and example. Nevertheless, it is something gained if the softening of the tubercles is in any way arrested by the use of alcohol, even though the arrest should be followed by a more general break-up in the end. But the use of alcohol in the treatment of disease, as contradistinguished from its prevention, belongs to the next section, where the subject shall be resumed.

The dietetic use of small quantities of alcohol tends to ward off morbid conditions which are secondary upon atonic dyspepsia; and therefore it may be credited with the prevention of anæmia, emaciation, premature old age from worry, melancholia, sleeplessness, and any other possible consequences of that form of indigestion in a considerable section of the population. Our weaker brethren (γάστερες ἀργοί, "slow stomachs," as the poet with anatomical precision calls them) owe it to alcohol, and often to alcohol only, that their weakness does not become disease. Very likely the *dura messorum ilia* are better without it, so long as they continue tough; but how long is that in this age of feverish cultivation of the soul at the expense of the body? And ill indeed can the world spare many of those weaker brethren, who without alcohol would be wailing invalids instead of the pillars of the country.

The form in which the product of alcoholic fermentation is presented to the digestive organs is far from being indifferent. The various methods of making it agreeable, or convenient, or cheap, offer an article varying from one little distant from a poison to one where all the advantages are obtained with an inappreciable risk of disadvantage. The most wholesome of these beverages are the immediate results of fermentation in its simple integrity, or with the addition of water only. But it is very seldom that the saccharine matters used are good enough, or the makers skilful enough, and honest enough, so to manufacture their liquors as to be agreeable to the palate, or capable of remaining agreeable for a sufficient time. And, therefore, the pure but nasty liquor has to be distilled; and the distillation partly concentrates, but principally newly develops the noxious "fusel oil," the bugbear of the distillers. All new, and therefore all cheap, spirits are made by it nauseous to the unaccustomed palate, and deleterious to the health.

Of pure wine the ripening is soon effected, and after a year or so, the palate only has to be consulted as to the time of consumption. But it is not so with vintages whose peculiar flavors it is desirable to preserve by retaining unfermented a portion of the sugar, and the oil of the grape-stone. Unless these wines are "fortified" with spirit, they decay prematurely, grow mouldy or "bitter," rancid, and sour. Now, if pure alcohol or old French brandy were used for this purpose, little harm would be done, but the manufacturers commonly employ for cheapness new potato or corn spirit, loaded with fusel oil; and then the only chance for the consumer getting a wholesome beverage is to keep it at the risk of deterioration, till it has become a costly luxury. Woe to the drinkers of cheap Peninsular wines and their imitations.

There has been some question lately as to how "purity" in wine is to be defined in legislative or other literature. Is it an adulteration to add alcohol to wine? I do not see how it can be considered so; and if it were, the proof of the addition of one of the normal constituents of the liquid would be difficult. Just as it is difficult to decide, except by a conventional standard, that the water in our milk does not come from the cow. And conventional standards are not strict justice. But if the offence were held to consist in "fuselling" wine, there would be but little impediment to our learning from an analyst whether we had an article fit for human consumption or not, and also its proximate commercial value.

What principally concerns us here is that not all the evil of strong drinks is to be attributed to alcohol; and that a man may by over-indulgence in "pineapple lozenges," "pear drops," or "fruity hardbake," made as usual with the products of the semi-decomposition of fusel oil, possibly do his health more harm than by visits to the tap-room.[1] The whole evil of drinking does not

---

[1] In the Fourth Annual Report of the State Board of Health of Massachusetts (January, 1873), mention is made by Dr. H. K. Oliver of two children

arise out of fusel oil; for we see the consumers of the very soundest beer and wine often suffer; but that an incalculable addition is made to it thereby is beyond doubt, and great gratitude is due to every one that helps us to exclude it from our beverages.

There are also doubtless other qualities besides the exclusion of fusel oil which may render fermented liquors wholesome. Four years ago there was living at the village of Menidi, near Athens, a priest 90 years of age, who from early manhood had consumed a dozen bottles of Greek wine a day, partly at meals, and partly at odd times.[1] It would be interesting, no doubt, to know the structure of this venerable toper's kidneys, as specimens of typical development, but the consul is probably right in attributing much to the quality of the wine, which is preserved from chemical change by means of rosin instead of fuselly spirit. Rosined wine *is* an acquired taste; and the British artist, when he first quenches his thirst at a Tuscan farm or rustic inn, is apt to exclaim that the landlord has drawn the wine in a varnish-pot, and to sneer at the balsamic and wholesome *Vino Vermuth*. But it is a taste well worth acquiring by thirsty souls in warm climates, and to be patronized by philanthropists. I understand that the flavor of *Vino Vermuth* is partly given by wormwood; and if so, the acknowledged wholesomeness of the wine would seem an answer to M. Magnan and others who have been trying to fix upon the innocent herb the blame of what they call "Absinthisme," and which is stated to contribute so largely to the degeneracy of the French nation.[2]

Water, again, and salt are directly antagonistic to alcohol in their physical action on the animal body. They both, especially the former, augment metamorphosis, as the previously quoted researches of Dr. Boecker fully show. Their presence therefore in the wine, especially if they are thoroughly incorporated with it from its first origin, cannot but be beneficial. On this ground,

---

being seized, after sucking a candy anchor, with alarming sedative symptoms, requiring active medical treatment. The flavoring was pine-apple essence (butyric ether).

[1] See a letter from the United States consul at Athens in the Second Report of the State Board of Health of Massachusetts, p. 258.

[2] Les diverses formes du délire alcoolique, par le docteur Magnan. Paris, 1874: Delahaye.

the addition of sea-water to the must (the marriage of Ampelos and Amphitrite of the Greek gem) is a custom to be highly commended.

Are there, against the ill consequences traceable to alcohol, any other preservatives, besides these found in the liquor itself?—any guardians, innate or acquired, resident in the body of the consumer?

There are probably few readers of British literature who have not smiled at Addison's light banter of the amateur self-doctor, who says he commits an excess on the first Monday in every month for the good of his health.[1] But perhaps all are not aware that this is a strict carrying out of the opinion deduced by the ingenious Sanctorius from careful observations made upon his own person. He says that "they who use regular diet want the benefit of those who debauch once or twice a month."[2]

He does not give it as a popular notion, and it does not appear to have been so, till attention was drawn to his series of self-denying and generally conclusive experiments, with its portrait of the author dining in his scales opposite the title. But if the previous aphorisms be read, it will be seen that he grounds his opinion on the fact, that the weight of the body is less the next day "without any sensible evacuation."[3] This merely implies that less food has been assimilated, an advantage only to those who habitually eat too much. But the author seems to think that "the corruptible body weigheth down the soul" in direct proportion to its pounds and ounces, a most unphilosophical superstition. All the possible benefit of the monthly debauch might be gained, in certainly a more dignified manner, by a day's fasting.

The only justification for an aphorism pregnant with so much danger is the observation, which is true enough, that occasional excesses, however much they may disturb the functions at the time, are infinitely less noxious than habitual dram-drinking. The most certain victims to alcohol are those who are always sipping, though "never drunk in their lives;" whereas we have all of us known many a toper who has attained a green old age by letting an interval of recovery elapse between his bouts of riot.

---

[1] Spectator, No. 25.
[2] Medicina Statica, Aph. 100. Quincy's translation.
[3] Id., Aph. 99.

A powerful collateral preservative is the kidney. If this organ
acts well and quickly, above the average of human kidneys, a
large proportion of the alcohol swallowed passes away unchanged
and harmless in a short space of time, as is shown by the known
researches of Dr. Percy, Mr. Masing, and the French physiologists, Perrin and Duroy, who have repeated their experiments.
This peculiarity probably distinguishes those strange mortals who
never seem the worse for anything they drink.

A further preservative against the wiles of alcohol is to be
found in a full aeration of the blood. Those who can bear much
liquor without constitutional effects have generally a high vital
capacity of chest in proportion to their stature; whereas those
whose health is injured by it fail to raise the spirometer to the
proper measure of cubic inches; so that by this means constitutions damaged by drink can be detected at insurance offices. It
may be observed also, that a free supply of fresh air during meals
modifies considerably the results of imprudence.

§ 2. Moral Effects of Alcohol.

As man consists of mind as well as body, it is impossible in
this essay to omit all mention of the moral effects of the consumption of alcoholic drinks, though the limitation of space prevents
a full discussion of the subject. All nations that have led the
van in the march of civilization, have been addicted to drink—
aye, and addicted to drunkenness. The Jews, the Greeks, the
Romans, the Germans, the Swedes, the Danes, not to mention the
English all round the globe, are amply attested by literature to
have been distinguished above their contemporaries in this way.[1]
It is true that some reactionary races, famous as conquerors, have
been abstinent, but they and their faiths are dying out, and the

---

[1] Dr. Bowditch contends that proneness to drunkenness is governed by climatic law, becoming stronger in proportion to distance from the equator.
But as he allows of the influence of hereditary disposition, the law does not
guide us to much inference. The widely dispersed races are not appreciably
affected in their habits by isothermal lines. Facilities for growing the grape
depend doubtless on climate, but I think Dr. Bowditch's own map and data
exclude its influence on man in respect of alcohol.—*Third Annual Report to
State Board of Health of Massachusetts.*

coloring they have given to civilization is even now fainter than that left by the robuster races a thousand years before they were heard of. Yet we learn at the same time from the same impartial witness, that vice, and vice too of a sort which is destructive to the social wellbeing of a community, is conspicuously traceable on a vast scale to strong drink. We read of it in patriarchal times as leading to a breach of the instinctive reverence due to parents, then to the vindictive cursing of a son, then to incest, then to murder, and so on in a continuous series up to the wife-beatings in yesterday's police reports.

Then, side by side with the innumerable judges' charges and prison reports, establishing irrefragably the fact that by far the greater proportion of crime brought under notice is due to drink, one reads with astonishment such unexpected revelations as those contained in the following letter from the United States consul at Copenhagen to the Massachusetts Board of Health. Mr. Yeaman says, after citing statistical evidence:

"Here is a people evidently more prosperous than formerly, evidently using more brandy than formerly, and evidently less given to intoxication than formerly. . . The people here appear so very sober, that I have been simply astonished to find how much brandy they really use."[1]

It is superfluous to notice that as intoxication is less, crimes of violence are less also.

Mr. Yeaman goes on to add:

"There is perhaps no country in Europe, except Prussia, where the average standard of education and intelligence is so high as in Denmark."

How shall this conflict be reconciled? How is it that ignorance, poverty, and vice, which daily are degrading the individual drinker, spare the community? Have the statisticians erred by collecting evidence only for the prosecution, and omitting a counterbalancing class to the criminals? I think so. The temperate man, who takes, in proportion to his needs, fermented liquors with his meals, has been put out of sight altogether. Yet is he the pillar of his country's prosperity, prolific, prosperous, and contented. It is probably his increase in number and power

---

[1] Second Annual Report, 1871.

which balances the amount of vice and misery caused by excess in drink. It is not safe, then, for the political economist to neglect his interest, as would be done by legislative measures tending to make the beverages he requires more difficult of attainment, by impediments to their purchase or the enhancement of their price. Still less is it safe by social pressure to lead him to banish the beer jug or the bottle from his table, and hide it in secret cupboards, or to drive him to sly grog-shops between meals. He should rather be induced by the cheapness of wholesome fermented liquors to employ them for domestic use, and to forego the dangerous product of the still.

In Nineveh of old (as Mr. Smith reads to us from the arrow-head records) Ishtar, goddess of carnal love, behaved at one time so scandalously, that the reigning sovereign persuaded the heavenly powers to let him banish her to the nether world. Few that read the published translation in the *Daily Telegraph* last year (and who did not?) can forget the picturesque monotone in which her gradual decay into silence is narrated at solemn length. But no sooner was she put down, than matters essential to the future continuance of the race went all wrong, and the frightened mover of this early repressive bill was obliged to petition for her return; and back again she came in the same tedious way that she went. If this is history, it is a valuable experience; if it is a myth, a prescient knowledge of man's nature is shown, pregnant of warning to the legislator.

We need not hesitate to transfer to the regulation of matters connected with alcoholic excess experience amply confirmed by the history of all ages, derived from another cognate temptation. The crimes attendant on both are developments of selfishness, and of the want of that self-restraint which we were sent into the world to learn, and which a man may fairly complain of a so-called paternal legislation depriving him of the opportunity of learning.

But the removal of the means of excess, out of false tenderness to the exceeders, and at the expense of the temperate, is one thing, and the repression of their excesses is quite another—nay more, it is antagonistic. Pagan and Christian moralists have, perhaps, rightly held that of the three main incentives to crime, malice, brutality, self-indulgence, the last is the more pardonable, and involves the least sinfulness.[1] And so long as law was considered to represent a vengeful Deity, and to punish in proportion to the moral enormity of the sinner, so long was drunkenness rightly pleaded in extenuation of offences. Even now in Italy it legally mitigates punishment. But I believe I am correct in saying that a higher political education has in our age changed this view, and the province of law is daily more and more considered to be confined to the protection of society, not of God—"*Deorum offensa Diis cura*" is said not in scorn, but in reverent humility. Now society, without a doubt, runs greater risk of damage by violence at the instigation of drunkenness than at the instigation of all other incentives whatever; and therefore justice has advanced so far as in theory to hold it in aggravation of an offence. She feels that she ought to be most a terror to those from whom she is in greatest danger. In theory, yes—but she has not reduced her theory to written rules; and in practice we daily read instances of the mediæval prejudice prevailing, and of criminals lightly treated because they are drunk. I should like to see this barbarous squeamishness pass away, and to see society protected by adding some deterrent degradation to crime committed during intoxication. I should like, for instance, to see the brute who kicks his wife and children when in liquor, well flogged, in addition to the penalty for assault. And drunken exposure of the person, and insults to young girls, might fairly be treated with the same indignity. The less innate turpitude there is in a criminal, the more possible it is to deter and reform him. Severity is worth practicing when there is a chance of its being successful; just as

---

[1] Dante's angelic guide is described as quoting Aristotle with approval:

Non ti rimembra di quelle parole,
Con le quai la tua Etica pertratta
Le tre disposizion, che'l Ciel non vuole;
Incontinenza, malizia e la matta
Bestialitade? e come incontinenza
Men Dio offende, e men biasimo accatta?

Inferno, xi, 79.

Reference is intended to the seventh book of the Nicomachean Ethics. "Incontinenza" is a translation of ἀκρασία, and does not mean "incontinence" or "intemperance" in their modern narrowed senses, but includes all varieties of the recklessness of unclean living.

when a patient has a good constitution, a vigorous treatment of accidental ailments is to be commended.

To sum up the influence of alcohol upon a healthy man, it would seem to be as follows:

(1.) When taken in moderate quantities, and with meals, it often somewhat increases the appetite and the digestive powers. And the *modus operandi* by which it effects this object, is by curbing the deleterious consequence of excessive energy of the nervous system.

(2.) When taken in excess its first noxious influence is on the nervous system, and hence we find in predisposed women "hysteria," and in men "alcoholism" or "chronic delirium tremens."

(3.) When taken in excess it so far impedes the renewal of the nerve tissue, that the vital activity of the bloodvessels is diminished over-much and over-long.

(4.) Of this diminution of the circulation, one result is local congestion of various parts, temporary at first and quite capable of relief, but by long continuance tending to produce permanent lesions.

(5.) If too long deprived of the nervous energy which presides over interstitial growth, the parts fall into atrophic degeneration; and instead of elastic, fibrous, areolar, muscular tissue, and whatever else of subtle structure may be needful to animal life, there is deposited only a mass of friable, lifeless substances, midway in chemical composition between albumen and fat.

(6.) It is probable that these noxious influences are continued so long as the alcohol is retained in the blood.

(7.) The principal lesions induced by steady regular excess of alcohol are cirrhosis of the liver, Bright's disease, condensation of the lungs, thickening of the larynx and bronchi, ossification and thickening of the arteries. And it is difficult to say what secondary lesions may not follow upon these.

(8.) The increased risk to life from injury or acute disease, incurred by the intemperate, may be accounted for, by some amount of the afore-mentioned atrophy having already taken place in their bodies, and so diminishing their force of recovery.

(9.) For there appears to be a likelihood that the immediate anæsthetic power of alcohol will tend to spare the nerves the shock of the external causes of disease, and so render it less injurious to a previously sound man.

(10.) Alcohol cannot be trusted to prevent tuberculosis, though it retards its progress.

(11.) Taken in still greater excess at once, in fact so as to produce drunkenness, it is probably not nearly so fatal in its consequences. The poisonous drugging of the nervous system is followed by an attack of indigestion and general pain, during which the blood purges itself of the alcohol.

(12.) Even should the blows to the nerves inflicted by drunken bouts be continually repeated, the consequences follow in crises, and may be recovered from, which cannot be said of atrophic lesions. These crises usually consist of the well-known phenomena of *delirium tremens*, which often so alarm the patient as to lead to a reform.

(13.) Some forms of alcoholic drinks are much more deleterious than others, especially those which contain a large proportion of fusel oil.

Hence we may deduce these rules for the use of alcohol by healthy persons.

(A.) Let it be taken never as a stimulant or preparative for work, but as a defence against the injury done by work, whether of mind or body. For example, it is best taken with the evening meal, or after toil.

(B.) Let the increase in the desire for and power of digesting food be the guide and limit to the consumption of all alcoholic liquids.

(C.) Let the forms be such as contain the least proportion of fusel oil.

(D.) Let all with an hereditary tendency to hysteria or other functional disease of the nervous system, refrain from its use altogether, even though as yet they are in good health. Among the hereditary tendencies must be classed a proclivity to delight in drunkenness, which remarkably runs in families. Children with such an heirloom had best be kept as late as possible without tasting strong drink.

The legislative authorities have the opportunity of materially advancing the good of their country:

(*a.*) By cheapening through fiscal regulations all wholesome fermented drinks, such as good beer and wine, and by laying the burden of taxation on the retail trade in spirits.[1]

(*b.*) By declaring fusel oil to be a poisonous adulteration of alcoholic beverages, so that the mother of invention may teach distillers to eradicate it.

(*c.*) By severely punishing intoxication.

(*d.*) Habitual intoxication might justly be made a ground for the dissolution of marriage, in that a wrong is done to the nation, as well as to the innocent partner, when it is burdened with members incapable of contributing to the common weal, which is so likely to be the case with the progeny of drunkards. When the wrong is involuntary, we pardon it; but alcoholism cannot plead that excuse, unless it take its stand on the ground of insanity, which removes the question into another category altogether. Those who claim the rights of free agents must answer for all the consequences of their acts.

Were the law to express such a marked disapproval of drunken husbands and wives, the benefit would extend much wider than the few whom it might free from the chain of matrimony; for it would teach men and women what a grave offence is committed by risking the parentage of an idiot, and lead them to take means to avoid it.

Nevertheless the most powerful engine for securing to mankind the beneficial influences of alcohol without its attendant dangers,

---

[1] Encouragement may be taken from what has happened in California. Mr. James Morrison writes:

"Before the introduction of native wines, when the stronger alcoholic stimulants were used, drunkenness was very common in California. The native wines are now found in nearly every houshold; they have supplanted to a great extent the use of the stronger alcoholic drinks. It is a fact, which no one familiar with California for the last twenty years will deny, that drunkenness is much less common now than formerly. This change is undoubtedly due to the substitution of the wine of the country for the stronger alcoholic stimulants." The very decided effect which financial legislation can have upon the comparative consumption of different sorts of liquors is shown by the effect of the budget of April 6, 1872, which laid a heavy tax upon absinthe in France. Within a year from that date M. Bergeron was able to report to the temperance society that the consumption of the said liqueur had diminished by nearly one-half. See Revue des Deux Mondes, March 15, 1874, p. 476.

is to be sought in the habits and silent pressure of social life. As our race becomes more humanized by training and experience, it is more to be trusted to make use of Heaven's gifts without abusing them; and the degradation of mind and body which alcohol is capable of entailing will become hateful at the same pace as its true value becomes understood. Such a progress is worth a hundredfold all possible legislation. The man who prefers law to custom announces a principle which would set the means above the end, would stereotype imperfect efforts, and substitute cork jackets for the power of swimming. Towards the progress of temperate habits every man can contribute—and who more powerfully than the medical man?—by the example of the moderate use of fermented beverages of the right sort at proper times, and by the discouragement of indulgence and excess. There is a difficulty in sending a drunkard to an asylum, but he is easily sent "to Coventry," and made to feel that he is a degraded animal. There may possibly be some rare cases to be found of true "dipsomania;" where, without any other mental disease, the patients are carried off by an uncontrollable impulse to drink; but they certainly are very rare indeed, and every alleged instance that I have investigated has always exhibited also some other form of insanity, sufficient to justify the imposition of restraint, or else proved to be using the cant of the day as an excuse for self-indulgence. The exceptional cases may be separately dealt with, when they occur; but as a rule I think it is better to give men the education of being their own gaolers than to let them lean on the weak crutch of state inebriate asylums.

Voluntary association for the purpose of mutual encouragement in temperance, for the discouragement of drunkenness and of the use of ardent spirits, or for the prosecution of the sellers of adulterated beverages, stands on a different footing from legislative control. It is a step in self-education. But unhappily it has been stopped from useful employment by the fatal snare of exaggeration. The denouncement of alcohol as always a poison, and its use for good fellowship as a sin, is now inseparably united in our minds with societies for the promotion of true temperance; and thus the latter virtue has been retarded by the fatal prejudice of being tied up in the same bundle with a fallacy. The society formed in France about two years ago, under the title "*L'Asso-*

ciation française contre l'abus des boissons alcooliques," though it may not make such a noise as our teetotalism, has a better basis of action. The proposed object is to encourage the substitution of inoffensive and salutary drinks for spirits. It accepts and recommends the use of beer, of natural wine, as well as of tea and coffee at meals, after the day's work is done, but sets its face sturdily against distilled liquors, facilities for liquoring between meals, and all excess. For my part I have faith in the continuous improvement in the human race; and I think I can see, not so very far ahead in the future, the influence of alcohol upon health perfected in its good, and deprived of its evil, without the dubious aid of any prohibitive legislation.

Some well-meaning persons think to discourage intemperance in drink by affecting a cynical carelessness as to the quality of that which is consumed. They allow their guests "a little wine for their stomachs' sake," but serve it out as like a drug as possible.

It is nothing to them if it be adulterated or not—"wipe away all poetry, all refinement from the cup," say they, "and it will cease to tempt." This reasoning is utterly false. As long as a sensual pleasure is coarse and rude, it can be attractive only by its quantity, and all indulgence can be only excess. But let it dress itself in the charms of a higher æstheticism, let the enjoyment of it be shared with others, and be associated with noble and beautiful memories, and moderation becomes a law of its existence. The sensuality of the savage is selfish, shameless, and unbridled. With the advance of manners it first parts with its selfishness, then it acknowledges the evils of excess, and then curbs itself more and more till it ceases to be intemperance at all. True of other gratifications of instinctive desire, this is especially true of drinking. The better the wine is, and the more attractive its aspect, the less likely are people to take too much of it. When Anacreon and Horace, Phidias and Cellini, Etruscan potters and Venetian glass-blowers, conspired to wreath the goblet with a halo of romance, they substituted an elegant appreciation for a degrading animalism, and led society a step onwards in morality. And it may be noticed that our Divine Exemplar first manifested forth his glory by giving to his village neighbors not only wine, but a pleasanter wine than they were accustomed to. I, therefore, hold it to be a duty of every one to see that whatever sort of beverage is provided for his household should be the best of its sort, the most agreeable to the educated palate, and adorned with the most refined surroundings that circumstances allow. However little a man's purse allows him to drink, let it be good.

PART III.

DIETETICS IN SICKNESS.

CHAPTER I.

THE DIET AND REGIMEN OF ACUTE FEVERS.

THE principal acute fevers enumerated in the Nomenclature are small-pox, chicken-pox, measles, scarlatina, typhus, enteric (or gastric), relapsing, yellow, remittent, cholera, mumps, influenza, weed. Some varieties of these, characterized, some by more severe, and some by milder phenomena than ordinary, receive special names, which would needlessly lengthen a list cited simply for the purpose of explanation.

One of the most natural and practical divisions of diseases is into *acute* and *chronic*. We are not always quite sure in which class to place the special morbid process we happen to be considering; but in an overwhelming majority of instances there is no doubt at all; and when we have decided, the light thrown on the path to judicious treatment is clear and bright. By *"acute"* disease is meant such as has a tendency to progress in a circle towards the recovery of health: each process, however dangerous and abnormal it may be, being a step towards the final arrival at that result, if only the sick man's strength hold out. While of *"chronic"* the natural road is straight on from bad to worse, unless from the interposition of some extraneous agency of accidental or designed origin foreign to the process of the disease itself. Dr. Pierre Petit, in the preface to his "Commentary on Aretæus," compares the former to race-horses, which run round to the goal, unless they founder on the way: one might in the same strain liken the path of the latter to that fatal descent which leads to Avernus. This division of maladies is of the most essential importance when we undertake to estimate the value of remedial measures. Nearly all the fallacies which have overladen our druggists' shops, not to speak of a variety of theories ending in "pathy," which crop up from time to time, flow from watching the acute phenomena in disease, as an index of the effect of a medicine. A very moderate spice of medical logic will suffice to show that it is only from the numerical comparison of many public institutions for many years that an opinion can be formed in acute disease concerning that effect. Whereas by observing its action in chronic disease, one cautious man may, from a very small number of well-considered cases, come to a reasonable conclusion as to the value of any really active medication. I am used, therefore, in instructing pupils, to insist very much on the differences between acuteness and chronicity: I believe the idea to be of great importance in the formation of the medical mind. And in the limited department of therapeutics treated of in this volume it must not be passed over, for in the most chronic cases there are acute phenomena, and in acute cases there are sometimes elements of chronicity which require a just estimate to be formed of them, when we come to test the efficacy of our efforts at relief.

In acute fevers, it is evident that it is not our business to try and arrest the symptoms, any more than it is our business to arrest the rising tides. We must only take care that they do not swamp our boat, and the safest way to do that is to keep the vessel as sound as possible and as capable of resisting external agencies as we can make it. The value of nutrition in fevers has been observed from the earliest times: Hippocrates thought so much of it that the point of his treatise "On the management of acute diseases" lies in his recommendation of the use of wine, and of a ptisan of barley, which we now call "gruel." The proper preparation of the latter he considers so important, that he condescends to tell us how to make it, so that it may be "thin, but not too thin; thick, but not too thick," as Miss Austen describes the perfection of this dietetic article. But during the dark ages of medical science a bugbear called Inflammation was set up, and acute disease grew to be considered a devouring flame, which must be starved out by removing fuel. This scare culminated in the French physicians of the last generation, who actually *deprived the sick of all food whatsoever*, as a mode of treatment, and called it *diète*. That was going too far, and the results terrified the neighboring islanders, who in the person of Dr. Graves found a voice, and maintained, not only in deeds but in words, that the "feeding of fevers" was the most essential feature in their cure. In following his master, Dr. Graves goes a step beyond the cautious Hippocrates; for though he sticks to gruel, barley water, and whey, the first three or four days, he quickly after that proceeds to chicken broth, meat jelly, and strong soup. It would be a platitude to point out the difference in the mortality since this change of practice. Everybody acknowledges its wisdom, except a very few eccentrics.

The great art of duly nourishing fever patients consists in giving a frequent, almost continuous, supply of liquid nutriment containing very soluble aliments in a dilute form. Perhaps I may be allowed to quote myself on this subject, and reproduce a clinical lecture I delivered at St. Mary's Hospital thirteen years ago.

"The physician sees that a large supply of nitrogenous material must be wanting. The nitrogenous tissues are devitalized, are drained away in a disproportionate excretion of urea and other organic compounds, and nothing is taking their place. Shall he act antagonistically, and try to stop the passage of urea by the kidneys? I do not know exactly how he would set about it; but I do know that if he succeeded, he would do positive harm; for the very worst cases of fever are those in which metamorphosis is active (as shown by the heat), while the excretion of urea is arrested (as shown by the lightness of the urine); they resemble cases of uræmic poisoning from diseased kidneys. The other principles of treatment which I noticed in my introductory lecture would not, perhaps, be so directly injurious, but common sense would still allot the palm to restoration here. Let it be your chief aim to supply that which you clearly see is passing away—nitrogenous tissue.

"But how will you supply it? Solid food would in all probability be vomited, from the unbearable loathing it excites. If not vomited, it would lie for some time a mere foreign body outside the mucous membrane of the digestive canal, and then pass away by diarrhœa, with much flatus and fetor, and much disengagement of gas during putrefaction. Your beefsteak might as well have been at once thrown down its final destination, the water-closet, to which it passes putrid, though undigested. Neither is it wise to fill the stomach with large quantities of victuals, for the same result follows. No 'meals,' therefore, must be allowed; and prudence suggests the giving in their place very small doses of nitrogenous aliment very frequently. These pass over the irritated stomach unconsciously, and are taken up gradually by the intestines, requiring but very little to make them fit for absorption. The suitablest food is that which is naturally supplied to the weakest stomach. The feeble digestive organs of babies can assimilate milk, and milk forms the most appropriate nourishment for the debilitated viscera of the fever patient. By giving two or three ounces every hour, you may get down a quart and a half per diem. But under ordinary circumstances every two hours is often enough. If there is sufficient acid left in the stomach to coagulate the casein into clots, and cheesy lumps are rejected by vomiting, as happens sometimes in milder cases, you may guard against this by adding liquor calcis or soda-water to the milk, or you may supply its place by beef tea. But it is the lumping of the cheese into solid masses that it is desirable to avoid, not the acidification, which is beneficial. If the patient takes thus a good supply of milk and beef tea, not only is the imminent danger of death by starvation avoided, but the emaciation which follows during convalescence is much less extreme, and the dangers in its wake less formidable.

"Eggs are a highly nutritious food; if taken raw, and diluted with milk or water, they are quickly absorbed. But should they be delayed and putrefy, the products of their decomposition are peculiarly injurious: the sulphuretted hydrogen and ammonia evolved are poisons to the intestines. I should recommend you to avoid eggs till convalescence has restored the gastric powers. The same objection does not lie against milk, the lactic acid arising from whose decomposition assists in the solution of the casein. Sour buttermilk is by no means to be despised as a food."[1]

When a patient cannot be raised in bed without risk of exhaustion, a crockery or glass feeder is a convenience, but the same vessel, or even one of the same appearance, should not be used for food and for medicine.

---

[1] Lectures Chiefly Clinical, London, 1864, lect. vi of 4th ed.

If the patient's mouth be foul, as in small-pox or putrid fever, it should be cleansed when he is fed. The administration of nutriment should then be so frequent that it is not allowed to become again foul. In fact the cleaner it is kept in the intervals, the better.

Food taken by the mouth should, as a rule, be as near the natural temperature of the body as possible. But when the febrile heat is very high, or there is much nausea, some of it may be iced with advantage.

When the stomach is inflamed, as we are taught by morbid anatomy to know it is in severe scarlatina, or when there is vomiting, only a few teaspoonfuls should be drunk at once and everything cold and dilute. The admixture of pepsin, in quantity not exceeding 20 grains a day, is also beneficial to these patients. If it purges, it may be guarded with a few drops of laudanum.

In obstinate vomiting and other instances where the ordinary paths of absorption refuse to fulfil their functions, it is necessary to get nutriment taken in by unusual routes. This is often the case at the height or later end of acute fevers, and then life may be sustained a long time by nutritive injections into the lower end of the bowel. Physiological experiments undertaken with the view of testing the power of absorption in these parts have demonstrated that many soluble drugs may be introduced into the blood through their mucous membranes. Opium, for instance, acts quickly when so given. And if drugs, why not aliment? Food thus administered in enema must be dilute, and it must be warm, and then very likely the still secreted intestinal juices which descend from above may dissolve a considerable amount of its starch and animal fibre. Yet a surer way of securing the solution is to add a small quantity of pepsin to take the place of gastric juice, and also of diastase (in the shape of malt) to supply the deficiency of saliva.

When life seems passing away under their eyes, the friends will often shrink from tormenting (as it seems to them) the sick man with food. Let them not despair: many a one has recovered after the doctor has taken his leave with a sad shake of the head, and without making a fresh appointment. And let them also be stimulated by this fact—namely, that the pains of death are aggravated, if not mainly caused, by the failure of nutrition. Even when apparently insensible, the dying suffer much increased distress from want of food, though they cannot express their sufferings.

The use of alcohol in fevers is regulated by the condition of the nervous system. If there is great prostration of strength, or tremulousness of the hands, or quivering in the voice and respiration, if there is delirium of a low muttering character when the patient is left quiet, then it is required. Or if the patient is habituated to a full allowance, it is well to continue to give a little. A sharp, weak, unequal beat of the heart is a warning that some of these symptoms are likely soon to come on. All these indicate that the nervous system is feeling very sensitively the destructive metamophosis going on, and has its power lowered by its sensitiveness. Then is the opportunity for the strong anæsthetic we are speaking of, which I order without scruple, though I do not view it, like food, as part of the necessary cure of fever. Above all, the friends must be warned not to employ it as a substitute for food: it may be useful as an adjunct, but can never take the place of true restoratives.

The form of alcoholic liquors must depend a good deal upon the purses of those who pay for it. Sound port, burgundy, and champagne, are the best, but I do not despise cheap gin if I can get nothing else.

It contributes greatly to the appetite of a sick person if the nurse is of fresh, clean, and cheerful aspect. Her dress should be of some washing material which does not rustle, of a soft warm color, and not stuck out with crinoline or any other sort of "dress improver." Black is always nasty to the sensitive olfactory nerves of an invalid. Above all, she must have a quiet decided manner, and never fidget.

Cookery for Fever Patients.

White Wine Whey.

(All Wheys are Sudorific and Nutritive.)

Put two pints of new milk in a saucepan and stir it over a clear fire till it is nearly boiling; then add a gill of sherry, and simmer for a quarter of an hour, skimming off the curd as it rises. Then add a tablespoonful more sherry and skim again for a few minutes.

Rennet Whey.

To a quart of new milk, either warm from the cow, or heated up to the same temperature, add a large tablespoonful of rennet.[1] Keep up the heat a little higher till the curd separates, and take it off with a spoon.

Lemonade.

Pare a lemon very thin and put the paring in a jug with an ounce of sugar candy. Squeeze the lemon into it, and pour on a pint of boiling water. Orange or pineapple may be used instead of lemon as a variety.

(Other aqueous drinks, good in both health and sickness, have been already mentioned at the end of Chapter II, § 10, in the First Part, p. 85.)

Linseed Tea (Demulcent and Diuretic).

Whole linseed, one ounce;
White sugar, one ounce;
Liquorice root, half an ounce;
Lemon-juice, four tablespoonfuls.

Pour on the materials two pints of boiling water, let them stand in a hot place four hours, and then strain off the liquor.

Do not give this to patients taking lead, iron, or copper.

Barley Water.

Wash two ounces of pearl barley with cold water. Then boil it for five minutes in some fresh water, and throw both waters away. Then pour on two quarts of boiling water, and boil it down to a quart. Flavor with thinly cut lemon rind, and sugar to taste, but do not strain unless at the patient's special request.

Water Gruel.

Mix one large tablespoonful of oatmeal into a smooth paste

---

[1] See end of chap. ii, sec. 7, in first part, p. 67.

with a little cold water. Pour in, mixing all the time, a pint of boiling water. Boil for ten minutes, stirring as before, and strain. It may be eaten with salt or sugar, according to taste.

Another Gruel.

Grits ½ oz.; Water ⅔ pint; Milk ⅓ pint; Sugar ½ oz. (*Children's Hospital*).

Arrowroot.

Mix one large tablespoonful of arrowroot into a smooth thin paste with a little cold water. Pour in a pint of boiling water, and flavor as required.

For "milk gruel" and "milk arrowroot" milk is substituted for water.

Rice Gruel (*somewhat astringent, in cases of diarrhœa*).

Take of
         Ground rice, two ounces;
         Cinnamon, a quarter of an ounce;
         Water, four pints.

Boil for forty minutes; then add a tablespoonful of orange marmalade.

Bael Drink (*in dysentery and diarrhœa*).

The liquid extract of unripe Bael fruit (*Liquor Belœ*), one or two tablespoonfuls to a pint of water.

Alum Whey (*in diarrhœa*).

Use alum in place of rennet to curdle the milk, in the proportion of about a quarter of an ounce to the first. Otherwise make it as rennet whey.

Almond Drink (*softening and nutritive in chest cases*).

A useful and pleasant drink may be quickly prepared by rubbing up two ounces of the "compound powder of almonds" (to be got at any chemist's) with a pint of water.

Claret Cup for Invalids.

Half a bottle of claret to a bottle of soda-water. Half a dozen drops of sweet spirits of nitre (*spiritus œtheris nitrosi*) put into the jug first gives a fruity flavor.

Egg Nogg of the British Pharmacopœia.

Best French brandy, four ounces;
Cinnamon water, four ounces;
Yolks of two eggs;
Sugar, half an ounce.

Rub the sugar and egg yolk together, then add the rest.

Cleansing Wash for the Mouth before Food.

A tablespoonful of Condy's solution[1] in a pint of tepid water.

Egg Soup.

Water, one pint;
The yolks of two eggs;
Butter, a lump as large as a big walnut;
Sugar, according to taste.

Beat them up together over a slow fire, gradually adding the water. When it begins to boil, pour it backwards and forwards between the saucepan and jug till quite smooth and frothy.

Panado.

Bread-crumb, one ounce;
Mace, one blade;
Water, one pint.

Boil, without stirring, till they mix and turn smooth, then add a grate of nutmeg, a small piece of butter, a tablespoonful of sherry, and sugar, according to taste.

Whole Beef Tea.

Make the cook understand that the virtue of beef tea is to contain all the contents and flavors of lean beef in a liquid form; and

---

[1] Or two tablespoonfuls of the "liquor of permanganate of potash" of the British Pharmacopœia.

that its vices are to be sticky and strong, and to set in a hard jelly when cold.

When she understands this, let her take half a pound of fresh-killed beef for every pint of beef tea required, and remove all fat, sinew, veins, and bone. Let it be cut up into pieces under half an inch square, and soak for twelve hours in one-third of the water. Let it then be taken out and simmered for two hours in the remaining two-thirds of the water, the quantity lost by evaporation being replaced from time to time. The boiling liquor is then to be poured on the cold liquor in which the meat was soaked. The solid meat is to be dried, pounded in a mortar, freed from all stringy parts, and mixed with the rest.

When the beef tea is made daily, it is convenient to use one day's boiled meat for the next day's tea, as thus it has time to dry and is easier pounded.

A wholesome flavoring for beef tea is fresh tomato, now so commonly grown on a south wall in gardens. A piece of green celery stalk, or a small onion, and a few cloves, may also be boiled in it. Leeks give it a fusty flavor, and mushroom ketchup, which some cooks introduce, is of doubtful composition.

While this is a-cooking, some more hastily prepared in the usual way may be used, or even Liebig's Extract, provided care be taken that the latter is sufficiently diluted, so as not to turn the stomach.

Quicker but less economical Beef Tea.

One pound of raw beef, minced, for each pint of water. Stir up cold and let it stand one hour. Then place the vessel in which they are mixed in a pan of water and heat them up to about 180° Fahr. for another hour over a slow fire. If boiled up to 212° it becomes gluey, and is not equally nutritious or digestible. Run the tea through a coarse strainer, and flavor at discretion.

Beef and Hen Broth (Gouffé).

Still quicker and still less economical is the following:

Take 1 lb. of lean beef;
½ a hen, boned;

Pound together in a mortar; add ¼ oz. of salt; put in a stewpan with 2½ pints of water; and stir over the fire till boiling; then add carrots, onions, leeks, and celery, cut fine; boil for half an hour; strain, and serve.

Beef tea and broth should not be kept hot, but heated up as required. It may be warmed, but never prepared in the sick-room, for nothing sets an invalid against food so much as cooking. If a nurse can rig up a fire or a gas stove for her own use in an adjoining room, well and good; but if not, she had better descend to the kitchen. In summer, when a fire is undesirable, she may employ Silver's patent Norwegian cooking box, in which cooking is done by hot water. But it is rather an expensive mode of attaining a simple object, unless the nurse were to keep one as part of her stock in trade.

Chicken Broth.

Skin, and chop up small, a small chicken, or half a large fowl, and boil it, bones and all,[1] with a blade of mace, a sprig of parsley, and a crust of bread, in a quart of water for an hour, skimming it from time to time. Strain through a coarse cullender.

Chicken broth poured on sippets laid on the bottom of the dish, makes a good sauce for boiled chicken or partridge, when the invalid is well enough to be allowed solid food.

Mutton Broth.

Lean loin of mutton, one pound, exclusive of bone;
Water, three pints.

Boil gently till very tender, throwing in a little salt and onion, according to taste. Pour out the broth into a basin, and when it is cold, skim off all the fat. It can be warmed up as wanted.

If barley or rice is added, as is desirable during convalescence, it must be boiled first separately, till quite soft, and added when the broth is heated for use.

---

[1] Some people add the feet, which contribute a peculiar, and not always acceptable flavor.

Eel Broth.

Skin, clean, and chop up into pieces an inch long, half a dozen small eels. Boil them in a pint and a half of water, skim, and then cover over and stew for forty minutes. This makes a capital stock for a *souchée* of flounders, sole, or perch, and a very good change for convalescents from fevers. The object is to avoid giving butter sauce, which is apt to turn rancid.

Boiled Pigeon or Partridge.

Clean and season the bird, inclose it in a puff paste, and boil. Serve in its own gravy, supplemented by the liver rubbed up with some stock, and do not forget the bread sauce.

Bread Sauce.

The crumb of a French roll;
Water, half a pint;
Black pepper, six to eight corns;
A small piece of onion and salt to taste.

Boil till smooth; then add a piece of butter about as big as a walnut, and mix for use. It is good hot with hot birds, cold with cold birds, and is an excellent food for the sick.

Nutrient Enema.

(*In cases where the stomach rejects food.*)

Take of beef tea half a pint, and thicken it with a teaspoonful of tapioca. Reduce 1¾ oz. of raw beef to a fine pulp, pass it through a fine cullender, and mix the whole up with 20 grains of acid pepsin (Boudault's *poudre digestive*) and 4 grains of diastase (or a dessertspoonful of malt flour)[1] (Fonssagrives).[2]

It should have a bright rose tint, and exhale a rich meaty odor. Not more than a quarter of a pint at once should be used, and that slowly. Thus, frequent repetition is facilitated.

---

[1] Where malt flour is used, the tapioca may be omitted.
[2] Journal de Méd. et de Chir. Prat., 1862.

Pending the arrival of the pepsin and malt, the other portion of the liquid may be administered alone.

Malt Tea.

(*Nutrient in cases where the mouth is very dry.*)

Boil three ounces of malt in a quart of water.

Biscuit and Milk.

Soak for about eight hours, till it is quite pultaceous, a hard captain's biscuit in milk or in water. Pour off what it has not absorbed, and mix it up in a pint of new milk.

Bread Pudding.

Pour over a French roll half a pint of boiling milk, cover it close, and let it stand till it has soaked up the milk. Tie it up lightly in a cloth, and let it boil for a quarter of an hour. Turn it out on a plate, and sprinkle a little sugar candy over it. The addition of burnt sugar or tincture of saffron will give it the established yellow color.

Rice Pudding.

Boil two ounces of rice in a pint of milk, assiduously stirring till it thickens. Take it off and let it cool. Then mix in well two ounces of butter, a quarter of a nutmeg grated, and sugar in moderation, according to taste. Pour it into a buttered dish and bake.

Batter Pudding.

Flour, three teaspoonfuls;
Milk, one pint;
Salt, a pinch;

Of powdered ginger, nutmeg, and tincture of saffron, each a teaspoonful. Boil.

The point in the last three recipes is the avoidance of eggs, which, when baked or even when boiled so long as it is necessary to boil puddings, are quite insoluble in a weak stomach. If, however, they are dressed separately, and either kept raw or lightly heated, they may be made into a custardy sauce, which is quite digestible in small quantities. To this a little wine may be added as a flavoring.

The Invalid's Tea.

Pour into a small china or earthenware teapot a cup of quite boiling water, empty it out, and while still hot and steaming, put it in the tea. Add enough boiling water to wet it thoroughly, and set it close to the fire to steam for five or six minutes. Then pour in the quantity of boiling water required, from the kettle, and it is ready for use.

The Invalid's Mashed Potato.

Boil one pound of potatoes with their jackets on till they are tender or brittle. Peel them, and rub them through a fine sieve; when cool, add a small teacupful of fresh cream and a little salt, beating the *purée* up lightly as you go on, till it is quite smooth, and warming it up gently for use.

CHAPTER II.

DIET OF CERTAIN OTHER INFLAMMATORY STATES.

INFLAMMATION arising from the congestion of any part of the body, whether in immediate consequence of injury, cold, heat, or other external accident, or from some previous disease or lesion, gets well soonest and safest when the blood is kept in as normal a condition as possible. The healthier and more efficient for nutrition the blood is, the more freely it circulates, the less are the throbbing, the swelling, the heat, the formation of mucus, pus, or fibrin, and the risk of ulceration or gangrene. Bruises, wounds, sores, and the necessary gaps made by the surgeon's knife, all heal the better for the blood being kept red and fluid.

Now starvation has precisely the opposite effect: it augments the proportion naturally borne by the fibrin to the red blood-globules; it makes the circulating fluid more readily coagulate, more adhesive, and consequently more prone to stick to the sides of the minute vessels through which it ought to flow, and more liable to block them up, and thus to favor their throbbing and swelling, and sometimes their rupture. Also, by letting effete material remain to form the bulk of the blood, it renders that fluid unfit to supply new tissue, and pus, mucus, etc., are supplied instead. Pus is, in fact, imperfect flesh strangled in its conception, bioplasm which has lost its plasticity.

Our object then should be to keep the blood in the highest health, by giving it all the nutriment it is capable of duly absorbing. The arguments used regarding general fevers apply more forcibly here, in that, the whole body not being so universally smitten with disease, the stomach (as we are often told by the appetite) retains much of its assimilative power, and is more likely to make a good use of what is given it than in the former case.

When the desire for food is genuine and keen, there is no objection to the continuance of the ordinary allowance of animal and vegetable diet at the usual times. But care must be taken to notice that the appetite is real, and not a wish dependent on whim, habit, or the aim at appearing less ill than people suppose.

The above rule, however, does not apply to stimulants. Beer, wine, or spirits, if allowed at all as a food or with food, must be restricted to smaller quantities than are taken in health. The patient's nutrition goes on better without them altogether, unless the state of the nervous system, as in fevers, should require the adventitious aid. (See last chapter.)

When loss of appetite indicates that the powers of the stomach are in partial abeyance, then the nutrient administered should be liquid, easily convertible into chyle, and in frequent small doses, for the reasons before detailed.

After surgical operations, and in many other cases where there is a tendency to a copious formation of pus, health is promoted by the use of fresh green vegetables and of lemonade, and of digestible fruits, such as recently gathered grapes, currants, blackberries, barberries, or whortleberries. Oranges and pomegranates are cooling to the mouth, and occupy agreeably many a dull minute in the eating. They are all wholesomer before food than after.

Hitherto common, or non-specific, inflammation has been spoken of. When the inflammation has a specific character, a new element is introduced into the treatment, and the general rules above alluded to do not apply.

In ACUTE RHEUMATIC FEVER experience demonstrates that a nutrient analeptic diet retards recovery, and will even bring on a relapse during convalescence. It is a painful thing to restrain a patient from wholesome food when he craves for it, but in this instance it is our duty. If meat in any form, solid or liquid, be eaten, it seems to turn into lactic acid, which many think is already in excess in the rheumatic blood—at all events it adds to the quantity of organic acids in the body. The power of fully converting it into living flesh is wanting, and until this power is regained, a semi-conversion into the substance named takes place. The smell of sour milk in the skin of rheumatic fever patients is well known, and seems to support the theory alluded to. Meat augments it, and adds also to the acidity of the urine. The redder and more muscular the meat is, the more it disagrees.

Vegetable food does not turn so copiously or so quickly acid;

so that water gruel of oatmeal, grits or rice, soup meagre, jellies, spiced rice, plain boiled rice, panado and other various preparations of bread, mashed potatoes, porridge, arrowroot and semolina puddings, Oswego corn, and the like, must be used to satisfy the mouths which often loudly complain of starvation. The cases which prosper are those where the appetite is keen, but has been denied any food beyond what is absolutely necessary to barely sustain life. The aqueous drinks should be alkaline.

Even when the pains are gone and nature calls for the replacement of the lost flesh, the diet which promises most readily to replace it will sometimes bring on a relapse, and you must very cautiously go back to "ordinary diet," else you run a risk of eventually losing time by overmuch haste.

But it cannot be said of acute rheumatism as of other cyclical fevers that the diet is the most important part of the treatment. The action of variations of temperature, especially in the direction of cold, is so extremely injurious in prolonging the disease, increasing its painfulness, and endangering life by causing pericarditis, that indubitably a more paramount value attaches to a scrupulous protection of the body against such changes. Freshness of air, so valuable in typhus and its allied states, is in this case a minor, very minor, consideration compared with its steady warmth. You must be careful, even when making your daily stethoscopic examination, not to expose the chest to chill. A careless and dangerous doctor may always be known by his neglect of precaution in these cases. Chilling the skin of a rheumatic fever patient not only turns the rheumatic inflammation into fibrinous inflammation (as in pericarditis), but it also still further deteriorates the process of nutrition. After a chill you will see the urine become turbid from the deposit of lithates, the effete products of a deranged assimilation. Dieting the sick is useless unless appropriate means are adopted for securing the digestion of the diet. And the most essential means in rheumatic fever is attention to the temperature.[1]

In GOUT, during the acute access, the same restricted diet conduces to shortening the inflammation; but it is not so imperative,

---

[1] I have enlarged more fully on this important subject in Lectures Chiefly Clinical, lectures xi, xii, and xiii, 4th edition.

and when the general health has been weakened by previous attacks it cannot often be borne. The treatment of chronic gout and the gouty diathesis requires separate consideration.

Soup Meagre.

Take of butter half a pound. Put it in a deep stew-pan, place it on a gentle fire till it melts, shake it about, and let it stand till it has done making a noise; have ready six medium-sized onions peeled and cut small, throw them in and shake them about. Take a bunch of celery, cut it in pieces about an inch long, a large handful of spinach cut small, and a little bundle of parsley chopped fine; sprinkle these into the pan, and shake them about for a quarter of an hour; then sprinkle in a little flour and stir it up. Pour into the pan two quarts of boiling water, and add a handful of dry bread-crust broken in pieces, a teaspoonful of pepper, and three blades of mace beaten fine; boil gently another half hour. Then beat up the yolks of two eggs with a teaspoonful of vinegar, and stir them in, and the soup is ready.

The order in which the ingredients are added is very important.

Bread Soup.

Take the crust of a stale roll, cut it in pieces, and boil it well in a pint of water with a piece of butter as big as a walnut, stirring and beating them till the bread is mixed. Season with celery and salt.

Alkaline Drink.

Cut the rind of a lemon very thin, and put it in a jug with a tablespoonful of powdered sugar-candy. Pour on it a little boiling water, and when it is dissolved, half a pint of Vichy water, and half a pint of common water.

The White Drink (*Decoction Blanche*).

Burnt hartshorn, powdered, two ounces;
Gum arabic, an ounce and a half;
Water, three pints.

Boil down to a pint, strain, and add sugar.

Isinglass Jelly.

Boil an ounce of isinglass and a dozen cloves (if liked) in a quart of water down to a pint. Strain hot through a flannel bag on two ounces of sugar-candy, and flavor with a little angelica root, or two or three tablespoonfuls of *Liqueur de la Grande Chartreuse*, if cloves are not relished.

Hartshorn Jelly.

Boil half a pound of hartshorn shavings (not "raspings," which are adulterated with bone-dust) or an equal weight of ivory turnings, in three pints of water down to a pint, strain, and add three ounces of white sugar-candy, and an ounce of lemon-juice. Heat again up to the boiling-point. As a variety in flavoring, white Capri, Moselle, or champagne may be used in quantity not exceeding two tablespoonfuls.

The gelatin derived from these sources claims no advantages over that obtained, as directed in cookery books, from calves' feet. But it is more soluble and digestible than if made from the ordinary "gelatin" of the shops, which is manufactured from old bones, probably after maceration in acid. There is something gained by using for invalids dishes differing in name and substance from those usually set before the robust, as it is thus easier to secure their being made properly, and not according to the traditions of the kitchen. A cook whose "calves' feet jelly" has been commended, will demur to receive a recipe for its ingredients from the doctor, but will allow him to be in his legitimate province when he orders hartshorn or isinglass—they are, or were lately, in the Pharmacopœia.

Bread Pudding (with egg).

Take of crumbs of bread, 2 ounces;
New milk, ⅓ of a pint, boiling hot.

Pour the hot milk on the bread, and let it stand about an hour covered up; then add the yolk of an egg, well beaten; then a teaspoonful of rose or orange-flower water, a little nutmeg, and half an ounce of sugar. (Instead of sugar, some prefer salt.)

Beat all up together. Tie up and boil or steam or bake three-quarters of an hour.

Bread Pudding (without egg).

Pour half a pint of boiling milk over a French roll, and let it stand, covered up, till it has soaked up the milk. Tie lightly in a cloth, and boil twenty minutes.

Mutton Broth.

¾ lb. of neck of mutton, free from fat, 1 oz. of carrots, 1 oz. of turnips, ½ oz. of onions, ½ oz. of barley. Boil first the barley in ½ pint of water, for half an hour, and throw that water away; then put the boiled barley and the meat in 1½ pint of water; boil and skim for a quarter of an hour; then add the vegetables, and boil till the meat is tender. To make 1¼ pint of broth.

Rose Tea.

Take of red rose-buds (the white heels being taken off) half an ounce;

White wine vinegar, three tablespoonfuls;

White sugar-candy, one ounce.

Put them in two pints of boiling water, and let them stand near a fire for two hours, then strain.

Similar acid drinks may be made of apple jelly, guava jelly, damson cheese, or syrup of gooseberries, barberries, etc., and a variety is always agreeable.

Sage Tea.

Take of leaves of green sage, plucked from the stalks and washed clean, half an ounce;

Sugar, one ounce;

Outer rind of lemon peel, finely pared from the white, quarter of an ounce.

Put them in two pints of boiling water, let them stand near the fire half an hour, then strain.

When the sage is dried, it must be used in rather less quantity than abovementioned.

In the same manner teas may be made of rosemary, balm, southern-wood, etc., and are convenient to prevent a thirsty patient taking too much tea and coffee, when not good for him. The use of acid is also avoided.

Oatmeal Tea.

Take of oatmeal a handful;
Boiling water a gallon.

Mix in a deep vessel. Let the oatmeal subside, which it does in half an hour, and pour off the tea. By this process hard water is made digestible.

The main difference between these recipes and those given at the close of the last chapter is, that nutrition is not made an object. When it is desired, the same means are available as have been already advised in lower fevers.

CHAPTER III.

THE DIET AND REGIMEN OF WEAK DIGESTION.

CHRONIC deficiency of power in the stomach, atonic dyspepsia, is evidenced by a sense of uneasy distension in the front of the waist, coming on an hour or so after food, and often lasting till it is temporarily relieved by the next meal or by sleep. There is no sharp pain on pressure—indeed, gentle rubbing seems to give relief. The sensation hardly ever amounts to pain, except in highly nervous or hysterical persons, or where a great formation of air gives rise to colicky spasms. Portions of undigested food are apt to regurgitate, and are rancid with butyric acid. This occurs at a later period after food than the more familiar "acidity," or sensitiveness to the normal acidity of the stomach, which is found in nervous hyperæsthetic persons.

Constipation, or what is called "sluggish liver," is often a symptom which attracts the notice of a patient and brings him to a physician. And most cases have been already much aggravated by the attempt to obviate the inconvenience with the aid of purgative drugs. The constipation really arises from the extension of the want of tone in the stomach to the secretions and propulsive movements of the intestines, and the habitual indulgence in opening medicine increases this. From want of tone, also, the ilia and colon dilate, and are blown up with retained air. And the dilatation causes pain, especially in the waist and under the shoulder-blades.

If regular, the evacuations are pale, scanty, and unformed. Should any excess in eating nitrogenous food be committed, there is sometimes a temporary diarrhœa and fetor in what comes from the bowels. Sometimes the looseness is almost constant.

This condition of the digestive canal is well demonstrated by the aspect of the tongue, the lips, and the gums. They are pale, and retain the impress of anything pressed upon them, and easily bleed. So that the tongue bears marks of the teeth along its edge, and the efforts of swallowing elongate the uvula. Sometimes streaks of blood which have exuded are hawked up, and the tickling of the glottis by the lax pendent uvula causes a cough, and the patient is alarmed by the fear of pulmonary consumption.

Palpitation of the heart, and sometimes intermission of the pulse, are noticed, especially where the flatulence is great. The pulsation is rapid on excitement, but is soft and slow during repose.

The urine, as a rule, is pale and watery, with a low specific gravity, indicating thus, by the deficiency of urea, an imperfect nutrition. The phosphates are also deficient, but still form a cloud when the fluid is subjected to heat, showing its abnormal deficiency of acid. Care must be taken that these characteristics of the renal secretion do not arouse a fallacious fear of Bright's disease in the kidney.

The intellectual faculties do not appear to be impaired, but there is a heaviness and disinclination for exertion which induces the patient to say they are, as an excuse for his sluggishness. If sleep be indulged in during the day, it is unrefreshing, and often leaves headache and weariness behind it.

In the case of atonic dyspepsia, I have departed from the general plan of this little volume in giving details of the symptoms of the disease, because these symptoms have each a decided bearing on the dietetic treatment. The first-named group of sensations appear to depend on the quantity which is taken at once, and are much less perceived when small and frequent meals are made the rule. These moderate meals during the day should be reinforced by a light supper on going to bed. And the period of waiting before breakfast may be made less exhausting by a cup of milk with a teaspoonful of rum in it.

Fluids should be drunk in moderation, and never at the commencement of the meal. For overdilution weakens the gastric juice, which is scanty at first; but when it is poured out more copiously, fluids assist the onward motion of the chyme through the pylorus, and obviate that delay which gives rise to the feeling of uneasiness.

The nature of the food is of less importance than its quantity; but still it has an influence. Rich, greasy sauces are especially to be avoided; for, meeting in the stomach with the acidifying remnants of the former meal, they rapidly undergo the butyric fermentation and turn rancid. No acid is so disagreeable to the stomach and œsophagus as the butyric; and therefore, when this forms, relief is often experienced on stopping that form of fermentation by another acid. A few drops of diluted phosphoric or muriatic acid, a teaspoonful of lemon-juice, or even a piece of sour apple, or a little Chili vinegar, often palliate the distress. But the wisest course is to avoid the cause.

The bill of fare should be varied from day to day, but as simple as possible at each meal. Considerable comfort is often gained by dividing the animal from the vegetable food, taking one at one meal and the other at another. Vegetable food is much less likely to cause flatulence if taken alone.

Vegetables should not be omitted from the dietary, or the future health will suffer. But such should be selected as cause least inconvenience, and the mode of dressing them adjusted and experimented upon till the patient finds how much and how many suit his peculiarities. Spinach can almost always be borne either chopped up, or served as cabbage, or in soup. And so generally can tomatoes, squashes, vegetable marrows, Elector's caps, salsify, scorzonera, beet root, and French beans. Peas must be very young and soft. Potatoes must be old, and cooked according to the recipe in the last chapter but one. Of cauliflower only the friable white flower is fit for eating.

Raw fruit should never be eaten at the end of a meal. But roast apples, if taken with a little cream and hardly any sugar, are digestible enough. Between meals or as a separate meal, a bunch of grapes, a few strawberries, currants, or raspberries with a crust of bread, aid rather than impede digestion.

All bread should be stale or toasted. The best is that which is closest grained and most friable. If the teeth are sufficiently sound, crust is preferable to crumb. Biscuits should be made with water, and not sweet. Of other farinaceous food it may be remarked that almost all depends on the cooking. Pastry, from its mechanical texture, is very slow of solution, and should be cut out of the bill of fare altogether; but the same flour made into bread pudding can generally be tolerated without inconvenience. Semolina, tapioca, and arrowroot are a safe variety.

Cheese after dinner always disagrees. But it is nutritious, and is often missed much by the patient. If so, let him try it toasted as a separate meal. It should be quite new, cut into thin slices, buttered and basted while toasting with a little cream. Let it be brought up on a hot-water plate, and not allowed to become hard and tough.

Meat should be eaten twice a day. It ought not to be cooked a second time. A table of precedence in comparative digestibility is given in a former part of this volume,[1] which, as far as the butcher's meat and game are concerned, is applicable to the selection of food in atonic dyspepsia. If any one can extend his acquaintance as far as "roast beef," he may consider that he is not an atonic dyspeptic.

The wholesomest fish for weak digestions are boiled flounders, whiting, haddock, sole, plaice, brill, perch, among which a free choice may be made. Cod must be taken with more caution. It, however, bears cooking by dry heat better than the aforenamed, and if well done in cutlets suits delicate stomachs often better than the others. Rich and oily-fleshed fishes, such as eels, herrings, pilchards, sprats, salmon, mullet, and those of very firm texture, such as turbot, tunny, pike, had best not be attempted by invalids at all. The condiment should be a squeeze of lemon or Duke of York's pepper sauce.

It must be observed that so powerful is the influence of the mind over the body that nothing is more common than for persons to exhibit wonderful idiosyncrasies respecting their capability of digesting peculiar articles of food. Those who have a strong will seem able by perseverance to use themselves to assimilate almost anything assimilable. The gratification of desire puts the nervous system into a state most favorable to the secretion of gastric juice, and the thwarting of a prejudice, however whimsical, upsets temper and stomach at the same time.

Some dyspeptics get into a bad habit of erasing from their future bill of fare everything that has once seemed to be followed by inconvenience. The result is an unwholesome monotony of wrongly selected victuals, and a despairing resignation to a needless abstinence. Let them on the other hand take the more hopeful course of *adding* to their dietary everything that they have

---

[1] Pp. 114, 115.

once found to agree, and they will have a choice nearly as extensive as their robust brethren could wish. If one cook cannot make a coveted article digestible, let them try another.

Sugar at the latter end of meals certainly generates an excess of organic acids, and is to be avoided. But yet, in moderation, it promotes (as observed by Blonlot) the flow of gastric juice; so that the custom adopted by Oriental nations[1] of taking a few bonbons and sweeties is in accordance with reason, and might prudently be experimented on in this country. An excellent dietetic fillip to the appetite is a couple of pepsin lozenges, or the same quantity made into pills with sugar.

Fermented liquors in atonic dyspepsia are apt to cause a congestive flushing of the brain, face, and neck, and a throbbing of the arteries. This may be viewed as an external evidence of what is going on unfelt in the abdomen. If these symptoms are prominent, a state of things is indicated not at all favorable to digestion, and we may be sure that alcoholic stimulants, *in the quantity taken*, are injurious. But that does not prove that in less quantity, or more dilute, they may not be beneficial. There is a quantity, capable of being arrived at in each separate case by experiment, *par la voie d'exclusion*, which just succeeds in stimulating the appetite without flushing the face. Very small, indeed, that quantity is sometimes; yet it is an actual measurable quantity, not "infinitesimal." It has a genuine physiological effect, to be accounted for by its observed action on the nervous system, and missed when it is not taken. I have known, for instance, the small sip of wine drunk at the Sacrament of the Lord's Supper to make a decided difference for good in the digestion of an immediately following meal.

If a patient has the common sense to regulate himself according to the forementioned guide, there is not much need for interfering with the nature of the vintage he patronizes. Port, Burgundy, and dry sherry agree if limited to tablespoonfuls. But if he persists in neglecting the warning, then we should persuade him to take claret, hock, or Capri. Not that these are really better for him, but they are more dilute. Beer, sweet champagne, and sparkling wines in general, almost always give rise to fer-

---

[1] See Rawlinson's notes to Herodotus, b. i.

mentation. Some persons find it suits them not to have any alcoholic drink during the day, but to take a glass of hot whisky toddy on retiring to rest.

Tea is most refreshing to the dyspeptic if made in the Russian fashion, with a slice of lemon on which a little sugar candy has been sprinkled, instead of milk or cream. One small cup of an evening is enough, and at breakfast its place is well taken by cocoa made from the nibs.

Constipation is effectually obviated by the habitual employment of green vegetables, porridge, brown bread, charcoal, and a few other articles of diet containing a good deal of unirritative matter which yet is not dissolved by the solvents. Olive oil and mustard also assist in effecting the same object. But it is better to allow the continuance of constipation than to administer purgatives, and it is wise from the first to try and disabuse your patients of the fallacy that the bowels not open every twenty-four hours are in mischief.

Where there is a looseness, from the food passing away undigested, a few drops of laudanum at night will often give it a chance of solution by detaining it in the bowels.

"Stomach cough" and "Stomach sore throat," described at the beginning of the chapter, are best treated by not sitting down to breakfast without having gargled the throat with alum water, or sprinkled the back of the fauces with dilute hydrosulphurous acid. If the mouth is foul and the breath heavy of a morning, a cleansing with a weak wash of Condy's solution is also a good preliminary to eating.

Palpitation of the heart and of the abdominal aorta are almost always made worse by special remedies directed to them, because the thoughts are thereby induced to concentrate themselves on the suffering organs. They do not lead to any organic affection, and are useful as an index to the improvement or decline of the patient.

A low specific gravity of the urine is an indication for the use of tonic medicines. An abnormally high specific gravity shows that more vegetables should be eaten.

When the medical attendant thinks fit in a case of atonic dyspepsia to order quinine and strychnine, alone or in combination, or iodide of potassium, care should be exercised not to eat within an hour before or after, as the mixing of the drug with food in the stomach diminishes the efficiency of both. But rhubarb or aloes, in the exceptional cases where they are advisable, may be used with advantage along with the meal; while pepsin is beneficial only in that combination, and iron should be taken by itself after food. It is right, however, to say iron rarely suits well for long together. Cod-liver oil, a good dietetic tonic in some instances of this form of disease, agrees best the last thing at night, and made up with a little soda and milk. During the day, it is apt to turn rancid, and return unpleasantly. Aromatics (such as teas or infusions of cloves, cinnamon, horse-radish, canella, and calumba), as also bitters (such as quassia, hops, and wormwood), are useful only just before meals. I have seldom found mere bitters do good.

The employment of alkaline mineral waters containing soda, especially Apollinaris and dilute Vichy water, is sometimes a great comfort. The only way in which we can explain this is that alkalies, as observed by Corvisart, by a sort of antagonism augment the secretion of gastric juice. They should be drunk, then, at the commencement of the meal, and not afterwards, when they would neutralize what is already secreted.

Mineral acids, the most digestible of which is the phosphoric, are best on a full stomach.

As respects the general regimen, it should be of a bracing character. As much time as possible should be spent in the open air, but exercise should not be carried to the point of producing over-fatigue. Riding in company, especially the company of ladies, is perfection; but bowls, archery, croquet, and lawn tennis, are not to be despised as wholesome amusements. Along with moderate exercise, sound rest, especially at night, in a quiet bed, should be secured; but coddling in hot rooms, sleeping by day, and giving way to slight feelings of discomfort, or the avoidance of society under pretence of fearing fatigue, must be discouraged.

An excellent device for invigorating the circulation, especially for women, is to take the morning bath by sitting in warm water and having a bucket of cold water poured down the spine from the nape of the neck, and then being rubbed dry immediately with a rough dried towel.

The quality of the water used by atonic dyspeptics is a matter of serious consideration. Hard chalky water produces flatulence, and frequently a grinding neuralgic pain in the epigastrium, so long as its use is persisted in; and the brighter and more sparkling it is, the worse it seems to agree. Chalybeate waters, as a daily drink, though the taste of iron be very slight in them, in a few months produce a peculiar debility and anæmia; even though at first, and when taken as a medicine for a short course, they had exhibited quite an opposite effect.

The best places of residence are high, dry, and bracing. The air of the Surrey and Sussex Downs, or such a combination of heather, fir-woods, and a sandy soil, as is to be found on Ascot Heath, seem the best specimens one can quote. It may be added that the last-named enjoys the advantage of a chalybeate spring, a short course of which will benefit those cases where atonic dyspepsia has induced scantiness of the evacuations peculiar to women, or strongly marked anæmia. It agrees better than other forms of steel; which, as previously noticed, is seldom well tolerated in pure atonic dyspepsia.

The cookery in cases of this complaint can hardly be too simple. The appetite should not be tempted by savory compounds, but by so preparing the article that its peculiar taste should be made prominent and attractive, and at the same time its texture made more soluble and digestible. The chapter on cookery in the first part of this work may supply a few hints, and the few recipes that follow are efforts in the same direction.

Duke of York's Universal Sauce.

(Pepys's Diary, February 10, 1668–9.)

Pound hard dried toast and a tablespoonful of peppercorns in a mortar, and boil them with an equal quantity of chopped parsley (or any other herb liked) and a dessertspoonful of salt in a small teacupful of water. Add a teaspoonful of white vinegar, Tarragon vinegar, or lemon-juice.

Relish for Fish.

Fish is made more digestible and has its flavor brought out by a few drops of lemon-juice squeezed over it.

White Mayonnaise Sauce.

Put the yolk of a hard-boiled egg in a small basin; break it up fine with a wooden spoon, adding a pinch of salt and a pinch of pepper. Then keep on stirring briskly, while you pour in, drop by drop, the best sweet olive oil to the extent of about two dozen teaspoonfuls. With each eighth teaspoonful of oil, add a teaspoonful of white vinegar. Tarragon vinegar, elder vinegar, or otherwise seasoned vinegar can be used by those who like them; or a tablespoonful of French mustard may be added.

Boiled Flounders.

Put the flounders in a stewpan with a moderate quantity of boiling water, seasoned with a little salt; take off the scum, and continue the boiling ten minutes. Drain the fish on a fish-plate before the fire, and serve.

Rice Milk (Gouffé).

Blanch in plenty of water, $2\frac{1}{2}$ ozs. of best Carolina rice;
Cool with plenty of cold water, and drain;
Boil 3 pints of milk in a 2-quart stewpan;
Mix the rice in the milk, and stir on the fire till boiling;
Add a lump of sugar and a small pinch of salt—say $\frac{1}{4}$ oz. of each;
Boil for an hour; serve.

Other milk soups—vermicelli, semolina, tapioca, etc.—are prepared in the same way, and form an excellent light supper.

Potato Surprise.

Scoop out the inside of a sound potato, leaving the skin attached on one side to the hole, as a lid. Mince up fine the lean of a juicy mutton chop, with a little salt and pepper, put it in the potato, pin down the lid, and bake or roast. Before serving (in the skin), add a little hot gravy, if the mince seems too dry.

CHAPTER IV.

GOUT AND RHEUMATISM.

There can be no sort of doubt that good cheer indulged in, either by himself or his ancestors, is the principal cause of every man's gout. Even where it is hereditary, the patient will generally be found to have assisted in bringing it on by his own love of evil things. The cases where it is traceable to lead poison, peculiar exposure to cold and damp, and a few other rare causes, need not trouble us much, as when they do occur their pathology is very obvious.

But if it is due to good cheer, shall we not be able to detect the identical article or articles out of whose pleasant substance is made the scourge to lash us? Shall we not be able, by excluding this, to enjoy the cup without the bitter drop at the end of it?

Those of our fellow-citizens who drink champagne, port, Madeira, sherry, Burgundy, London porter, are the most, of all the population, disposed to have gout. These liquors contain a peculiarly large proportion of alcohol; and so, exclaim the growers of light wines and total abstainers, alcohol is the cause of gout. But wait—observe the Glasgow laborer and the Irish peasant, and the Swede and Russ, who consume in the form of spirits six times the amount of alcohol that we do. They are quite free from gout.

Is it the acid in these drinks? Hardly—for they really do not contain any marked excess. Or the sugar? No—or sweetened tea would be equally noxious.

Or is it the peculiar combination of all those? Possibly. But I think a more rational explanation is that the classes who are rich enough to enjoy these luxuries are rich enough to enjoy other luxuries also. In fact, the cases of gout originating in the individual himself, without his being able to throw any blame upon his pedigree, occur in those who daily consume a larger quantity of nutritious food than their bodies require, or than can, by the assimilative powers they possess, be converted into useful muscle. Hence there is an accumulation of oxidizable matters which fail to attain the end of their chemical transformations, which do not become urea. So that there remains in the system an excess of them, especially of uric acid, convertible into urate of soda, the characteristic deposit in gout. Other contributors to the production of gout run hand in hand with the over-eating of meat, rather than with indulgence in fermented liquors only. Laziness can hardly be dissociated from gluttony—picture the latter, and you picture the former. And laziness can easily be understood to act in the same direction, by arresting the oxidation which is the result of due exercise. Immoral self-indulgence of another sort aids the development of gout by the exhaustion of the nervous system which it induces without corresponding muscular action.[1] And the same may be remarked of intellectual pursuits, a love for which has been noticed frequently to go along with gout, and which tempts the possessor to use his brain and let his limbs languish. Damp cold also, by weakening and retarding the circulation, helps in a minor degree the evil effects of gorging the digestive organs with nitrogenous food. Such seems to me the history of the origination of gout in the constitution.

The part played by fermented liquors seems rather the bringing on of the acute attack, of which almost all habitual sufferers can cite an instance. One man takes some port wine, which he has left off for years, and the same night is woke up with the horribly familiar stab in the small joints. Another drinks a few glasses of champagne, and feels his foot twinge and swell before he leaves the table.

This consequence of an unhabitual excess is of the greatest importance in diagnosis. For we may lay it down as universally true that when a few glasses of wine or beer are followed soon by the inflammation of a joint or pain in a trunk nerve, the inflammation or pain is of a gouty nature.[2]

The views above stated as to the origin of gout can suggest but one line of preventive treatment. The children of gouty families should be brought up to a life of strict abstemiousness

---

[1] Λυσιμελοῖς Βάκχου καὶ λυσιμελοῦς Ἀφροδίτης
Γέννᾳται θυγάτηρ λυσιμελὴς Ποδάγρα.—*Anthol. Græca.*
[2] Cyr, de l'Alimentation, part ii, chap. ii, article 1.

and muscular activity—"to scorn delights and live laborious days." They must not compound for temperance in alcohol by indulgence in dainty meats, sweet pastry, soft beds, or idleness. At the same time I would not encourage them to an ambitious athleticism, to glorying too much in their strength. The result of this very often causes a reactionary depression at the period of middle life, which persons hereditarily healthy can resist, but which, in the case of those with ancestral tendency to gout develops the disease.

From the earliest years, vegetables should form a considerable portion of the dietary. They should be furnished in great variety, so that the young may acquire a taste and a digestion for them. The various sorts of "meagre" (or meatless) soups found in cookery books, and others which an inventive mind will suggest, should be habitual. Porridge for breakfast may be taken to any amount—what men are grown upon it in Northumberland and the North of Yorkshire! And in buttermilk will be found the best quencher of the thirst and nourishing digester of other victuals as well as itself. As they grow up, the young people should be impressed with the reason of this temperance, and urged to persevere in it, not only for their own sakes, but as a moral duty for the sake of their descendants. A man has a right to choose for himself one of two paths, but he has no right to lead involuntary followers along that which seems the pleasantest at the moment.

If the tackling with the disease by diet and regimen be undertaken only after the development of it has taken place, the task is much more difficult, but not a fair subject for despair. Dr. Garrod, than whom none has sought more diligently for medicinal remedies for gout, or more advanced our knowledge of its pathology, says that "if a gouty man could entirely lay aside his usual habits, and follow in all respects the dictates of nature, there would probably be little need to seek relief from medicine."[1] By "nature" must be meant the higher nature, reason or common sense; for man's desires are so perverted by the prejudices of education, and by moral and physical inheritance, that one cannot well trust to their guidance in the treatment of disease. But by

---

[1] Reynolds's System of Medicine, vol. i, p. 875.

rational management, it seems pretty certain that gout may be so far checked as will enable the patient to enjoy life, to live to his full term, and to hand down an untainted constitution to his successors.

An honest trial must be made, for three months at least, of an entire abstinence from all alcohol, except a little weak claret and water at dinner. If in that time weight is gained, or not lost, let the rule of abstinence be established for good and all.

If it is found that alcohol cannot, in the judgment of a competent temperate medical man, be left off, it will be best to take, as the habitual form of it, hollands and soda water. The quantity of spirit should never exceed a small wineglassful, and should be taken either at dinner, or, in cases of sleeplessness, at bedtime, but not at both.

Meat should be eaten but once a day. In other respects, the advice given in the last chapter as to food and cookery may be held valid for gouty persons as well as atonic dyspeptics. The obedience of the cook to orders is of paramount importance; and it will be well also to warn our patient not to be trying this or that dish as good for gout, at the bidding of officious friends, but to use his own judgment, usually sounder than theirs and also strengthened by self-interest.

The digestion of gouty persons is usually much improved by a winter sojourn in the dry warm air of the islands and highlands which get the benefit of Mediterranean breezes. I am unwilling to injure one place by mentioning another, and am disposed rather to enlarge upon the general qualities by which that great nurse of the civilization of the human race, the Mediterranean, can earn still that gratitude, though we are out of her leading strings.

Without seeking for recondite cosmic influences, take the map of Europe, and consider the obvious characters of the waves and winds. The supply of fresh water brought by rivers to the Mediterranean is very much less than that which comes by the Gulf Stream from the Polar glaciers to the Atlantic. It is therefore salter, and would be salter still, and in course of generations all salt, were not a vast stream of Atlantic water flowing in through the Straits of Gibraltar. Being salter, it is more ready to remove moisture from the air, and less ready to part with it to the air under varying temperatures than the ocean which surrounds us at home. Its temperature also is more equalized throughout; there are not, as with us, hot and cold streams the vapors from which meet in the air and precipitate their watery burden on the earth. So that the atmosphere is, in the first place, clearer, or has a minor degree of hygrometric saturation, than in the British Isles, and in the second place is less rainy.

Of course there is rain, and of course there is east wind and north wind in Italy as elsewhere; but there is not that continuous depressing influence of the combination of a cold wind with a saturated air which we have. Then the avoidance of these black un-Italian days does not involve an imprisonment of a week or two in the house; and, in travelling, a change of residence of a few miles will often put you completely under shelter. Those who go abroad with an old-fashioned guide-book, or take their ideas of Italy from the queer little pictures in blue body color that used to adorn our nurseries, may be found grumbling, but those who have ever tried to travel in England, and have the faculty of comparison, will be well content. When they see the deep indigo horizon, the mark of extra saltness in ocean, which so generally bounds the traveller's landscape in Mediterranean lands, they will bless it for the clear air of which it is not only the indication but the cause.

The high temperature of the great body of water, so intimately intermingled with the land, preserves that of the air from the variations which we experience. This is of course most especially the case on the coast, but even inland it may be experienced. The sole exception is that of certain spots in the immediate vicinity of snow-clothed mountains, the air blowing over which in the early spring is much chilled. So that in such places as Florence, for example, a change of wind from a sea-breeze to a Tramontana may cause a very sudden fall of the thermometer. But yet the even temperature of Italy as a whole, compared with England as a whole, must be allowed; its exposed climates are less variable than our exposed climates, and its sheltered nooks more completely sheltered than ours.

There is a higher amount and degree of sunlight than in our skies, so that all the chemical actions in the animal body must be intensified. A conspicuous evidence of this is the readiness with which the skin of the face becomes tanned even in the short and fresh winter days.

Again, the nervous system is much more awake to the effects of alcohol; so that instinctively less quantities are taken to produce the required effects. There are south of the Alps very few, if any, water drinkers; but there are also very few indeed who indulge in strong drink. One does not "feel to want it." A single glass of Orvieto or Capri there seems to produce as much exhilarating relief as an allowance of the domestic port or sherry containing five times the quantity of spirit. And not only is it felt superfluous, but positive discomfort is felt as the immediate effect of what appear to a stranger moderate doses of alcohol.

A corresponding observation may be made in respect of animal food. Less is hankered after by the invalid, and, moreover, a less rich or fat meat than in England. Without a failure of appetite, he yet feels satisfied with a smaller quantity.

In Italy and Spain sleep is less profound, and easier shaken off and dispelled by slight disturbances. This light and short repose is soon found to be the custom of the country, and the mind being set at rest is satisfied with it.

There are no lands where there is so much variety to divert the mind from preying on its own thoughts as these. It may be doubted whether the past, the present, or the future offers most to attract the attention. If none of the strata of historical relics interest the traveller, he can still hardly help feeling their present picturesqueness, and the loveliness of the forms and colors by which they are surrounded; while to the social or political observer the spectacle of a naturally intellectual and industrious people, long restrained by untoward circumstances, and now at length bursting out for good and for evil into new life, is unique. He desires to live, were it only to see what becomes of them.

Now reflect how these various agencies may act on the human body. The dryness of air, without excessive heat or cold, renders it unnecessary for mucous membranes to put on their slimy coats of mucus. They are in a more active condition for the work of absorbing oxygen, digesting, extracting nutriment or water, or whatever else they are required to do. They are filled with blood, and pass it on rapidly with its fresh burden of renewed life to the tissues. So that the disintegration and removal of the effete products that cause gout is actively promoted.

The evenness of temperature renders it possible for all, invalids included, to be a great deal in the open air, and take the exercise which has been described as so essential to those of gouty constitution.

The action of sunlight in reddening the blood is familiar to even the poets. The etiolated plant of northern climes is quickly by it rendered crisp and hardy, and the same effect may be inferred to take place in the etiolated animal. It is of great importance in atonic gout.

Spite of all the arguments of philanthropists, alcohol will always remain an evil necessity to brain-working populations in northern climes. I do not know but that its occasional use is a necessity in all climes. But certainly a little goes further in the south, and therefore less is requisite, if any. Thus is avoided the temptation to the slight daily excess, which is the most dangerous temptation to the gouty.

It is not alone on the passing traveller that Italy's special (I will not call it "*fatale*) dono di bellezza" acts. There is scarce a peasant who is not proud of the external aspect of his country,[1] and I am not at all sure that this feeling is not as powerful an assistant as any material agency in enabling him to live and flourish contentedly on what would starve or drive into melancholy madness an English convict or pauper. Do not set down this as sentimentalism or artistic whim. Consider how continually the converse of the experiment is impressed upon us by bluebooks and sanitary reporters, who are constantly repeating that a monotonous colorless life, such as finds its exaggerated type in solitary confinement, renders absolutely requisite an excessive quantity, and under other conditions an unwholesomely stimulating quality, of food, to preserve in health the mind and body, although no complaint is made by the sufferer. Why then should

---

[1] A genuine Italian will often prefer his country's praises to his own in whimsical ways. At Salerno we commended a soufflé to a waiter, who was the author of the delicacy: "Call that well made?" said he; " why every little lad in Salerno can make as good a one—Si, signori, ogni ragazz' di dieci anni nella città—and there are thirty thousand inhabitants!" Untrue of course, but patriotic.

it not be true that the daily life among picturesque and cheery scenes saves the victuals, although the benefited are ignorant of their blessings?

But it is indubitably for the stranger, as Filicaja complains in his touching sonnet, that this banquet of sweets is principally spread. The temporary resident profits more than the native by the fair scenes and their associations, by their ennobling and rousing effect on the mind, and through the mind on the body. They have for him the additional charm of variety and novelty.

I have dwelt somewhat lengthily on the effects of a warm tonic climate in this chapter, not because its benefits are exclusively or even principally confined to the gouty, but to finish up the subject as a whole, and be enabled, when speaking of other diseases, merely to refer to it.

The most expedient form of exercise for a gouty person is riding, as he will not be deterred from it by a little tenderness of foot or ankle so often as if he only walks.

Gouty persons suffer frequently from relaxed, sluggish piles, especially when travelling. The best regiminal treatment for them is to take some cayenne pepper with food as an astringent to the mucous membrane of the rectum, and to not only wash the external orifice, but to dash quite cold water against it till a glow of reaction is felt in the bowels, every time after the bowels have been opened. A full supply of green vegetables in the dietary should be secured.

I cannot recommend that the malaise which gouty persons often feel for a considerable period, no better and no worse, should be put an end to by a debauch productive of an acute attack. It possibly may hasten the advent of the attack, but there is no proof of its bringing the end of it any nearer. Whereas other treatment, which it is not the place to discuss here, accomplishes both objects, and gives a chance of escape altogether.

Timid sufferers will sometimes demand spirits "to keep the gout out of the stomach," as they say. Now real gouty gastritis is a very terrible malady. I had once the opportunity of examining the stomach of a debauchee who died of it, and found the mucous membrane in a state exactly resembling that produced by oxalic acid. This, however much to be dreaded, is not likely to be kept off by alcohol. But in point of fact such affections of the stomach are very rare, and when they do occur, do not appear to be retrocedent, or to arise from repulsion of the external inflammation into the viscera, but to be an extension of the morbid process, and to bear the same relation to gout that pericarditis bears to rheumatic fever. So that one does not see how anything but harm can be done by alcohol.

In spite of its rarity, it is curious how much patients talk about "gout in the stomach." I have often thought that medical men must have called by that name attacks of acute dyspepsia from indigestible food, or of windy spasms, or of gastric neuralgia. These last are likely enough to happen to the gouty, but differ nowise in their requirements from the same ailments occurring in others.

The regimen of RHEUMATIC GOUT does not vary except in the details, which must be suited to each individual case, from that of Gout.

CHRONIC RHEUMATISM, or *the rheumatics*, seldom need lay a patient up, if he will adopt the simple dietetic expedient of eating a certain quantity of fresh mustard with every meal, much if he is worse than usual, a little if he is pretty comfortable. He should avoid bathing in cold water, which will tend to derange his digestion.

We must, however, carefully make the diagnosis in each instance between chronic rheumatism, which is a disease of the fibrous tissues, and neuralgic pains, which run along the lines of the nerves. The latter are not unfrequently benefited by cold water, and are aggravated by warmth, especially by a hot bed. Medicinal treatment is here essential.

CHAPTER V.

GRAVEL, STONE, ALBUMINURIA, AND DIABETES.

URINARY DISEASES form a separate class for the consideration of the dietician, because in them we have to keep in mind not only what is likely to affect the general health, but also what will affect the kidneys and bladder specially, by producing a secretion capable of exercising a locally deleterious action.

Uric acid forms the nucleus of the great majority of urinary concretions, and many entirely consist of it. The small red grains, like cayenne pepper, and the brown lumps of gravel or stone, are all of considerable hardness and of smooth exteriors, so that their presence is often unfelt, till they make themselves evident, either by passing out into daylight, or by concreting together and becoming large enough to be a mechanical impediment. The great object in treatment must be to prevent the deposit of uric acid in the renal secretion; and when the tendency to it is once proved, not to await the advent of symptoms, which only arise at late periods of the disease.

Uric acid is deposited only when it is in excess. Excess of uric acid in the system is, if not absolutely identical, at all events closely related to the diathesis of gout. So that all that has been said about the regimen of that disease may be applied without any reservation to this contingency also. The strict observance of it is the more urgent in the degree that stone is the more painful and dangerous of the two.

Oxalate of lime is a substance whose very presence in the urine is decidedly abnormal and morbid. It may be derived by occasional accident from some article of food, such as tomatoes, or rhubarb, or sorrel, without inconvenience accruing; but when not to be accounted for in this way, it evidences a departure from health, and the persistence of its deposit leads to a most painful form of stone. Its shape is sharp-pointed crystals, and these grow together into knobby masses, which, whether from their form or other unexplained causes, give rise to great agony in the parts they come into contact with. It is of great moment, therefore, to detect its existence in the urine at an early date, so as to anticipate the evil results.

Oxalate of lime is often found in children who live in the country and have been under-fed. In this case it would appear to arise from their eating an excess of acid fruits and bad vegetables, and drinking hard unboiled water. Chemists tell us that citric, malic, and other organic acids distributed through the vegetable world, are liable to conversion into oxalic; so that it is not necessary that it should be naturally a constituent of the food eaten, but may arise during its fermentation or digestion.

It also is formed in the urine of adults in consequence of exposure to damp cold, deficiency of fresh air, and a low monotonous diet. The debility arising from frequent attacks of gout also induces the formation, so that it appears mingled with urate of ammonia or soda, easily distinguishable from that amorphous deposit by the pointed octahedral shape of the crystals under the microscope. The persons in whom I have found oxalate of lime have been of a nervous sensitive constitution, and those of the male sex have frequently suffered from nocturnal seminal emissions. They always feel much the irritation of the neck of the bladder induced by the morbid urine.

The indication afforded by the formation of oxalate of lime is to secure a generous animal diet, warm clothing, fresh air, gentle exercise and rest; but, above all, that repose of mind the want of which is a prominent cause of the indigestion. Rhubarb, sorrel, apples, pears, and in short any raw fruit and vegetables, should be avoided. Water-cresses and lettuce are the least objectionable, as they are sedative to the urinary organs and prevent a scorbutic condition of the blood. A small quantity of old spirits and water at meals does not in general disagree.

Phosphatic salts are a natural ingredient in the urine. The healthier a man is the more of them does he excrete. In the normal condition of acidity of the urine they are quite dissolved. But when from the irritating presence of a foreign body, such as a uric acid calculus for instance, the surface of the mucous membrane of the urinary apparatus becomes so alkaline that it makes the water alkaline also, then the phosphatic salts are concreted into crystals and solid masses. Thus the uric acid calculus, or other foreign body, becomes coated with a layer of a nature quite different from its own. Idiopathic or constitutional inflammation of the bladder and kidneys may also deprive the urine of its healthy acidity, and cause phosphatic gravel to concrete without any visible nucleus to concrete upon. But I cannot say I have seen the same happen from the temporary alkalinity dependent on taking soda, magnesia, or potash in the medicine or diet. However, where there is a fear of this result, in case of an alkali being required, it is wise to use the more diffusible one, ammonia.

The treatment of the morbid condition here alluded to is wholly surgical. The mechanical removal of the cause of irritation is imperative where it is possible. Where it is impossible, the special treatment of the local inflammation causing it is the only rational procedure.

Gravel formed of the strange greenish substance called "*cystin*" is very rare. I have met with only one person who has ever passed it, and he seems to be nowise deteriorated in health. From the large quantity of sulphur it contains (26 per cent.) it would seem to be formed by the decomposition of some animal matter, perhaps solidified mucus.

Chronic Albuminuria, or Bright's disease of the kidneys, is a complex state, the complication of which involves several conditions with separate bearings on the choice of the most appropriate diet and regimen.

The essential feature in Bright's disease is an imperfect nutrition of the smaller bloodvessels, by reason of which they lose their elasticity and tone, become thick and friable, are obstructed, and are unable to supply the glandular structure of the kidney with the blood for depuration. Hence a retention of the urea which should be cleansed out from the circulating blood. The blood thus fouled fails to absorb sufficient oxygen, is imperfectly aerated, and cannot renew its red particles. Hence, with the retention of urea there is anæmia also. By the force of the heart the fluid is still driven along, but it meets obstruction, instead of help, in its course from the action of the capillaries; and there ensues an exudation of the easiest exuded, that is the serous, part of its substance into the nearest open space. Hence the urine, at the same time that it lacks the urea which it ought to carry away, is loaded with albumen abnormally abstracted from the circulation.

The object aimed at by the dietician should be to supply the nutrient organs with aliment which is the most readily and rapidly convertible into blood and tissue, namely, digestible animal food in frequent moderate quantities. He will also feed the anæmic blood with iron, and in his use of drugs will choose such as contribute to rectify the deficiencies of the digestion. Bearing in mind the fouling of the blood with effete matters, he will advise a copious use of aqueous drinks, which flow out readily by the kidneys, carrying with them such constituents of the diseased body as are soluble in water. Especially will he carefully disabuse the mind of his patient of the notion that drinking water leads to dropsy, and that the thirst should be thwarted on this ground. The warning is very necessary; for an albuminuriac hardly ever escapes at least some subcutaneous collection of fluid in the face or legs; and if his doctor has previously been urging on him the use of diluents, he accuses them of having caused the symptom. In point of fact the best and only thoroughly safe diuretic for the relief of albuminurious dropsy, is water, and it is also a valuable preventive.

In disease of the structure of the kidneys, and especially in albuminuria, it is obvious that harm must be done by alcohol. There is no thoroughfare through the kidneys for the natural amount of urea and phosphates, still less for the alcohol. Thus the ordinary safety-valve is shut up. In this condition coma is often brought on prematurely by neglect of the warning voice of the sensations. Albuminuriacs will always remark that they feel stupid and sleepy after their accustomed amount of wine, and this should be a hint to their medical attendant to leave it off, instead of encouraging them to persist on the plea of weakness, as is often done.

An additional reason in behalf of abstinence is that albuminuria, at least in its chronic form, indicates a degenerative diathesis, which has been already pointed out as one of the bad influences upon health of alcohol. Whether alcohol or cold be the sharper blade of the shears so apt to cut short life's thread by Bright's disease, I care not greatly, inasmuch as it cannot be questioned that both of them have that tendency. It is their union which is so irresistible a foe to the poorer working classes who come to hospitals, and it is the avoidance of that union which protracts the lives of our wealthy patients. I cannot repent of including albuminuria with its attendant evils among the diseases caused by alcohol, and I very much object to the gin and whisky punch, the hot grogs and the spirituous diuretics which I have known albuminuriacs advised to drink. If they must have the liquors, let them at any rate indulge in an occasional bout and be done with it, and not consume their poison in its most fatal form of small repeated doses. Warm diluents are just as good diuretics without alcohol as with it.

There is considerable difficulty in persuading albuminuriacs to persist in a systematic milk diet; but if it could be managed, I am persuaded nothing would give a better chance of putting a stop to the progress of the lesion.

In *Diabetes* certainly life is prolonged, and the risk of the intercurrent maladies diminished, by a diet from which sugar and articles which form sugar are, as far as practicable, excluded.

M. Bouchardat, Professor of Hygiene at the University of Paris, gives the following list of eatables whose chemical composition makes them injurious to diabetics:

Sugar;

Bread of any kind, or pastry;

Rice, maize, and other starchy grains;

Potatoes, arrowroot, tapioca, among root products;

Sago, among piths;

Among manufactured starches, macaroni, vermicelli, and semolina;

Of vegetable seeds, peas and beans of all sorts, and chestnuts;

Radishes, turnips, beetroot, carrots;

All preserved fruits, apples, and pears;

Honey, milk, beer, cider, sweet and sparkling wines, lemonades, and such-like sweetened acid drinks.

Happily the list of permissible articles of diet is somewhat longer, and might easily be extended by the introduction of nutritious eatables kept out of European markets by want of demand. The diabetic may eat without fear—

Meat of all kinds, brown or white, boiled, roast, or grilled, and seasoned with any sauce pleasing to the palate, provided there be no flour or sugar in it;

To the curious gourmet, affected with diabetes, it may be mentioned that Professor Bouchardat especially commends the flesh of carnivorous animals, and advises their trying cat, dog, and fox. Probably larks, robins, and ducks will be more to their taste. The liver, however, should not be eaten, and by removing the fat a great deal of the rank flavor of the carnivora is avoided;

All sorts of fish, shell-fish, and lobsters;

Eggs;

Cream and cheese;

Spinach, endive, lettuce, sorrel, asparagus, hop-tops, artichokes, French beans, Brussels sprouts, cabbage (the last " very good with pickled pork or bacon," says the good-natured professor);

Salads of cress, endive, American cress, corn salad, dandelion, lettuce, with a full allowance of oil and hard-boiled eggs;

Fresh vegetable gluten, i. e., dough with the starch washed out, may be made into an agreeable dish with grated Parmesan or Gruyère cheese and butter;

Anchovy and Ravigote butter (see Gouffé);

For dessert, olives. On high days and holidays, when the patient has begun to improve, some fresh summer fruit, of course without sugar.

The wearing hunger may be much appeased by chewing cocoa beans.

For drink, a bottle and a half of good claret or Burgundy may be taken in the day. Those who prefer it may take instead brandy and soda-water, one part of the former to nine of the latter. Fresh beef tea is a capital quencher of thirst.

Coffee with cream.[1]

It may be observed that several of these last-named victuals are not entirely devoid of sugar, starch, or inosite; but the quantity is so small that they may be conceded as a variety, and are not considered dangerous by our chief authority on the subject who has been here quoted. The *menu* has not been improved by any of the subsequent writers on the subject, whose efforts have been mainly directed to the pharmaceutical treatment of the disease.

---

[1] Bouchardat, Du Diabète sucré, Paris, 1852.

Diabetic patients should always chew slowly, eat often but moderately; and drink should be taken in the same fashion.

They will assist digestion much by wearing a flannel belt round their waist next the skin.

They should take a great deal of gentle exercise in the open air.

The digestion will also be improved by quinine and strychnine, to which may be added, with the approbation of the dietician, cod-liver oil. The action of soda, also, in augmenting the secretion of gastric juice, may be brought into play by the dietetic employment of Vichy water.

Among the substances forbidden to diabetics, is unhappily included that which is proverbially the one thing needful, "the staff of life." Bread is the article of diet most prominently starchy and deleterious, and at the same time the most difficult to deny. To get over this difficulty, M. Bouchardat devised a sort of sham bread made of gluten. But it is agreeable neither to the eye nor to the taste, and the patients on whom I have tried it have preferred going without bread altogether.[1] A much more palatable substitute is proposed by Dr. Pavy in his well-known work on Diabetes—namely, cake made of powdered almonds and eggs.

However, my own opinion is that we do not act wisely in enforcing a diet which is really intolerable to the patient. The object to be gained is to conciliate the stomach, appetite, and fancy into taking the greatest possible amount of animal food and oleaginous matter—in fact, to assimilate the patient, as much as possible, to the Esquimaux and Pampas Indians, who have nothing but water and beef, beef and water, from the cradle to the grave. And if he eats the heartier for having a biscuit, or crust, or glass of porter, or even a forbidden vegetable with his meals, it is better to give him his way than to tempt him to break through all rules altogether by playing the tyrant.

There is a doubt felt how far the patient's unnatural thirst ought to be gratified. Abstinence diminishes the quantity of urine of course. But is that an advantage? It would seem better, if the blood gets loaded with sugar, as analysis proves it to be, for the foreign ingredient to be washed out by an ample outflow, than for it to remain at the risk of poisoning the tissues. It

---

[1] Lectures Chiefly Clinical, lect. xxxviii.

will be found that the thirst is closely proportioned to the amount of sugar requiring expulsion. When the diet is changed from starchy to meaty food, much less is drunk, and much less evacuated by the kidneys, although no restriction has been laid upon the appetite for liquids. I do not, therefore, check patients in drinking as much diluent as they wish.

In all diseases which affect the mucous membrane of the urinary apparatus, such as the irritation of a stone in the kidneys, ureter, or bladder, an enlarged prostate, gonorrhœa, etc., malt liquors are peculiarly deleterious. They increase the secretion of pus, and often bring it back again when it has stopped for several days. This is not to be accounted for by the alcohol therein contained, since neither weak spirits and water nor sound claret has the same effect; but it must arise from some peculiar condition of the transformed malt forming an irritating ingredient in the water.

CHAPTER VI.

DEFICIENT EVACUATION.

COSTIVENESS must be regarded as a disorder of the whole system, and not of the intestinal canal alone. The first object of treatment should be to relieve the body of the presence of excrementitious matter, the predominance of which in the blood is shown by the dingy, feculent tint of the complexion, the sluggish thoughts, the headache, and the slow circulation. Purgatives, then, may judiciously be used to commence with; for the immediate relief they afford to the feelings of discomfort is great. But let not the relief be set down to the mere "clearing out the bowels;" it is the cleansing of the blood which is the real aim of the remedy, and the real cause of the relief. An inspection of what comes away shows that it has been newly made; it is fresh bile and other constituents of recent fæces, not of those which have rested long in the canal.

Nothing is easier than thus with a vigorous pill and draught to drive away, as by a charm, the patient's discomforts; and he is ready enough to cry out that no more is wanted. But what is the consequence of leaving off treatment? The renewal of the blood and tissues not having had time to regain its original activity—there not being enough new-made blood to carry on vigorous life—the effete materials again collect, and the disease takes a fresh start. Again and again the coarse expedient is called for, and at last fails to effect its object of giving relief.

To avoid this evil consequence it is prudent to administer no quickly acting purgatives which completely empty the abdominal canal, but rather such as cause a gradual increase in the solid matter of the stools. Aloes and rhubarb are the best evacuative; and I find it also beneficial to combine with the chosen drug some resin, which acts as a tonic to the mucous membrane, and prevents the exudation of serum and mucus. Two or three grains of aloes-and myrrh-pill, every night, will in a week produce all the good effect of strong purgation; and it will produce the good permanently instead of merely for a time.

All superfluous food of the sort that has the property of arresting evacuation must be left off. Wine, beer, tea, and coffee must, on this account, be excluded from the dietary; and milk, cocoa, whey, soda-water, Seltzer-water, etc., substituted for them.

Water is a very ready remedy, and certainly a very rational one, when the evacuation by destructive assimilation is deficient. The experiments of Dr. Böcker and of Dr. Falck[1] show the increase of interstitial metamorphosis by this agent to be in close proportion to the quantity taken; and all who have heard or read of the agreeable sensations of patients during their submission to the "Water-cure" cannot question its capability of removing matter from the tissue. Herein lies its strength; for, as Dr. Böcker observes, "the demand for new tissue, as expressed in the sensation of hunger, keeps pace exactly with the extent of the metamorphosis."

A much less amount of water-drinking than is involved in a hydropathic course will often be of great service. Let the patient take, the first thing of a morning, a tumbler of water made spicy by a few cloves which have been placed overnight in the tumbler and had boiling water poured on them—a weak clove-tea in fact.

Warm hip-baths are also of great service, and can be borne from the first even by those reduced to extreme anæmia and lifelessness. Afterwards a cold douche bath, administered by sitting in warm water and having a bucket of cold water poured down the spine, or squirted against the loins with a garden syringe, is of essential use, by actively promoting the life of the skin and capillaries. Raising the specific gravity of the water by the addition of salt prevents the chill which fresh water is apt to impart.

Green vegetables, of which spinach is the completest specimen, should be freely eaten, and summer fruits. Porridge is an excellent breakfast; and brown bread, so that it be not new, is beneficial as supplying a good quantity of material for the bowels to act upon. Watercresses and dandelion are useful eaten raw, and the watery extract from the root of the latter forms a suitable

---

[1] Zeitschrift der K. K. Gesellschaft der Aerzte zu Wien, April, 1854, and Vierordt's Archiv, i, p. 150.

drug to assist the digestive organs. Horace recommends lettuces, and with some reason. Both they and spinach may be made into soup.

Roast apples and stewed prunes are much better suited than pastry as a second course for the patients we are now considering. Stewed prunes can also be taken with meat. Figs can be eaten either cold or in a pudding.

Bacon, the most soothing of fats to the digestive canal, eaten at meals, or two teaspoonfuls of salad oil taken at bedtime, prevent that drying and hardening of the contents of the bowels which causes much of the inconvenience, and they also augment the activity of the liver.

Red wines should be avoided by the costive; they cause piles. The best of the white sorts are Chablis, Sauterne, and Capri.

A careful diagnosis should be made between deficient evacuation (as above described) from costiveness, and that which depends on anæmia or old age.

In anæmia there is almost always joined with it (in the female sex) a deficiency of catamenial discharge. It is scanty, pale, or even altogether absent. This is an indication for the use of iron, phosphatic salts of lime, and cod-liver oil. A diet drink of cinchona is especially useful in such cases.

Porridge.

Always use the coarsely ground Scotch oatmeal. Mix two tablespoonfuls of it with a small teacupful of cold water till it is of uniform consistence. Then pour in a pint of boiling water, and keep boiling and stirring it for forty minutes. It is then fit to eat, but may be kept simmering till wanted, if a little more water be added as the other steams away. It should be served in a soup plate quite hot, and cold milk added to reduce it to an eatable temperature.

Brown Bread.

One of the most useful purposes to which to apply brown bread is to make it into bread sauce, in the manner before mentioned for white bread. Mixed with tomatoes, it makes a capital sauce for costive persons.

Brown bread biscuits are also eligible.

Oatmeal Flummery.

Take crushed Embden grits in proportion to the quantity required. Put them into a broad pan, cover with water, stir up together, and let stand for twelve hours. Then pour off the water so long as it runs clear. Add fresh water, mix, and let stand twelve hours more. Repeat the same process a third time. When the oatmeal has been thus macerated thirty-six hours, strain it through a hair sieve and boil it, stirring vigorously till it is quite thick. Pour it, to cool, in a dish, and eat it cold with milk, or a little wine and sugar.

Spinach Soup.

Pick all the stalks from 1½ lb. weight of fresh spinach; wash it and chop it; put it in a 3-quart stewpan with four ounces of butter; stir it over the fire for five minutes; add an ounce of flour and stir again for three or four minutes. Then stir in two quarts of chicken broth till it boils. Simmer it on a cool stove for half an hour, and add a small teaspoonful of cream. Serve with it some pulled bread fried or baked.

Endive or lettuce soup may be prepared in the same way.

Lettuces with Gravy (Gouffé).

Take 8 round and full cabbage-lettuces; trim off all the outside leaves; wash and blanch for 10 minutes; cool them well; squeeze the water out; cut them in two; lay them open on a dish, and season them with 3 pinches of salt; tie the halves together, and put in a 2-quart stewpan; cover them with broth and add 2 gills of stock pot fat, 1 fagot,[1] and 1 onion with 2 cloves stuck in it; place a round of paper on the top, and simmer for 2 hours; when cooked, drain on a cloth: untie and open the lettuces again; cut the stalks out, and fold the leaves round, giving to each piece an oval shape, about 3 inches by 2 inches, and dish them in a circle.

Reduce 1½ pint of household gravy to half the quantity; pour it over the lettuces, and serve.

Thin slices of crumb of bread, cut to the shape of the lettuces, can be put in the intervals between the lettuces.

---

[1] A fagot should contain a small handful of parsley, a sprig of thyme, and a bay-leaf. It should be tied up with string, so that none of the leaves break off in cooking. It is a perfectly wholesome seasoning.

CHAPTER VII.

NERVE DISORDERS.

§ 1. HYSTERIA.

IT has been authoritatively stated that "the physician can do but little for one who is born hysteric; *i. e.*, for one whose disease is but an exaggeration of her habitual, constitutional state."[1] I cannot agree in that sentence, for I am sure that a judicious medical adviser may very often give, even to such a born sufferer as above described, advice which will enable her not only to be comfortable herself, but to be a useful member of society. And of such advice one point of dietetic regimen will always form a part in every single case—"Let no fermented or spirituous liquid ever cross your lips." I have not a shadow of a doubt that at least one-third of the subjects of hereditary hysteria we come across, would never have developed into the declared disease had they observed this abstinence.

Hysteria has already been spoken of, in a chapter concluding the second part of this volume, as kept up, if not originally started, by alcohol. This is most especially true (according to my observation) of those irregular chronic forms of the ailment where there are few, if any, stormy paroxysms or convulsions. I have been surprised to find how often a total abstinence from fermented and spirituous liquors is followed by a restoration of the invalid's strength. That incapacity for exertion, which made a witty hysteriac say the other day that she should take "*Non possumus*" as her motto, vanishes as if by magic, and your patient cheerfully adopts all the further means for her cure.

It is not here meant to be implied that hysteria in all cases, or even in any large proportion of cases, is brought on by tippling. Such an idea is quite negatived by the age and social condition

---

[1] Reynolds's System of Medicine, vol. ii, p. 325.

of most of its victims. Compared with hereditary and educational influences, with mental depression, with starvation, with tea-drinking, with pelvic irritation, with dyspepsia, the experience of the ailment (as collected by M. Briquet[1]) shows the evil power of alcohol to be inconsiderable in the number it thus immediately affects. Still I cannot but think that if they had been made from clinical records in this country, his statistics would have assigned more importance to it as an originating influence. What we are concerned with now, however, is its injurious action on the disease already established. Of that I have no doubt. And I find a full explanation of the fact in the innutrition of the nerve tissue which was illustrated by experiment in the former chapter on " Alcohol."

There is in the treatment of hysteria a difficulty of diagnosis which especially concerns the use of alcohol, namely, that arising from the frequent occurrence in the same person, and at the same time too very often, of some variety of neuralgia, such as intermittent tic, brow ague, and stitch in the side, which when they occur alone are benefited by high living and port wine. A trimming treatment is necessary here, but if the hysteria predominates, I strongly object to the employment of alcoholic stimulants.

The momentary relief afforded by the poison puts a mountainous difficulty in the path of those who would give their clients advice grounded on the truths established by science. We feel ourselves so weak, that some have swallowed the fallacies of teetotalism, so as to be able to "answer a fool according to his folly." Others of us, to keep one another up by mutual moral support, banded together two years ago, to sign a protest against the misuse of alcohol in medicine. And I do think that the latter movement has had considerable effect in rousing the consciences of medical men to the meanness of conciliating patients, especially nervous patients, by allowing them to drown pain in what is a final poison.

The best way of breaking off the habit of yielding to the perverted sensation which so insidiously cries for alcohol, is immediately and altogether to relinquish it. Terrible sometimes is the struggle, yet it is a bracing and ennobling conflict; whereas the long-continued daily annoyance of giving it up little by little is

---

[1] Briquet sur l' Hystérie.

on the whole quite as painful, and is often enfeebling to the mind. Moreover, courage is likelier to give way in a month than in a day.

A generous nitrogenous diet is essential in the treatment of hysteria. Meat should be eaten at breakfast, and it will be found to be less repulsive and easier digested if the internal viscera be previously braced up by a shower bath, or a cold douche to the spine. The best drink is milk and soda water, in the proportion of about half of each.

The hysteric temperament will in the female sex sometimes cause functional derangement of the periodical evacuation, either in excess, defect, or painfulness. As a rule, all local treatment is to be avoided, and the attention of the patient withdrawn from what is, in point of fact, a mere accidental symptom to a due regulation of the diet and regimen. And the same rule will apply to irritation of the urinary organs.

Pollutio nocturna in men also often arises from the same cause. Excluding the cases where this annoyance depends on worms or piles in the rectum, it will be found that nine-tenths of the subjects of it come of hysterical families, and may be relieved by anti-hysteric treatment. If it is associated with *impotentia coram fœmina*, the diagnosis of the temperament is the more easy. Here again local interference is most pernicious.

§ 2. Delirium Tremens, Alcoholism.

Nobody but a whimsical doctrinaire would propose to relieve physical phenomena really similar in essential matters to such as are caused through a drug, by administering an additional similar stimulant. No man of plain, unclouded common sense does so in words; but in act, the vulgar drunkard looks upon himself as performing a feat of wisdom, when he takes "a hair of the dog that bit him."[1] What this does (if it is strong enough to do any-

---

[1] *Si nocturna tibi noceat potatio vini,*
Hoc tu mane bibas iterum, et fuerit medicina.
"If overnight's debauch does hurtful prove,
A glass next morning will your pains remove."
Sanctorius.

thing) is to stop the painful process of eliminating the toxic agent. It is equivalent to an attempt to soothe retching when a man has swallowed an ounce of laudanum. The retching may be soothed certainly, and by an additional dose of the laudanum itself; but the sufferer is all the worse in the end.

As a remedy for drunkenness some have employed the agency, supposed to be antagonistic to alcohol, of sulphuric ether. It has been useful in dead-drunkenness (coma), and those who have indulged in secret will sometimes steady themselves and cover the airy evidence of their brandy dram with an odor which hopes to pass as an article of materia medica. I suspect the proceeding differs in nowise from that above-named. The safest and quickest, though not the pleasantest, cures for the remorse of a guilty stomach, are patience and diluents. Let the sick man, if he still on principle adheres to his " hair," make it as little a one as possible, and drowned, as completely as possible, in the antagonistic fluid, water; and let the patience be seasoned with repentance; and then he may calculate safely on being no worse in the end for his temporary folly. In this case what happens, probably, is a retention of effete matters, especially in the nervous tissue, proportionally greater than that noted in Experiments I, II, and III,[1] in the ratio of the amount of alcohol held in the body at once—a retention followed by a copious evacuation, not only of the retained matters, but of the results of an increase in destructive assimilation. Hence a feeling, not of satisfaction, as when retained effete matters are got rid of by a timely purge, but of painful exhaustion. The body is actually lighter, as noticed by Sanctorius, but, from weariness, feels heavier. After the evacuation has gradually ceased, there is no reason to doubt that the body quite recovers itself, for temporary functional disturbances, when time is allowed for full completion of the crisis, leave no organic traces.

So that the occasional (say " monthly " in deference to Sanctorius) drunkard, if he also takes a monthly bout of penitence and temperance, probably does not shorten his days. But that he thereby lengthens them or gladdens them, implies a condition of things that never need voluntarily be submitted to in Europe.

---

[1] See pages 200, 206, 208.

It implies a monotony of surroundings which will thank a misery for a change, and clock-like habits unbroken by the due mental and bodily stimulants of natural life. It implies an existence approaching to that of the Pitcairn Islanders, the descendants of the crew of the "Bounty," who are described by their rare visitors as having eliminated from their little community all that could lead to sorrow or irregularity, without diseases and without doctors; and yet they rapidly grow old, and soon after forty are calmly gathered to their forefathers, simply because they have nothing else to do. Their last sexagenarians were some survivors of the original mutineers. This is an extreme case, but we sometimes come across examples of too monotonous an existence even in civilized life, as, for instance, in the families of respectable persons, engaged in safe trade. Amusement and ambition, from want of use, fail to attract them, and considerable ingenuity is often required to make them care for anything. Dramatic entertainments, either on stage, platform, or pulpit, afford the readiest resource; but of course any hobby in which the patient may himself act a part, is much more valuable. All these are much healthier stimulants than alcohol; and moreover, whatever sanatory merits intoxication might have, the moral objections to it would outweigh them all, and in spite of the authority of Sanctorius and our gratitude to him as a physiologist, I hope I shall never again hear any one advised to get drunk for the good of his health.

Delirium Ebrietatis, "mad drunkenness," is the systematic name given to a form of intoxication, whose peculiarity depends on the diathesis or temperament of the sufferer's brain. While most grow stupid, quarrelsome, or maudlin under drink, some (with probably a latent taint of insanity in them) have a dangerous tendency to grow furious, and to have delusions which still further excite their fury. The affection is, in fact, a sort of mania, exhibited only during the presence of alcohol in the blood. Indeed, in Dr. Darwin's days, it was apparently considered true mania; and as bleeding was usual in mania then, it was practiced for *delirium ebrietatis* also, though not with the approbation of that far-sighted physician, who, in this and many other things, was much in advance of his generation. He says, "Venesection, I believe, sometimes destroys those who would otherwise have re-

covered in a few hours."[1] He recommends emetics and watery
liquids, which are antagonistic to alcohol, and therefore I presume
he would join me in classing this among the diseases where alcohol is to be avoided. It is important to distinguish it from madness, in which alcohol is sometimes useful, as will be afterwards
noticed. A few hours will certify the diagnosis, as *delirium ebrietatis* is soon recovered from, unless the patient be a constant
tippler.

Delirium Tremens—or "the horrors"—is the familiar and distinct form of disease of the nervous system arising from the long-continued deleterious influence of alcohol on health. Its pathology
has already (p. 211, etc.) been assigned to an arrest of the renewal
of the tissues, especially those which contain phosphorus—an arrest
found by experiment when even small quantities of alcohol are
ingested, and apparently directly proportioned to the continuousness of the ingestion. It is distinct from *delirium ebrietatis*, not
only in its symptoms, which are too well known to need repetition, but in the fact that these symptoms seldom, if ever, follow
immediately on an unusual excess. It is true that a man may be
mad drunk, and after a time drivel off into the horrors, if he is
an habitual dram-drinker; but he does not, as a rule, get struck
with the horrors during his debauch. Indeed, this form of disease, as often as not, is the consequence of a sudden omission of
the customary stimulant. It is on the ground of this last-named
fact, apparently, that some practitioners have thought it best to
continue the use of alcohol, while the patient is still raving and
trembling from its effects. I do not understand them to suppose
that they are acting on any homœopathic principle: they aim only
at fending off the danger of the excessive crisis by restoring that
condition of alcoholic saturation the patient enjoyed previously,
and during which he did *not* fall into delirium tremens. I cannot agree with those practitioners. Truly enough the raving, and
the terror, and the trembling, and the sweating, and the awful
chaos of dreaming and wakefulness, in which the poor wretch is
plunged, are made somewhat more tolerable by the dose—the
pains are palliated. Nevertheless, there is a higher aim of treatment than palliation of pain; the danger to life does not depend

---

[1] Zoonomia, ii, 1, 7.

on these symptoms, but on the suspension of the controlling functions of the brain; and to that suspension and danger each additional drop of alcohol must contribute its share. I feel sure that life is risked by the alcoholic treatment of delirium tremens; and the risk of life is, in this disease especially, the paramount consideration. For the patient has brought it on himself by his own fault, and so long as he does not die I am very little anxious to spare him the pains and penalties of his self-indulgence, which pains and penalties may perhaps contribute useful lessons of future conduct.

My own plan of treatment has been to cut off the supply of alcohol—at once, and, so far as my power extends, forever. Where there is no vomiting, I give a mustard emetic, and then a purgative enema. If the patient is quiet enough, the vital temperature, which, as we have seen (p. 203), is lowered by alcohol, should be kept up by a warm bath, or at least by hot ablutions. And fresh material for new tissue, with assistance in getting rid of effete tissue, should be assiduously furnished in the shape of meat and water united, as beef tea with the pounded beef in it. A sift of pepsin mixed with grated Parmesan cheese will make it spicy for the jaded stomach, and promote its digestion. The addition is especially desirable if in consequence of vomiting the food has to be taken by enema. (See recipe for the enema, p. 242.)

It will be seen that the intention of the abovementioned treatment is to evacuate the alcohol, and to counteract its suspensory influence on the vital warmth and on the assimilation and renewal of the nervous tissue.

There is one other highly important measure which is rendered needful by the reduced power of the nervous system during its functional arrest. The functions that depend upon it must be spared work. Complete quiet, and shielding from whatever can excite the mind, is important, not only during the disease but during convalescence, lest a relapse should occur. The excitement may be so mild as scarcely to occur to us as a risk; yet by the frail machinery and unfed tissues it cannot be borne. I will quote an example from my note-book. An earnest young curate who came under my care for delirium tremens, contracted by drinking bad brandy to support him in his daily ministration to a large country parish, had got so well again that I consented to his breaking the intolerable idleness of a Sunday in London, by hearing a famous preacher. During the night (to my shame) he had a relapse, and acknowledged that the cause was the contagious enthusiasm of the orator. Again, in a former chapter (page 212) is mentioned a case strongly impressed on my memory by being the only young female patient I have lost by delirium tremens, and in which the disease was due to the mental excitement caused by a false accusation of theft after dram-drinking of by no means an excessive character. This weakness of the nerve tissue which gives way under such moderate pressure as it could easily resist in health, is explained by the arrest of its vitality indicated by the non-excretion of phosphates during the after-effects of alcoholic anæsthesia, as before described. We must not forget to allow for it in the treatment of patients.

It is rather beyond the scope of the subject in hand to speak of the influence of other anæsthetics in disease; yet I cannot pass on without saying that if an ordinary dose of opium or its alkaloids, or of hydrate of chloral, gives a patient a few hours' sleep, he seems to me none the worse for it. But it is most dangerous to look upon them as specific, and to administer them in continuous reprises till sleep follows, or in the enormous doses that have been employed by some practitioners. The reason for those enormous doses being tolerated is that absorption has been arrested, and that they have passed away harmlessly. But supposing that absorption should be suddenly restored, the enormous dose can hardly fail to cause sudden death. No advantage is gained, and a great risk is incurred.[1]

In *chronic alcoholism* I allow the same argument to hold good. I have never found harm to arise from leaving off alcohol alto-

---

[1] It is surprising to find such a generally judicious physician (Dr. Todd) an advocate of the homœopathic treatment of alcoholic poisoning. In the last of his Clinical Lectures on certain Acute Diseases (one on the therapeutical effects of alcohol) he commends the free administration of wine to a child of three years old, who was dying from convulsions and hemiplegia in consequence of having swallowed a large quantity of gin, and regrets that brandy had not also been poured in by the rectum. He relates also specimens of delirium tremens similarly managed, but their result cannot be said to justify the use of brandy except to a most bigoted adherent. He himself does not propose to act in the same way in respect of other articles of the materia medica.

gether and at once. I am not prepared to deny that delirium tremens may have been the result, in some cases, of such precipitation, but it has not happened in my experience; and I should be ready, if it did happen, to treat it in the usual way, confident that the patient, in spite of the accident, was still gaining a positive advantage by abstinence.

§ 3. Ague and Intermittents.

In ague, alcohol should not be spared as an article of diet; and taken more freely than usual it contributes to lengthening the interval and shortening the paroxysms, converting quotidian into tertian, and curing tertian altogether. And, in brow-ague, neuralgia, inconstant sciatica, and other intermittent disease of similar nature, it is sometimes equally efficacious. But, given during the paroxysms, I have not found so much benefit—perhaps it is not absorbed.

In ague the combination in which alcohol is offered is of considerable importance. The most generous red wines should be used, and the distance at which their bouquet may be smelt may be taken as a rough test of their utility. I remember learning a lesson on this point from a most unscientific source. I was chatting in the market-place at Dijon with a farmer's wife, when she incidentally mentioned that her husband was a great sufferer from ague, and was quite tired of swallowing quinine. I advised her to take home a good supply of Burgundy in her market-basket, and begged to contribute the few francs I had in my pocket. She tripped straight off to a grand wine-merchant's office; but instead of coming out fully laden, she bore only two bottles, to the price of which she had contributed out of her own purse. It was of a vintage such as is allowed to trickle slowly over the tongue at the table of a prince, and I promptly called her a prodigal. "No, no," said she, "I am not; a mouthful of this is worth to a sick man a bucket of commoner wine"—and yet the common wine of Dijon is not to be sneered at. She was quite right; there is no wine like Burgundy for ague, and the price (provided the merchant be honest) is a direct measure of its medicinal value.

§ 4. Other Diseases of the Nervous System.

Alcohol has been used with great advantage in *Sunstroke.* Dr. Maclean (in his article on the subject in Reynolds's "System of Medicine") strongly commends the employment of *Warburg's Tincture,* the only known ingredient in which is alcohol, and in addition a judicious quantity of dietetic stimulants. The treatment advocated by this gentleman does not seem to have yet taken hold of the Anglo-Indian medical mind, for the author speaks of venesection still being employed on a large and fatal scale quite recently; so that numerical evidence of success cannot be furnished. I should expect it when collected to be decisive; for there are reported in sunstroke several most important symptoms which connect its pathology with that of other morbid states benefited by alcohol, viz., a copious excretion of solids in the urine, a high temperature of blood, delirium, weakness of the sphincters, trembling of the muscles, and death by syncope or coma.

I have never had a case of genuine sunstroke under my care, the best imitation of one which this cloud-kerchiefed climate has afforded me having been caused by long sitting at a desk with a bald head close to an Argand gas burner. But if ever I were to practice in a hot climate, I should unhesitatingly adopt Dr. Maclean's treatment, both as to the use of ice and of alcohol.

In *concussion of the brain* from accident, stimulants are usually administered before the arrival of a surgeon, and no injurious consequences appear to follow. And blows on the head which happen to revellers produce, as a rule, much less cerebral lesion afterwards than is expected from their violence. The hint should not be lost, especially since some of the symptoms of severe jar to the brain-tissue much resemble those recorded as produced by sunstroke; indeed, so closely as to have suggested the insertion of this paragraph in its present place.

Neuralgia is often decidedly benefited by alcohol, as has been already said. Taken dietetically at meals, it involves no risk of tempting the patient to intemperate indulgence; and if good generous wine be the vehicle, a very moderate quantity often suffices.

Spirits are not nearly so effectual, and require to be employed in larger doses.

But this is not all that the sufferer asks; he wants to know if he may not have recourse to the remedy when his agonies are upon him; for neuralgia is not like ague, and is relieved in an unmistakable manner by alcohol during the paroxysm. It is hard to refuse, and, unless the patient surpasses the average in folly, I do not refuse. But at the same time I give this caution—that the intended dose should be fixed upon, and not exceeded even if the pain remain as severe as ever. The chief benefit is by no means immediate, but consists principally in shortening the paroxysm. An intimate friend of mine with an "irritable stump," often brought on (like Nelson's) by cold or mental annoyance, on the invasion of the spasmodic movements takes a tumbler of very hot spiced port (bishop) or of stiff whisky toddy, and then waits patiently till the trouble ceases. This it usually does in two or three, instead of lasting twenty-four, hours. But whatever happens, he strongly advises that the stimulant should not be repeated.

Hypochondriasis is too generally a constitutional and hereditary complaint to be a good test of the virtue of any remedy. The few hypochondriacs in whom I have witnessed any improvement have been persons who had been living a dull abstemious life, and have been persuaded on sanatory grounds to indulge the palate and the fancy from time to time with unwonted liberality. It is very possible that Sanctorius's monthly excess might be a sanatory measure here, though one is loath to give advice which is capable of misinterpretation. But the daily and habitual use of alcohol in hypochondriasis is worse than useless—in fact, deleterious. I have known several cases receive much benefit from entirely leaving it off.

There is a peculiar form of *hypochondriasis* which arises from eating too little vegetables and too much meat. It is distinguished by the high specific gravity of the urine (1.025 to 1.035) dependent on the presence of urea alone without sugar. There is in these cases often a remarkable lassitude and apparent paralysis of the limbs suddenly occurring after exertion. Sometimes there is emaciation. Both these symptoms usually lead the patient and his friends to attribute the morbid state to insufficient nutrition, and to increase more and more the proportion of meat in spite of the aggravation of the ailment. A rapid cure attends the diminution of the meat meals to one daily, and the supplying their place with plenty of porridge and green vegetables.

In *mania, melancholia,* and *dementia,* the most recent experience seems unanimous in recommending alcohol. Dr. Maudsley makes no distinction between them in speaking of the therapeutics of insanity, and indeed seems to proportion the strength of the liquor to the violence of the disease. He speaks of "wine" as a prophylactic against madness in general, and "brandy" as a remedy in maniacal excitement. I cannot find, as I should have liked, any warning bearing on cases which can be traced to drunkenness in the individual affected, though it can hardly be doubted that such do exist, and require special management. Even if the sequence of causation have been reversed, and the intemperance be the consequence of the madness, the association of the two must surely modify the treatment. It may be that Dr. Maudsley approves of the homœopathic therapeusis of alcoholism, and his evidence to its prudence would be very valuable; but it should be distinctly and boldly expressed.

Leaving this branch of the question open, it may be freely granted that his sketch of the current practice in curable disorders of the intellect is correct, and his confidence in its success is not misplaced. But here, as in health, I would limit the quantity, as a rule, to that which suits the stomach best and increases the appetite.

The advantage of an ample and nutritious diet in insanity is daily more and more forcing itself upon the proprietors of lunatic asylums, though their interest of course would tempt them to an opposite creed. It is good economy in the end to feed highly even pauper lunatics, unused as they may have been to such treatment; for by such means cures are effected, and the country relieved of the charge.

CHAPTER VIII.

SCROFULA, RICKETS, AND CONSUMPTION.

SECTION I.

SCROFULA is indubitably an hereditary disease; and some observations incline me to the suspicion that the development of the ancestral stain is more common when it is on the maternal side than when it is derived from the father. A faulty early nutrition would seem to have as much influence in drawing it out, as the source of the vital spark. For this reason a mother who is conscious of the existence of scrofula in her family, even should she never have been herself afflicted, should deny herself the privilege of suckling her infants; and a healthy irreproachable wet-nurse should be got to supply her place. On the other hand, when it is from the father's side that danger is feared, it is peculiarly incumbent on her to play her important part with assiduity, and to protract the time of lactation to its full period, observing accurately the suggestion made in the first chapter of the second part of this volume. I cannot too strongly urge the duty of impressing this upon patients.

The nursing mother will act wisely also if she takes after her meals small doses of the Syrup of the Phosphates, or Parrish's Chemical Food.

Lactation ended, the child's diet should be arranged so as to contain a rather larger proportion of animal food than recommended for ordinary nurseries and the instinct for a carnivorous nutriment should be encouraged, or at least not thwarted. Warm clothing, much sunlight, frequent exercise in the open air, and an annual breath of sea air will powerfully aid in keeping off the dreaded disease.

The "weariness to the flesh" induced by overmuch study must be sedulously prevented. It spoils the appetite and digestion, so essential to the object of our care.

The cold plunge-bath, well-ventilated rooms, moderation in the pursuits of pleasure and ambition, and a virtuous life, will usually carry our patients safely up to manhood and womanhood, and after that they may be considered safe.

Supposing the tendency is so strong that in spite of these precautions the disease appears, I do then think drugs are useful, but chiefly those which have the nearest relation to aliments, such as iron, the phosphates, cod-liver oil. Their effect on the appetite must be carefully watched; and the end must not be sacrificed to the means—if they spoil the appetite they must be left off immediately.

In this case also, I think alcohol is of service, employed with the same watchfulness as articles of the pharmacopœia.

Section II.

Rickets (*Rachitis*) is distinguished from scrofula by being very rarely, if ever, hereditary. It consists in a softening of the bones, especially of the back bones, in children who do not absorb enough bone-building materials in their food. I think I cannot do better here than quote the words of Dr. Trousseau, who, with his usual happiness of expression, has put before the world the striking results of M. Jules Guérin's researches:

"Of all causes, that most sure to produce rickets is improper food.

"In his first works M. Jules Guérin had adopted the idea generally admitted that rickets and scrofula are occasioned through deficient feeding; and by that term, according to vulgar prejudice, was meant suckling carried on too long. But his observation taught him soon that the direct converse of this proposition was true, and that the babies who became rickety were not those who had been kept too long at the breast, but those on the contrary who had been prematurely weaned. In fact, it appeared to be true enough that under the influence of an insufficient supply of proper food, the malady developed itself; but by proper food was to be understood something different from what was commonly meant. Experiments tried upon animals made the question quite clear. In these experiments M. Jules Guérin set himself to find out if it were feasible to produce rachitis at will. He took a number of puppies in equal condition; and having let them suckle for a time, he weaned suddenly half the lot and fed them on raw meat, a diet which at first sight would seem the most suitable for carnivorous animals. Nevertheless, after a short time, those who continued to take the mother's milk had grown strong and hearty, while those who had been weaned on an apparently more substantial diet pined, and were taken with vomiting; then their limbs bent, and at the end of four or five months the poor little beasts showed all the symptoms of confirmed rickets. From these experiments we must conclude with M. Jules Guérin that the rachitis depended in great measure on the derangements of nutrition which claimed improper diet as their cause. A diet which is taken to at a wrong season may fairly be called improper. For carnivora, it is flesh before the age of suckling has passed; for herbivora (and an experiment bearing on the point has been made on pigs) it is vegetable feeding, given them too soon, when they ought to be still at the teat. In the human race the same thing happens. Rickets is never so common as it is in babies weaned ere the teething is forward enough, and brought up on pap, vegetables, or even meat."[1] It would be waste of time to point the moral of these valuable sentences.

The daily addition to the dietary of a small quantity of the phosphatic salts is still more indicated here than in scrofula.

But still nothing can take the place of milk, an article of diet in which our laboring population, both in town and country, is sadly deficient. I have no fault to find with coal clubs and clothing clubs, but I am sure that as a preserver of infantile health a "cow club" is worth them all. Even the small weekly contribution, so difficult to extract from our reckless race, can be dispensed with, by arranging the price paid on delivery, so as to cover keep, insurance, contract for medical attendance, and replacement of superannuated animals. Charitable help is most valuable when it takes upon itself the duty of advice and superintendence. But still, O Lady Bountiful, do not give your spare and skim milk to the store pig, when there are half a dozen children growing up bow-legged and crooked within half a mile of

---

[1] Trousseau, Clinique Médicale, vol. iii, p. 484, 3me édition.

your lodge gate for lack of it. Do not *give* it at all, but bestow the much greater boon of selling it—as low as you like, but still sell it.

Section III.

Consumption.—Since the introduction into our materia medica of Cod-liver oil by Dr. Bardsley of Manchester, the profession has been growing gradually, but surely, convinced of the all-importance of the dietetic treatment of pulmonary consumption. Cod-liver oil is a typical aliment, representing what is the fittest of all known substances to supply the deficiency that constitutes the disease. To use the language of microscopic physiology, "the basis of molecular growth" is poisoned, so that instead of actively secreting gland, or elastic tissue, or bloodvessel, or epithelium, etc., fit for their various duties, being formed, only a useless cheesy substance is the product of nutrition. The due powers of life are lacking in it. We call it "tubercle," and look upon it, if we are thoughtful people, as an infant tissue strangled in its cradle. To save then the parts threatened with tubercle, we must anticipate the formation of the imperfect matter by supplying a groundwork for perfect tissue. This groundwork is laid by freshly assimilated oleaginous substance; oleaginous substance is what is furnished by nature for the primary growth and nutrition of all the higher tissue of animal bodies; and indeed many physiologists assert that without it there is no growth; so that in administering it, we are closely imitating the wisest teacher of medicine, mother Nature.

The reason, then, for giving cod-liver oil is to counteract the tendency to form morbid solids by supplying the most suitable material for healthy solids—in short, by overcoming evil with good. This is a much higher aim than the mere replacement of the plumpness of flesh whose waste gives the name to the disorder. The object is not so much to cure emaciation, as to cure the cause of the emaciation. Very necessary it is to understand this; for not a few tuberculous and many tuberculously inclined patients are well furnished with fat, and some are even corpulent; and they will demur to the prescription of our remedy, if they are allowed to retain the dominant idea that all it can do is to fatten.

Young women, especially, of consumptive families will often object to the risk of losing the elegant slimness of youth, and insist that they are as plump and round as they ought to be. What they say may very likely be true; but yet their store of fat can only keep up the temperature of the body, and cannot be used again for nutrition. For the purpose of nutrition, recently assimilated fat is requisite; and thus tuberculosis may go on unchecked for lack of it, even in a corpulent person, since his store of fat is not available for digestive purposes.

At present cod-liver oil is the most readily assimilated fat we know of: and so to it all other means of treating this class of diseases have a reference. Some substances ground their claims for notice on being substitutes for it, and drugs of various kinds act beneficially by preparing the stomach to digest it; but anything which does not look to it for a character is of only accidental utility.

I heartily wish that some of the efficient substitutes for fish-oil were more agreeable to the palate; for it must be confessed that the taste and smell are a serious impediment to its employment. Suet is apparently the best: in milk it is to some persons not repugnant, and its digestibility is increased by its being made into an emulsion with one of the gastric solvents, to be obtained in the shops under the name of "Pancreatic emulsion."[1] Milk, by itself, comes next in value, and a milk diet has from the earliest times been recommended in advanced phthisis. But the large quantity required to be swallowed, often deranges the stomach and produces repugnance; so that in recommending it to a patient it is necessary to warn him of this difficulty, and take example by our master Hippocrates, who, in advising a consumptive to drink a large quart jug ($\tau\rho\iota\chi\acute{o}\tau\upsilon\lambda o\nu$ $\varkappa\acute{\upsilon}\lambda\iota\varkappa a^2$) of mare's milk the first thing every morning, adds significantly "if he can."

Devonshire cream has been used as a substitute, but I cannot say it has proved in my hands an efficient one. If taken in sufficiently large quantities to be of service, it is apt to exercise a purgative action.

For my part, I think it is wisest to try and get habituated to

---

[1] The credit of introducing this to the profession is due to Dr. Dobell.
[2] Equal to two pints and a quarter.

the fish-oil, adopting some of the various little devices which are used to diminish its nauseousness. In the first place it is essential that the article should be the best of its sort, that is, as free from smell, taste, and color as possible, showing its careful and recent preparation. Though they do not always tell their customers, the importance of this freshness is well-known to manufacturers. I learnt so when prescribing it for a stranger a few years ago: her husband, who was present, stopped me when I was describing what cod-liver oil should be, saying he knew all about that, for he made it, and engaged that she should have it fresh prepared daily.

Some patients will take the oil easiest in milk. Some find the taste annulled by eating a piece of red herring, or anchovy, or sardine, before and after the dose. Some like the bitter aromatic of coffee to counteract the rankness; and this fancy I am glad always to hear of; for it gives an opportunity for recommending a bitter drug which assists powerfully in the assimilation of the oil. If coffee is approved of, I then advise the oil to be taken as a parenthesis in a mixture of quinine, or of strychnine, or of both together. A sip of the medicine is to be taken, then the oil swallowed, and washed down with the remainder of the draught. If there is persistent nausea afterwards, a few drops of hydrocyanic acid may be added each time. Nausea may also be avoided by taking the oil on going to bed. Where there is cough, the sheets should be previously warmed.

But the taking of cod-liver oil is not an exemption from all other care. The student of the literature of the subject will find his faith claimed by two extreme parties, one of which is all for beefsteaks, porter, and in short for as carnivorous a diet as possible, the other in favor of pulse and herbs, seeds and potatoes. Both are as far right and as far wrong as it is probably possible to be. They are quite right in what they include, and quite wrong in what they exclude. The food should be as near that of health as the digestion of the patient will allow, that is to say, mixed and varied, liberal in frequency, and moderate in quantity. The whims of appetite should not be thwarted, nor the prejudices of theory indulged: the demands of muscular exertion should be provided for by the starchy ingredients of the dietary, and the increased wear and tear of the machinery by the nitrogenous.

The full powers of digestion should be brought into activity by plenty of open air and exercise within the bounds of weariness. And this is the reason why English patients are advised to live during the winter months in more genial climes than their own. They get thus more sunlight and oxygen without damp. But also the change is good; for it has been found that to those living in parts of Britain where the atmosphere is soft and mild, as in the West, a colder and more bracing air, provided it be dry, is sometimes beneficial. From Cornwall the small trading vessels will sometimes carry a consumptive miner to pass the winter in Greenland; and patients have been sent with advantage from such localities as Leghorn to the exhilarating sharpness of a snow-clad Alpine upland. But this plan requires great caution.

Much more confidently can the recommendation be given that the advantages of a sea-life should be favorably put before the youths of families, where there is an ancestral tendency to pulmonary tubercle. I do not mean that they should be driven into it against their will, but that it should be made as attractive as possible. The healthy state of a young sailor's digestive organs is the best guarantee that can be obtained against consumption.

As to the use of alcohol in threatened cases, and in the early stages of tubercle, I have no hesitation in pronouncing an opinion adverse to it. But if the morbid matter has broken down, and there is either nocturnal perspiration, copious purulent expectoration, diarrhœa, extreme emaciation, or depression of spirits, wine, especially port wine, in quantity equal to the occasion, is often of decided use. When the demand for it has passed away, it may be left off.

CHAPTER IX.

DISEASE OF THE HEART AND ARTERIES.

Section I.

In disease of the heart the most remarkable change in respect of digestion is the slowness with which liquids are absorbed by the stomach. If much is drunk at once, it will remain a long time gurgling about, and, instead of assisting digestion, will offer an impediment and cause inconvenience. The reason of this is the slackening of the circulation of the blood, which is directly or indirectly effected by all the various cardiac lesions in common. Whether the heart is too thick or too thin, too large or too small, whether the openings of valves are obstructed by warty growth, or whether their stretched curtains fail to control the flux and reflux of the current, the practical result is one, namely, that the vital stream is unnaturally sluggish. Now it is a well known law governing the passage of liquids through membranes that it is promoted by the rapid movement, and retarded by the sluggish movement, of the fluid towards which the endosmosis is tending. It is clear, therefore, that the absorption of potables from the digestive canal into the blood must be deficient in close proportion to the obstruction presented by the defective machinery of the circulation.

The observation of a dry diet contributes greatly to the comfort of the patients now being considered. The dinner should never be prefaced by soup, and the drink should be taken only in sips during the meal. And what is taken should not be cold, but at least have the chill taken off. When thirst occurs between meals, the same rule of sipping should be observed. In this way quite a sufficiency to supply the requirements of the kidneys may be introduced—indeed, more is probably got into the blood than when larger quantities are gulped down at once, chilling and paralyzing the absorbent powers of the mucous membranes.

Independent of the inconvenience which the indigestion of fluids causes, it may be feared also that the distension of the stomach accompanying it interferes with the already laborious action of the heart, and aggravates the organic source of the malady. This is especially the case in Dilatation with Thinning of the heart's walls, and is least noticeable in Valvular Disease.

Where Heart-disease is complicated with obesity, especially if the fat is accumulated in the chest, the enforcement of a dry diet is still further to be viewed as imperative; inasmuch as it contributes powerfully to the reduction of the hypertrophied adipose tissue.

The only symptom which may perhaps render it a doubtful policy is the occurrence of dropsy. Should this be very decided, it is of importance to keep the kidneys active to a degree that can be accomplished only by the ingestion of a considerable supply of watery drink. I think, however, this should be always followed up by purgative medicines, often even by mercurials. And when the necessity has ceased, the dry regimen should be resumed.

The dietary of persons with imperfect hearts should be at least as nitrogenous as if they were completely sound. What we have to dread is the atrophic degeneration of the cardiac muscle, for till this degeneration occurs the original lesion is not aggravated; and the constitution often gets so used to the state of mechanism of the heart that no inconvenience of any kind is felt. And atrophic degeneration is warded off by keeping up the redness and fluidity of the arterial blood.

The importance of the existence or non-existence of valvular disease lies not in the injury it directly inflicts, as in the likelihood of the induction of other lesions of the heart. If the muscular structure remains healthy, injured valves do not appear capable of causing death. But very surely are they fatal when they are followed by dilatation or thickening or degeneration of the cardiac walls, with the sad train of dropsies, apoplexy, pulmonary hemorrhage, etc.

In 2161 post-mortem examinations at St. George's Hospital in ten years, the cardiac valves were diseased without the walls of the heart being affected 113 times; but in every case there were other lesions amply sufficient to account for death quite independent of the valves, such as accidents, surgical complaints, cancer, low fever, etc. In one alone, where anasarca from granular kidneys was the immediate cause of death, could any symptom be debited to diseased valves with healthy heart, and that symptom was pulmonary hemorrhage.[1]

In the classes of people represented by hospital patients, the probability that valvular disease will be followed by its unhappy consequences is very great. When a patient thus affected leaves our hospital wards, we expect to see him again shortly, and on each fresh admission with a more severe complaint. But the same expectation must not be applied to the more comfortable classes of society. Persons in easy circumstances have valvular lesions for years and years, perhaps through the greater part of a long life, and not only continue to live, but even fail to experience symptoms bad enough to consult a medical practitioner.

In my volume of Clinical Lectures[2] I gave some two dozen examples of persons found, when examined for insurance, to have valvular lesions, but who, during periods varying from 50 years downwards, had not in anywise suffered in consequence. This I attribute to their not being dependent for daily bread on daily labor—in fact, to ease of mind and body.

The renewal of the destroyed tissue being impossible, and equally so any mechanical compensation for the arrested function, it is obvious that in the treatment of the disorganized valves themselves, restorative medicine, in the strictest sense, must be quite at fault. But indirectly it is almost as effective in prolonging life, as if it could put in a new valve, or make another muscle do duty for the resting ventricle. It may repair those reparable conditions which are injurious, and which by bringing on enlargement constitute the real danger in cardiac cases. Let us try and cure what is curable, and trouble ourselves as little as possible about bygone injuries.

While we bid our patients live generously, we must disabuse them of the notion that the advice includes a free allowance of alcohol. Alcohol is really the most ungenerous diet possible. Addiction to it impoverishes the blood, and is the surest road to that

---

[1] Decennium Pathologicum, MS. in library of Medico-Chirurgical Society, chap. x, sect. i.
[2] Lect. xxv.

degeneration of the muscular fibres which is so much to be feared. And in diseases of the heart it is especially hurtful, by quickening the beat, causing capillary congestion and irregular circulation, and mechanically dilating the cavities. Let the fermented drink be limited to that quantity which increases the appetite. In a great many instances this may be very shortly described as none at all, but often also, whether from the force of habit, or the nature of the constitution, a sip of wine enables a sufficient dinner to be eaten, and more fully digested than when not thus assisted. Burgundy and Champagne are the best adapted for the purpose, as a small quantity of these wines is inspiriting, with a minimum of alcoholic contents.

Section II.

In the treatment of *Aneurism* the science of diet plays a part different from that which it plays in any other diseases which we have been reviewing. In them it aims at restoring health by replacing what is deficient, and bringing the body directly into a state as near the full natural state as possible. Here its intention more resembles that of a drug; it would bring the blood temporarily into a morbid condition, to accomplish a certain definite object; it makes a sacrifice to realize a future advantage.

By *Aneurism* I mean the formation of a pouch in an artery through the yielding of the sides of the tube. The danger is lest they should go on yielding, burst, and let out the lifeblood. Therefore, if the artery be a small one, and capable of being spared till it is replaced by a collateral circulation, mechanical surgery can save life by tying, compressing, obliterating it, or by removing the limb where it is situated. In the largest arteries such trenchant treatment is obviously impossible; yet there are found from time to time instances of aneurism in even the largest arteries being rendered innocuous by a natural process. In patients who have died of other diseases, the pouch has been found plugged up, and leakage prevented by a firm caulk of fibrin, clotted out gradually from the blood. It is usually adherent to the roughened or torn sides of the artery.

Why is not this happy cure always found? In a great many cases because the mouth of the pouch is so wide and open that the blood cannot stagnate long enough to form a clot. But in others also, because the fluid is so rich in non-coagulating material (to wit, blood-disks) that it coagulates with difficulty, and under some circumstances not at all.

In such a case—that is to say in a person whose fresh color, muscular development, and good digestion show him to be healthy-blooded—if an aneurism forms, the best chances of restoration to safety are to be sought in making the circulating liquid—first, unnaturally stagnant, and secondly, unnaturally coagulable.

These objects can be attained by keeping the patient in bed and starving him. Thus is produced an artificial anæmia, a state of blood in which the red disks are deficient, and the fibrin is in excess, a state in which coagulation is at its maximum, and the force of the heart at its minimum. The method is commonly known as Valsalva's, and it has met with a sufficient measure of success to warrant a trial of it being made in every case where no impediments lie in the way. About fifteen years ago, I published a clinical lecture on several patients treated thus, and I concluded that "not merely is it the best mode, but the only honest mode, of treating aneurisms of the trunk vessels; because it is the only one we at present know consonant to reason and experience."[1]

The most successful cases are those where the sac is in the descending and abdominal aorta, as that of the stone-mason I mentioned in the lectures just quoted. Yet others are not to be despaired of. In the "British Medical Journal" (December 16th, 1865) is published a case by Dr. Waters of Liverpool, of thoracic aneurism successfully treated by rest and low diet, and Mr. Tufnell of Dublin has advocated its use in those cases of aneurism which would come under the care of a surgeon.

Bread and water, or pudding and water, as advised by Valsalva, and used as punishment diet at prisons, is the most effective. Valsalva gave as little as half a pound of pudding in the morning, and half a pound in the evening, till the pulsation of the tumor was arrested, and then he gradually increased the quantity till ordinary diet was resumed, carefully watching against a relapse, however, during this latter process. And I agree that it is better at once to try the experiment with extreme rigor, and give

---

[1] Lectures Chiefly Clinical, lect. xxiv.

it up if necessity will have it so, than to shiver and vacillate on the brink of the serious struggle for life, perhaps too late to be of use.

The confinement to bed must be absolute.

To the future conduct of those who have suffered from aneurism, and in whom a recurrence of the disease is, of course, to be feared, it is obvious that the principles above advocated do not apply. The formation of an aneurismal sac is a degenerative process; and starvation, bleeding, and confinement to bed would be most deleterious, as tending to induce further debility and degeneration. The anæmic dietary is for the sole purpose of bringing about the formation of a clot, and the prevention of further degeneration of the arteries must follow an opposite path. Iron, nutritious food, good dry air, moderate exercise, and suitable clothing should be perseveringly adhered to for the rest of the patient's life.

I would strongly advocate, also, in the interests of the digestive organs, a change from the climate of England, where degenerative disease is the rule, and acute disease the exception, to Italy, where degenerative disease is the exception and acute disease the rule—a change, if not for life, at all events for long enough to alter the constitution. It may be remarked that it is in the absence of aneurism that this peculiarity of the Italian climate is shown. In the statistics of Milan Hospital (*Rendiconto della Benificenza dell' Ospedale Maggiore, etc.*, 1862), I find but 4 instances of thoracic aneurism in 61,761 patients, or 1 in 15,440; whereas at St. Mary's Hospital, London, I find 29 in 7319 patients, or 1 in 252. Again, at Genoa I find the last published bills of mortality (for 1860, printed by Dr. Giovanni du Jardin) without any deaths by aneurism. The corresponding report for London contains 103. The differences are so great, that the limited range of the observation is unimportant, and I cannot but conclude that there is something in the climatic influences of London favorable to, and something in Milan and Genoa antagonistic to, the formation of aneurism.

INDEX.

Acids in atonic dyspepsia, 254
Adulteration, definition of, 218
Ague, 290
Air, as a defence against intoxication, 221
Albuminuria, 272
Alcohol, abstinence from, in gout, 263
    action (physiological) of, 107, 200, 273
    beneficial effects of, 216
    different forms of, 73, 217, 235, 290
    influence of nationality in consumption of, 221
    injurious to children, 137, 226
        hysterics, 282
        nurses, 126
    moral effects of, 221
    use of in atonic dyspepsia, 256
        consumption, 300
        fevers, 235
Alkaline drink, 248
Almonds, 51, 238, 276
Aneurism, 304
Arrowroot, 72, 238
Artichoke, 46
Artichokes, Jerusalem, 44
Asparagus, 43, 45
Athletic training, 155, 181
Athleticism, evils of, 263
Atonic dyspepsia, 217, 252

Bacon, 61, 280
Bael drink, 238
Barley, 71
    water, 85, 237
Bathing, 163, 258, 295
Batter pudding, 243
Bed, lying-in, 167
Beef tea, 131, 239, 275
Biscuits, 70
    and milk, 243
Boils, 166, 172
Brain, concussion of, 291

Bread, 68, 254
    brown, 280
    pudding, 243, 249
    sauce, 242
    soup, 248
Bright's disease of kidneys, 213, 272
Butter, 66
Buttermilk, 65

Cabbage, 43, 45
Cardoons, 46
Carnivora, flesh of, 275
Carrots, 44
Caviare, 41, 60
Celery, 49
Cellulose, conversion into sugar, 59
Change of air, 182, 265, 300
Chantarelles, 47
Cheerfulness, 116, 199, 229, 286
Cheese, 67, 254, 275
Chestnuts, 46
Chewing, 105, 108
Chicken and "Hen" broth, 240
Children, diet of, 134
Chocolate, 55, 96
Cirrhosis from alcohol, 213
Claret cup, 85, 238
Cleanliness, 48, 50, 87, 97, 163
Climate, effects of, 175, 221
    in atonic dyspepsia, 259
    in consumption, 300
    in gout, 264
    of Italy, 265, 306
Clotted cream, 66
Cocoa, 55, 96
Cod-oil, 297
Coffee, 55, 96
Commercial life, diet of, 140
Condy's wash, 239
Constipation, 252, 257, 278
Consumption, pulmonary, 297
Cookery, 52, 87, 106
Corpulence, reduction of, 166
Crabs, 42
Cream, 298

Cucumber, 43, 49
Cups, 85
Curaçoa, 85
Curds, 65

Degeneration, arterial, from alcohol, 215
   in meat, 32, 37, 38
Delirium tremens, 212, 284
Diabetes, 274
Diarrhœa, 171
Diet, dry, 301
Digestibility, degrees of, 114
Digestible, meaning of, 111
Digestion, 101
   time of, 99
Diluents, 50, 51, 85, 113, 250, 277
Disease, transmission of, in meat, 37
   in milk, 63
Dress, 136, 165, 180
Ducks, 39
Dynamic equivalents, 24

Eel broth, 242
Egg nogg, 239
   soup, 239
Eggs, 39, 59, 67, 98, 100
Emotion, action of, 110
Endosmosis, 111
Enemata, nutritive, 235, 242, 288

Fagot, for flavoring, 281
Famine, 176, 194
Fasting, 121, 195
Fevers, diet and regimen of, 231
Fish, 40
Flatulence, 252, 259
Flavorings, 49, 93, 97, 281
   poisonous, 218
Flounders, 260
Food and work, 23, 28
   natural, of man, 17
   quantity of, 22, 27, 28, 120
Fraud, 60 (note), 62, 69, 71, 78, 218
French beans, 47, 275
Fruit, 51, 246, 275
Fusel-oil, 81, 218

Game, 39, 40
Geese, 40
Gelatin, 249
Gluttony, 137
Gout, 247, 261
   in the stomach, 268
Grape sugar, 56, 73
Gravel, 270

Grits, 71, 238
Groceries, 52

Habit, force of, 18
Ham, 61, 170
Hartshorn drink, 248
   jelly, 249
Heart, disease of, 301
Heat, 113
Hypochondriasis, 292
Hysteria, 282
   in the male, 284

Ice, 85
Infants, diet of, 125
Inflammation, diet for, 245
Insanity, 293
Intemperance, dangers of, 213, 220
Iron in medicine, 82, 258, 259, 295

Junkets, 65

Kidneys, activity of, defence against alcohol, 221
   diseases of, 270

Leeks, 44
Lemonade, 85, 237
Lentils, 47
Lettuces, 50, 281
Liebig's food for infants, 130
Linseed tea, 237
Literary life, diet of, 144
Liver, sluggish, 252
   use of, 122
Luncheon, 49, 137, 166, 253

Macaroni, 57, 176
Maize, 71
Malt liquors, 80, 277
   tea, 243
Mayonnaise, 260
Meals, times of, 117
Meat, choice of, 29
   storing of, 100
Milk, 20, 62, 126
   boiled, 170
   diet, 274, 297, 299
   goats', 170
Mineral waters, 84
Mixed diet, 20
Morels, 47
Mothers, diet of, 126
Mould, blue, 64
Mushrooms, 47

INDEX.

Mustard, 61, 269
Mutton broth, 241, 250
    club, 34

Neuralgia, 291
Noxious trades, 152
Nurse, choice of, 236
Nutrition, 122
Nutritiousness, comparative, 123
Nuts, 51

Oatmeal, 71
    flummery, 281
    tea, 251
Obesity, 302
Oil, 59, 170
Old age, 189, 197
Olives, 59, 275
Oxalate of lime, 270
Oysters, 42

Panado, 239
Parsnips, 44
Partridge, 40, 89, 242
Peas and beans, 46, 47, 59, 274
Pepper, 60, 255, 259
Pepsin, 235, 256, 258
Phosphates in medicine, 294
    urine, 271
Physicking, 138
Pickles, 60
Pigeon, 212
Porridge, 280
Potato surprise, 260
Potatoes, 43
    for invalids, 244
Poultry, 39
Poverty, 91, 192
Preserved food, 94, 99
Professional life, diet of, 144
Puff-ball, 47
Purgatives, 278, 288

Raisins, 56
Rennet, 67, 237
Rheumatic fever, diet in, 246
Rheumatism, chronic, 269
Rice, 57
    gruel, 238
    in broth, 241
    milk, 260
    pudding, 243
Rickets, 132, 295
Riding, 268
Rose tea, 250
Rosin, use of, in wine, 219
Rot in sheep, 32

Rowing, 136, 157, 165
Rye, 71

Sage tea, 253
Salad, 48, 254
Salsify, 44
Salt, use of, in wine, 219
Salting, 94
Sauce, universal, 97 (note), 255, 259
Sausages, 61
Scrofula, 294
Seakale, 44
Sea-sickness, 173
Semolina, 57
Shell-fish, 42
Sleep, want of, 172, 198, 217
Soup for poor, 91
    meagre, 248
Sour crout, 45, 111
Spices, 61
Spinach soup, 281
Spirituous liquors, 80
Starch, conversion of, 106
Starvation, 184, 245, 305
Stomach cough, 257
Suet, 298
Sugar, 55
    action of, in digestion, 256
    formation in body, 106, 123, 274
Sunstroke, 291
Swiss condensed milk, 65, 130

Tea, 53, 119
    for invalids, 244
Teeth, 17, 108
Teething, 132
Temperance societies, 228
Temperature, influence of, in digestion, 247
Toast and water, 85
Tomatoes, 46, 49
Travelling, hints about, 169
Treacle, 56
Trichina, 36
Truffles, 47
Tubercle, softening of, arrested by alcohol, 217
Turkeys, 40, 89
Turtle, 41

Uric acid, 270

Variety, 114, 119, 135, 254
Vegetable food, uses of, 42, 49, 194, 254, 263, 279, 292
    marrow, 46

Vegetables, cookery of, 92
   storing of, 100
Vermicelli, 57
Vichy water, 248, 258
Vinegar, 57, 170
Voisin, on idiocy produced by alcohol, 216

Water, choice of, 81
   as a remedy, 219, 279
   gruel, 71, 237
Whey, 65, 236, 238

Wine, 72
   light, beneficial influence of, 217, 266
   in old age, 198
   red, 280
Winter greens, 45
Worms, round, 50
   tape, 84

Yams, 44
Yeast, 70

PAVY ON FOOD.—Just Issued.

A TREATISE ON FOOD AND DIETETICS,
PHYSIOLOGICALLY AND THERAPEUTICALLY CONSIDERED.

By F. W. PAVY, M.D., F.R.S.,
Physician to and Lecturer on Physiology in Guy's Hospital.

In one very neat octavo volume of nearly six hundred pages; cloth, $4.75.

No modern treatise on this subject having existed in the English language, Dr. Pavy's work supplies a want which has been very seriously felt, and in a manner which shows that the author is an extensive reader and has judiciously arranged the numerous facts and theories, together with the most striking experiments and the deductions drawn therefrom. It seems to us that he has truly conferred a great benefit upon all interested in the subject-matter of his work, and that nobody will study its pages without having derived valuable instruction therefrom, and without considering it not only useful, but next to indispensable.—*Amer. Jour. of Pharmacy*, Aug., 1874.

The present book is a result of his work in this direction, and is well calculated to do credit to his perseverance in collecting facts, and his judgment in arranging them in an entertaining, as well as a practical form. It is but rarely that we have had offered us so much practical information in so agreeable a manner as is done by Dr. Pavy in the present instance.—*New Remedies*, July, 1874.

Not pretending to contain much original research, it presents an admirably clear, full, well digested account of all important facts and theories on the supply of force and material in the human organism. The size of the work, the conciseness of its style, the extent of its information, the completeness of its scope, and the general accuracy of its statements reflect not less credit on the industry than on the ability of its author.—*London Lancet*, Oct. 10, 1874.

We can very cordially commend the book to our readers.—*Brit. and For. Med.-Chir. Rev.*, Oct., 1874.

The work will amply repay the reader, whether professional or general, and should find a place in the library of every physician, in the dispensary of every hospital, and would constitute a valuable addition to the household library.—*Chicago Med. Journ.*, Nov., 1874.

In one word, Dr. Pavy has favored his brethren with an admirable book on a most interesting and important subject. When the practice was to reduce the diet of the sick to the *minimum* compatible with existence, physicians might despise dietetics; but since one of the great points in practice has come to be the support of our patients by suitable aliment, the study of food has become one of the most important in medicine.—*Am. Practitioner*, Oct., 1874.

It is seldom, indeed, that we have the good fortune to read so admirable a book as this treatise of Dr. Pavy. A feeling of satisfaction is always felt in the reader's mind when the author writes clearly, instructively, and as master of his subject; and these three characteristics are found combined in the work before us—a work which is sure to become a standard authority on the subject of which it treats, and which we have great pleasure in recommending to the attentive study of our readers.—*Dublin Journ. of Med. Sci.*, Oct, 1874.

BY THE SAME AUTHOR—Lately Issued.

A TREATISE ON THE FUNCTION OF DIGESTION: Its Disorders, and their Treatment. By F. W. Pavy, M.D., F.R.S., Senior Assistant Physician to, and Lecturer on Physiology at, Guy's Hospital, etc. From the Second London Edition. In one very handsome octavo volume of about 250 pages; extra cloth, $2.00.

It is a model of its kind. The author has a happy faculty of entertaining as well as instructing his reader, combining, so to speak, pleasure with profit. Accustomed as a teacher to present salient points, he succeeds admirably in outlining his subject.—*N. Y. Med. Record*, Sept. 1, 1869.

This work is well worthy careful perusal, and should be in the library of every practitioner.—*St. Louis Med. Archives*, Sept., 1869.

In the field thus defined Dr. Pavy has given us the most satisfactory essay yet published.—*N. Y. Med. Gazette*, Aug. 28, 1869.

HENRY C. LEA, Philadelphia.

FOX ON THE STOMACH.—Now Ready.

THE DISEASES OF THE STOMACH.

Being the Third Edition of the "Diagnosis and Treatment of the Varieties of Dyspepsia."

By WILSON FOX, M.D.,

Holme Professor of Clinical Medicine, University College, London.

Revised and enlarged with illustrations, in one handsome octavo volume, cloth, $2.00.

A complete "Treatise on Diseases of the Stomach," which bears testimony not only to a vast amount of work, but also to a surprising research in the literature of the subject. As a work of reference, we consider the volume before us most complete, and one which every man engaged in thoroughly studying diseases of the stomach should consult.—*Dublin Journal of Med. Science*, Oct., 1873.

This work, by Dr. Wilson Fox, is a highly valuable one, representing very fully the most recent views relative to the pathology and symptomatology of diseases of the stomach, and offers an excellent digest of the principles and details of treatment advocated by the most eminent practitioners of the day.—*British and Foreign Med.-Chirurg. Review*, July, 1873.

For want of space we are compelled to omit any notice of the other subjects treated of in this admirable work; and in what we have written we have not been able, for the same reason, to give the author's remarks on the etiology, symptoms, and diagnosis of the various forms of dyspepsia, which to the student will prove the most interesting and useful part of his treatise.—*American Practitioner*, March, 1873.

We have more than usual pleasure in calling attention to this work by Dr. Fox, as well adapted to meet the troubles of practitioners.—*Cincinnati Lancet and Observer*, March, 1873.

BRINTON ON THE STOMACH.—Lately Published.

LECTURES ON THE DISEASES OF THE STOMACH, with an Introduction on its Anatomy and Physiology. By WILLIAM BRINTON, M.D., F.R.S., Physician to St. Thomas's Hospital. From the second and enlarged London Edition. With illustrations on wood. In one handsome octavo volume of about 300 pages; extra cloth, $3.25.

This is no mere compilation, no crude record of cases, but the carefully elaborated production of an accomplished physician, who, for many years, has devoted special attention to the symptomatology, pathology, and treatment of gastric diseases.—*Edinburgh Medical Journal*.

FLINT'S MEDICAL ESSAYS.—Now Ready.

ESSAYS ON CONSERVATIVE MEDICINE, AND KINDRED TOPICS. By AUSTIN FLINT, M.D., Professor of the Principles and Practice of Medicine in Bellevue Med. College, N. Y. In one very handsome volume, royal 12mo.; cloth, $1.38.

This little volume consists of a collection of thoughtful essays on important topics, which have appeared from time to time in various periodicals. The subjects treated of are as follows: I. Conservative Medicine. II. Conservative Medicine as applied to Therapeutics. III. Conservative Medicine as applied to Hygiene. IV. Medicine in the Past, the Present, and the Future. V. Alimentation in Disease. VI. Tolerance of Disease. VII. On the Agency of the Mind in Etiology, Prophylaxis, and Therapeutics. VIII. Divine Design as Manifested in the Natural History of Diseases.

A more suggestive collection of topics it would be difficult to conceive. The essays on conservative medicine are peculiarly valuable.—*Peninsular Journal of Medicine*, Oct., 1874.

TUKE ON MENTAL INFLUENCE.—Lately Issued.

ILLUSTRATIONS OF THE INFLUENCE OF THE MIND UPON THE BODY IN HEALTH AND DISEASE. Designed to illustrate the action of the imagination. By DANIEL HACK TUKE, M.D., Joint author of "The Manual of Psychological Medicine," etc. In one handsome octavo volume of 416 pages; extra cloth, $3.25.

Dr. Tuke's idea in writing this book is to begin work in this uninvestigated field. It is the first work on the subject extant, and practicing physicians will be amply paid for reading it carefully through, and studying the chapter on psycho-therapeutics. Almost every physician rejects the systematic, scientific employment of psychical aids in curing disease, and in so doing one means is thrown away more powerful than all the agents in materia medica.—*Chicago Med. Journal*, Jan., 1873.

The influence of the mind upon the body is without doubt one of the strongest therapeutical agencies which can be wielded by the physician, and he who passes by it as unworthy of his attention, who uses it carelessly and in ignorance of its true scope and value, is neglecting or misusing a valuable assistant in his battle with disease and death. In writing this book the author has placed the profession under many obligations to him, and as the subject is more fully investigated as time passes along, the reader of the future will feel with increased force his debt of gratitude to Dr. Tuke for placing in his hands a work filled with fresh and entertaining truths, stated in terms easily understood. We advise our readers to obtain it and study it with care.—*Buffalo Med. and Surgical Journal*, April, 1873.

This book of Dr. Tuke is one of intense interest, not only to the scientific man but also to the student of the rational school of metaphysics.—*Obstetrical Journal*, June, 1873.

HENRY C. LEA, Philadelphia.

HENRY C. LEA'S
(LATE LEA & BLANCHARD'S)

CLASSIFIED CATALOGUE
OF
MEDICAL AND SURGICAL PUBLICATIONS.

In asking the attention of the profession to the works advertised in the following pages, the publisher would state that no pains are spared to secure a continuance of the confidence earned for the publications of the house by their careful selection and accuracy and finish of execution.

The printed prices are those at which books can generally be supplied by booksellers throughout the United States, who can readily procure for their customers any works not kept in stock. Where access to bookstores is not convenient, books will be sent by mail post-paid on receipt of the price, but no risks are assumed either on the money or the books, and no publications but my own are supplied. Gentlemen will therefore in most cases find it more convenient to deal with the nearest bookseller.

An ILLUSTRATED CATALOGUE, of 64 octavo pages, handsomely printed, will be forwarded by mail, post-paid, on receipt of ten cents.

HENRY C. LEA.

Nos. 706 and 708 SANSOM ST., PHILADELPHIA, July, 1875.

ADDITIONAL INDUCEMENT FOR SUBSCRIBERS TO
THE AMERICAN JOURNAL OF THE MEDICAL SCIENCES.

THREE MEDICAL JOURNALS, containing over 2000 LARGE PAGES,

Free of Postage, for SIX DOLLARS Per Annum.

TERMS FOR 1875:

THE AMERICAN JOURNAL OF THE MEDICAL SCIENCES, and } Five Dollars per annum.
THE MEDICAL NEWS AND LIBRARY, both free of postage, } in advance.

OR

THE AMERICAN JOURNAL OF THE MEDICAL SCIENCES, published quarterly (1150 pages per annum), with
THE MEDICAL NEWS AND LIBRARY, monthly (384 pp. per annum), and
THE MONTHLY ABSTRACT OF MEDICAL SCIENCE (592 pages per annum), } Six Dollars per annum, in advance.

SEPARATE SUBSCRIPTIONS TO

THE AMERICAN JOURNAL OF THE MEDICAL SCIENCES, when not paid for in advance, Five Dollars.
THE MEDICAL NEWS AND LIBRARY, free of postage, in advance, One Dollar.
THE MONTHLY ABSTRACT OF MEDICAL SCIENCE, free of postage, in advance, Two Dollars and a Half.

It is manifest that only a very wide circulation can enable so vast an amount of valuable practical matter to be supplied at a price so unprecedentedly low. The publisher, therefore, has much gratification in stating that the very great favor with which these periodicals are regarded by the profession promises to render the enterprise a permanent one, and it is with especial pleasure that he acknowledges the valuable assistance spontaneously rendered by so many of the old subscribers to the "JOURNAL," who have kindly made known among their friends the advantages thus offered, and have induced them to subscribe. Relying upon a continuance of these friendly exertions, he hopes to be able to maintain the unexampled rates at which these works

(For "THE OBSTETRICAL JOURNAL," see p. 22.)

are now offered, and to succeed in his endeavor to place upon the table of every reading practitioner in the United States the equivalent of three large octavo volumes, at the comparatively trifling cost of SIX DOLLARS *per annum*.

These periodicals are universally known for their high professional standing in their several spheres.

I.

THE AMERICAN JOURNAL OF THE MEDICAL SCIENCES,

EDITED BY ISAAC HAYS, M.D.,

is published Quarterly, on the first of January, April, July, and October. Each number contains nearly three hundred large octavo pages, appropriately illustrated wherever necessary. It has now been issued regularly for over FIFTY years, during nearly the whole of which time it has been under the control of the present editor. Throughout this long period, it has maintained its position in the highest rank of medical periodicals both at home and abroad, and has received the cordial support of the entire profession in this country. Among its Collaborators will be found a large number of the most distinguished names of the profession in every section of the United States, rendering the department devoted to

ORIGINAL COMMUNICATIONS

full of varied and important matter, of great interest to all practitioners. Thus, during 1874, articles have appeared in its pages from nearly one hundred gentlemen of the highest standing in the profession throughout the United States.*

Following this is the "REVIEW DEPARTMENT," containing extended and impartial reviews of all important new works, together with numerous elaborate "ANALYTICAL AND BIBLIOGRAPHICAL NOTICES" of nearly all the medical publications of the day.

This is followed by the "QUARTERLY SUMMARY OF IMPROVEMENTS AND DISCOVERIES IN THE MEDICAL SCIENCES," classified and arranged under different heads, presenting a very complete digest of all that is new and interesting to the physician, abroad as well as at home.

Thus, during the year 1874, the "JOURNAL" furnished to its subscribers 85 Original Communications, 113 Reviews and Bibliographical Notices, and 305 articles in the Quarterly Summaries, making a total of about FIVE HUNDRED articles emanating from the best professional minds in America and Europe.

That the efforts thus made to maintain the high reputation of the "JOURNAL" are successful, is shown by the position accorded to it in both America and Europe as a national exponent of medical progress:—

America continues to take a great place in this class of journals (quarterlies), at the head of which the great work of Dr. Hays, the *American Journal of the Medical Sciences*, still holds its ground, as our quotations have often proved.—*Dublin Med. Press and Circular*, Jan. 31, 1872.

Of English periodicals the *Lancet*, and of American the *Am. Journal of the Medical Sciences*, are to be regarded as necessities to the reading practitioner.—*N. Y. Medical Gazette*, Jan. 7, 1871.

The *American Journal of the Medical Sciences* yields to none in the amount of original and bor-

rowed matter it contains, and has established for itself a reputation in every country where medicine is cultivated as a science.—*Brit. and For. Med.-Chirurg. Review*, April, 1871.

This, if not the best, is one of the best-conducted medical quarterlies in the English language, and the present number is not by any means inferior to its predecessors.—*London Lancet*, Aug. 23, 1873.

Almost the only one that circulates everywhere, all over the Union and in Europe.—*London Medical Times*, Sept. 5, 1868.

And that it was specifically included in the award of a medal of merit to the Publisher in the Vienna Exhibition in 1873.

The subscription price of the "AMERICAN JOURNAL OF THE MEDICAL SCIENCES" has never been raised during its long career. It is still FIVE DOLLARS per annum; and when paid for in advance, the subscriber receives in addition the "MEDICAL NEWS AND LIBRARY," making in all about 1500 large octavo pages per annum, free of postage.

II.

THE MEDICAL NEWS AND LIBRARY

is a monthly periodical of Thirty-two large octavo pages, making 384 pages per annum. Its "NEWS DEPARTMENT" presents the current information of the day, with Clinical Lectures and Hospital Gleanings; while the "LIBRARY DEPARTMENT" is devoted to publishing standard works on the various branches of medical science, paged

---

* Communications are invited from gentlemen in all parts of the country. Elaborate articles inserted by the Editor are paid for by the Publisher.

separately, so that they can be removed and bound on completion. In this manner subscribers have received, without expense, such works as "WATSON'S PRACTICE," "TODD AND BOWMAN'S PHYSIOLOGY," "WEST ON CHILDREN," "MALGAIGNE'S SURGERY," &c. &c. With Jan. 1875, was commenced the publication of Dr. WILLIAM STOKES'S new work on FEVER (see p. 14), rendering this a very desirable time for new subscriptions.

As stated above, the subscription price of the "MEDICAL NEWS AND LIBRARY" is ONE DOLLAR per annum in advance; and it is furnished without charge to all advance paying subscribers to the "AMERICAN JOURNAL OF THE MEDICAL SCIENCES."

III.
THE MONTHLY ABSTRACT OF MEDICAL SCIENCE.

The publication in England of Ranking's "HALF-YEARLY ABSTRACT OF THE MEDICAL SCIENCES" having ceased with the volume for January, 1874, its place has been supplied in this country by a monthly "ABSTRACT" containing forty-eight large octavo pages each month, thus furnishing in the course of the year about six hundred pages, the same amount of matter as heretofore embraced in the Half-Yearly Abstract. As the discontinuance of the "Ranking" arose from the multiplication of journals appearing more frequently and presenting the same character of material, it has been thought that this plan of monthly issues will better meet the wants of subscribers, who will thus receive earlier intelligence of the improvements and discoveries in the medical sciences. The aim of the MONTHLY ABSTRACT will be to present a careful condensation of all that is new and important in the medical journalism of the world, and all the prominent professional periodicals of both hemispheres will be at the disposal of the Editors.

Subscribers desiring to bind the ABSTRACT will receive, on application at the end of each year, a cloth cover, gilt lettered, for the purpose, or it will be sent free by mail on receipt of the postage, which, under existing laws, will be six cents.

The subscription to the "MONTHLY ABSTRACT," free of postage, is Two DOLLARS AND A HALF a year, in advance.

As stated above, however, it will be supplied in conjunction with the "AMERICAN JOURNAL OF THE MEDICAL SCIENCES" and the "MEDICAL NEWS AND LIBRARY," making in all about TWENTY-ONE HUNDRED pages per annum, the whole *free of postage*, for SIX DOLLARS a year, in advance.

The first volume of the "MONTHLY ABSTRACT," from July to December, 1874, can be had by those who desire to have complete sets, if early application be made, for $1.50, forming a handsome octavo volume of 300 pages, cloth.

In this effort to bring so large an amount of practical information within the reach of every member of the profession, the publisher confidently anticipates the friendly aid of all who are interested in the dissemination of sound medical literature. He trusts, especially, that the subscribers to the "AMERICAN MEDICAL JOURNAL" will call the attention of their acquaintances to the advantages thus offered, and that he will be sustained in the endeavor to permanently establish medical periodical literature on a footing of cheapness never heretofore attempted.

PREMIUM FOR NEW SUBSCRIBERS TO THE "JOURNAL."

Any gentleman who will remit the amount for two subscriptions for 1875, one of which must be for a *new subscriber*, will receive as a PREMIUM, free by mail, a copy of "FLINT'S ESSAYS ON CONSERVATIVE MEDICINE" (for advertisement of which see p. 15), or of "STURGES'S CLINICAL MEDICINE" (see p. 14), or of the new edition of "SWAYNE'S OBSTETRIC APHORISMS" (see p. 24), or of "TANNER'S CLINICAL MANUAL" (see p. 5), or of "CHAMBERS'S RESTORATIVE MEDICINE" (see p. 16), or of "WEST ON NERVOUS DISORDERS OF CHILDREN" (see page 21).

⁎ Gentlemen desiring to avail themselves of the advantages thus offered will do well to forward their subscriptions at an early day, in order to insure the receipt of complete sets for the year 1875, as the constant increase in the subscription list almost always exhausts the quantity printed shortly after publication.

☞ The safest mode of remittance is by bank check or postal money order, drawn to the order of the undersigned. Where these are not accessible, remittances for the "JOURNAL" may be made at the risk of the publisher, by forwarding in REGISTERED letters. Address,

HENRY C. LEA,

Nos. 706 and 708 SANSOM ST., PHILADELPHIA, PA.

DUNGLISON (ROBLEY), M. D.,
Late Professor of Institutes of Medicine in Jefferson Medical College, Philadelphia.

MEDICAL LEXICON; A DICTIONARY OF MEDICAL SCIENCE: Containing a concise explanation of the various Subjects and Terms of Anatomy, Physiology, Pathology, Hygiene, Therapeutics, Pharmacology, Pharmacy, Surgery, Obstetrics, Medical Jurisprudence, and Dentistry. Notices of Climate and of Mineral Waters; Formulæ for Officinal, Empirical, and Dietetic Preparations; with the Accentuation and Etymology of the Terms, and the French and other Synonymes; so as to constitute a French as well as English Medical Lexicon. A New Edition. Thoroughly Revised, and very greatly Modified and Augmented. By RICHARD J. DUNGLISON, M.D. In one very large and handsome royal octavo volume of over 1100 pages. Cloth, $6 50; leather, raised bands, $7 50. (*Just Issued.*)

The object of the author from the outset has not been to make the work a mere lexicon or dictionary of terms, but to afford, under each, a condensed view of its various medical relations, and thus to render the work an epitome of the existing condition of medical science. Starting with this view, the immense demand which has existed for the work has enabled him, in repeated revisions, to augment its completeness and usefulness, until at length it has attained the position of a recognized and standard authority wherever the language is spoken.

Special pains have been taken in the preparation of the present edition to maintain this enviable reputation. During the ten years which have elapsed since the last revision, the additions to the nomenclature of the medical sciences have been greater than perhaps in any similar period of the past, and up to the time of his death the author labored assiduously to incorporate everything requiring the attention of the student or practitioner. Since then, the editor has been equally industrious, so that the additions to the vocabulary are more numerous than in any previous revision. Especial attention has been bestowed on the accentuation, which will be found marked on every word. The typographical arrangement has been much improved, rendering reference much more easy, and every care has been taken with the mechanical execution. The work has been printed on new type, small but exceedingly clear, with an enlarged page, so that the additions have been incorporated with an increase of but little over a hundred pages, and the volume now contains the matter of at least four ordinary octavos.

A book well known to our readers, and of which every American ought to be proud. When the learned author of the work passed away, probably all of us feared lest the book should not maintain its place in the advancing science whose terms it defines. Fortunately, Dr. Richard J. Dunglison, having assisted his father in the revision of several editions of the work, and having been, therefore, trained in the methods and imbued with the spirit of the book, has been able to edit it, not in the patchwork manner so dear to the heart of book editors, so repulsive to the taste of intelligent book readers, but to edit it as a work of the kind should be edited—to carry it on steadily, without jar or interruption, along the grooves of thought it has travelled during its lifetime. To show the magnitude of the task which Dr. Dunglison has assumed and carried through, it is only necessary to state that more than six thousand new subjects have been added in the present edition. Without occupying more space with the theme, we congratulate the editor on the successful completion of his labors, and hope he may reap the well-earned reward of profit and honor.—*Phila. Med. Times,* Jan. 3, 1874.

About the first book purchased by the medical student is the Medical Dictionary. The lexicon explanatory of technical terms is simply a *sine qua non*. In a science so extensive, and with such collaterals as medicine, it is as much a necessity also to the practising physician. To meet the wants of students and most physicians, the dictionary must be condensed while comprehensive, and practical while perspicacious. It was because Dunglison's met these indications that it became at once the dictionary of general use wherever medicine was studied in the English language. In no former revision have the alterations and additions been so great. More than six thousand new subjects and terms have been added. The chief terms have been set in black letter, while the derivatives follow in small caps; an arrangement which greatly facilitates reference. We may safely confirm the hope ventured by the editor "that the work, which possesses for him a filial as well as an individual interest, will be found worthy a continuance of the position so long accorded to it as a standard authority."—*Cincinnati Clinic,* Jan. 10, 1874.

We are glad to see a new edition of this invaluable work, and to find that it has been so thoroughly revised, and so greatly improved. The dictionary, in its present form, is a medical library in itself, and one of which every physician should be possessed.—*N. Y. Med. Journal,* Feb. 1874.

With a history of forty years of unexampled success and universal indorsement by the medical profession of the western continent, it would be presumption in any living medical American to essay its review. No reviewer, however able, can add to its fame; no captious critic, however caustic, can remove a single stone from its firm and enduring foundation. It is destined, as a colossal monument, to perpetuate the solid and richly deserved fame of Robley Dunglison to coming generations. The large additions made to the vocabulary, we think, will be welcomed by the profession as supplying the want of a lexicon fully up with the march of science, which has been increasingly felt for some years past. The accentuation of terms is very complete, and, as far as we have been able to examine it, very excellent. We hope it may be the means of securing greater uniformity of pronunciation among medical men.—*Atlanta Med. and Surg. Journ.,* Feb. 1874.

It would be mere waste of words in us to express our admiration of a work which is so universally and deservedly appreciated. The most admirable work of its kind in the English language.—*Glasgow Medical Journal,* January, 1866.

A work to which there is no equal in the English language.—*Edinburgh Medical Journal.*

Few works of the class exhibit a grander monument of patient research and of scientific lore. The extent of the sale of this lexicon is sufficient to testify to its usefulness, and to the great service conferred by Dr. Robley Dunglison on the profession, and indeed on others, by its issue.—*London Lancet,* May 13, 1865.

It has the rare merit that it certainly has no rival in the English language for accuracy and extent of references.—*London Medical Gazette.*

HOBLYN (RICHARD D.), M. D.

A DICTIONARY OF THE TERMS USED IN MEDICINE AND THE COLLATERAL SCIENCES. Revised, with numerous additions, by ISAAC HAYS, M. D., Editor of the "American Journal of the Medical Sciences." In one large royal 12mo. volume of over 500 double-columned pages; cloth, $1 50; leather, $2 00.

It is the best book of definitions we have, and ought always to be upon the student's table.—*Southern Med. and Surg. Journal.*

NEILL (JOHN), M.D., and SMITH (FRANCIS G.), M.D.,
Prof. of the Institutes of Medicine in the Univ. of Penna.

AN ANALYTICAL COMPENDIUM OF THE VARIOUS BRANCHES OF MEDICAL SCIENCE; for the Use and Examination of Students. A new edition, revised and improved. In one very large and handsomely printed royal 12mo. volume, of about one thousand pages, with 374 wood cuts, cloth, $4; strongly bound in leather, with raised bands, $4 75.

The Compend of Drs. Neill and Smith is incomparably the most valuable work of its class ever published in this country. Attempts have been made in various quarters to squeeze Anatomy, Physiology, Surgery, the Practice of Medicine, Obstetrics, Materia Medica, and Chemistry into a single manual; but the operation has signally failed in the hands of all up to the advent of "Neill and Smith's" volume, which is quite a miracle of success. The outlines of the whole are admirably drawn and illustrated, and the authors are eminently entitled to the grateful consideration of the student of every class.—*N. O. Med. and Surg. Journal.*

There are but few students or practitioners of medicine unacquainted with the former editions of this unassuming though highly instructive work. The whole science of medicine appears to have been sifted, as the gold-bearing sands of El Dorado, and the precious facts treasured up in this little volume. A complete portable library so condensed that the student may make it his constant pocket companion.—*Western Lancet.*

In the rapid course of lectures, where work for the students is heavy, and review necessary for an examination, a compend is not only valuable, but it is almost a *sine qua non.* The one before us is, in most of the divisions, the most unexceptionable of all books of the kind that we know of. Of course it is useless for us to recommend it to all last course students, but there is a class to whom we very sincerely commend this cheap book as worth its weight in silver—that class is the graduates in medicine of more than ten years' standing, who have not studied medicine since. They will perhaps find out from it that the science is not exactly now what it was when they left it off.—*The Stethoscope.*

HARTSHORNE (HENRY), M.D.,
Professor of Hygiene in the University of Pennsylvania.

A CONSPECTUS OF THE MEDICAL SCIENCES; containing Handbooks on Anatomy, Physiology, Chemistry, Materia Medica, Practical Medicine, Surgery, and Obstetrics. Second Edition, thoroughly revised and improved. In one large royal 12mo. volume of more than 1000 closely printed pages, with 477 illustrations on wood. Cloth, $4 25; leather, $5 00. (*Lately Issued.*)
The favor with which this work has been received has stimulated the author in its revision to render it in every way fitted to meet the wants of the student, or of the practitioner desirous to refresh his acquaintance with the various departments of medical science. The various sections have been brought up to a level with the existing knowledge of the day, while preserving the condensation of form by which so vast an accumulation of facts have been brought within so narrow a compass. The series of illustrations has been much improved, while by the use of a smaller type the additions have been incorporated without increasing unduly the size of the volume.

The work before us has already successfully asserted its claim to the confidence and favor of the profession; it but remains for us to say that in the present edition the whole work has been fully overhauled and brought up to the present status of the science.—*Atlanta Med. and Surg. Journal,* Sept. 1874.

The work is intended as an aid to the medical student, and as such appears to admirably fulfil its object by its excellent arrangement, the full compilation of facts, the perspicuity and terseness of language, and the clear and instructive illustrations in some parts of the work.—*American Journ. of Pharmacy,* Philadelphia, July, 1874.

The volume will be found useful, not only to students, but to many others who may desire to refresh their memories with the smallest possible expenditure of time.—*N. Y. Med. Journal,* Sept. 1874.

The student will find this the most convenient and useful book of the kind on which he can lay his hand.—*Pacific Med. and Surg. Journ.,* Aug. 1874.

LUDLOW (J. L.), M.D.

A MANUAL OF EXAMINATIONS upon Anatomy, Physiology, Surgery, Practice of Medicine, Obstetrics, Materia Medica, Chemistry, Pharmacy, and Therapeutics. To which is added a Medical Formulary. Third edition, thoroughly revised and greatly extended and enlarged. With 370 illustrations. In one handsome royal 12mo. volume of 816 large pages, cloth, $3 25; leather, $3 75.
The arrangement of this volume in the form of question and answer renders it especially suitable for the office examination of students, and for those preparing for graduation.

TANNER (THOMAS HAWKES), M.D., &c.

A MANUAL OF CLINICAL MEDICINE AND PHYSICAL DIAGNOSIS. Third American from the Second London Edition. Revised and Enlarged by TILBURY FOX, M. D., Physician to the Skin Department in University College Hospital, &c. In one neat volume small 12mo., of about 375 pages, cloth, $1 50.
*** By reference to the "Prospectus of Journal" on page 3, it will be seen that this work is offered as a premium for procuring new subscribers to the "AMERICAN JOURNAL OF THE MEDICAL SCIENCES."

Taken as a whole, it is the most compact vade mecum for the use of the advanced student and junior practitioner with which we are acquainted.—*Boston Med. and Surg. Journal,* Sept. 22, 1870.

It contains so much that is valuable, presented in so attractive a form, that it can hardly be spared even in the presence of more full and complete works. Its convenient size makes it a valuable companion to the country practitioner, and if constantly carried by him, would often render him good service, and relieve many a doubt and perplexity.—*Leaven-*

The objections commonly, and justly, urged against the general run of "compends," "conspectuses," and other aids to indolence, are not applicable to this little volume, which contains in concise phrase just those practical details that are of most use in daily diagnosis, but which the young practitioner finds it difficult to carry always in his memory without some quickly accessible means of reference. Altogether, the book is one which we can heartily commend to those who have not opportunity for extensive reading, or who, having read much, still wish an occasional practical reminder.—*N. Y. Med. Gazette,* Nov.

GRAY (HENRY), F.R.S.,
Lecturer on Anatomy at St. George's Hospital, London.

ANATOMY, DESCRIPTIVE AND SURGICAL. The Drawings by H. V. CARTER, M. D., late Demonstrator on Anatomy at St. George's Hospital; the Dissections jointly by the AUTHOR and DR. CARTER. A new American, from the fifth enlarged and improved London edition. In one magnificent imperial octavo volume, of nearly 900 pages, with 465 large and elaborate engravings on wood. Price in cloth, $6 00; leather, raised bands, $7 00. (*Just Issued.*)

The author has endeavored in this work to cover a more extended range of subjects than is customary in the ordinary text-books, by giving not only the details necessary for the student, but also the application of those details in the practice of medicine and surgery, thus rendering it both a guide for the learner, and an admirable work of reference for the active practitioner. The engravings form a special feature in the work, many of them being the size of nature, nearly all original, and having the names of the various parts printed on the body of the cut, in place of figures of reference, with descriptions at the foot. They thus form a complete and splendid series, which will greatly assist the student in obtaining a clear idea of Anatomy, and will also serve to refresh the memory of those who may find in the exigencies of practice the necessity of recalling the details of the dissecting room; while combining, as it does, a complete Atlas of Anatomy, with a thorough treatise on systematic, descriptive, and applied Anatomy, the work will be found of essential use to all physicians who receive students in their offices, relieving both preceptor and pupil of much labor in laying the groundwork of a thorough medical education.

Notwithstanding the enlargement of this edition, it has been kept at its former very moderate price, rendering it one of the cheapest works now before the profession.

The illustrations are beautifully executed, and render this work an indispensable adjunct to the library of the surgeon. This remark applies with great force to those surgeons practising at a distance from our large cities, as the opportunity of refreshing their memory by actual dissection is not always attainable.—*Canada Med. Journal*, Aug. 1870.

The work is too well known and appreciated by the profession to need any comment. No medical man can afford to be without it, if its only merit were to serve as a reminder of that which so soon becomes forgotten, when not called into frequent use, viz., the relations and names of the complex organism of the human body. The present edition is much improved.—*California Med. Gazette*, July, 1870.

Gray's Anatomy has been so long the standard of perfection with every student of anatomy, that we need do no more than call attention to the improvement in the present edition.—*Detroit Review of Med. and Pharm.*, Aug. 1870.

From time to time, as successive editions have appeared, we have had much pleasure in expressing the general judgment of the wonderful excellence of Gray's Anatomy.—*Cincinnati Lancet*, July, 1870.

Altogether, it is unquestionably the most complete and serviceable text-book in anatomy that has ever been presented to the student, and forms a striking contrast to the dry and perplexing volumes on the same subject through which their predecessors struggled in days gone by.—*N. Y. Med. Record*, June 15, 1870.

To commend Gray's Anatomy to the medical profession is almost as much a work of supererogation as it would be to give a favorable notice of the Bible in the religious press. To say that it is the most complete and conveniently arranged text-book of its kind, is to repeat what each generation of students has learned as a tradition of the elders, and verified by personal experience.—*N. Y. Med. Gazette*, Dec. 17, 1870.

SMITH (HENRY H.), M.D., and HORNER (WILLIAM E.), M.D.,
Prof. of Surgery in the Univ. of Penna., &c.   *Late Prof. of Anatomy in the Univ. of Penna., &c.*

AN ANATOMICAL ATLAS, illustrative of the Structure of the Human Body. In one volume, large imperial octavo, cloth, with about six hundred and fifty beautiful figures. $4 50.

The plan of this Atlas, which renders it so peculiarly convenient for the student, and its superb artistical execution, have been already pointed out. We must congratulate the student upon the completion of this Atlas, as it is the most convenient work of the kind that has yet appeared; and we must add, the very beautiful manner in which it is "got up," is so creditable to the country as to be flattering to our national pride.—*American Medical Journal.*

SHARPEY (WILLIAM), M.D., and QUAIN (JONES & RICHARD).

HUMAN ANATOMY. Revised, with Notes and Additions, by JOSEPH LEIDY, M.D., Professor of Anatomy in the University of Pennsylvania. Complete in two large octavo volumes, of about 1300 pages, with 511 illustrations; cloth, $6 00.

The very low price of this standard work, and its completeness in all departments of the subject, should command for it a place in the library of all anatomical students.

HODGES (RICHARD M.), M.D.,
Late Demonstrator of Anatomy in the Medical Department of Harvard University.

PRACTICAL DISSECTIONS. Second Edition, thoroughly revised. In one neat royal 12mo. volume, half-bound, $2 00.

The object of this work is to present to the anatomical student a clear and concise description of that which he is expected to observe in an ordinary course of dissections. The author has endeavored to omit unnecessary details, and to present the subject in the form which many years' experience has shown him to be the most convenient and intelligible to the student. In the revision of the present edition, he has sedulously labored to render the volume more worthy of the favor with which it has heretofore been received.

HORNER'S SPECIAL ANATOMY AND HISTOLOGY. Eighth edition, extensively revised and modified. | In 2 vols. 8vo., of over 1000 pages, with more than 300 wood-cuts: cloth, $6 00

WILSON (ERASMUS), F. R. S.
A SYSTEM OF HUMAN ANATOMY, General and Special. Edited by W. H. GOBRECHT, M. D., Professor of General and Surgical Anatomy in the Medical College of Ohio. Illustrated with three hundred and ninety-seven engravings on wood. In one large and handsome octavo volume, of over 600 large pages; cloth, $4 00; leather, $5 00.

The publisher trusts that the well-earned reputation of this long-established favorite will be more than maintained by the present edition. Besides a very thorough revision by the author, it has been most carefully examined by the editor, and the efforts of both have been directed to introducing everything which increased experience in its use has suggested as desirable to render it a complete text-book for those seeking to obtain or to renew an acquaintance with Human Anatomy. The amount of additions which it has thus received may be estimated from the fact that the present edition contains over one-fourth more matter than the last, rendering a smaller type and an enlarged page requisite to keep the volume within a convenient size. The author has not only thus added largely to the work, but he has also made alterations throughout, wherever there appeared the opportunity of improving the arrangement or style, so as to present every fact in its most appropriate manner, and to render the whole as clear and intelligible as possible. The editor has exercised the utmost caution to obtain entire accuracy in the text, and has largely increased the number of illustrations, of which there are about one hundred and fifty more in this edition than in the last, thus bringing distinctly before the eye of the student everything of interest or importance.

HEATH (CHRISTOPHER), F. R. C. S.,
Teacher of Operative Surgery in University College, London.
PRACTICAL ANATOMY: A Manual of Dissections. From the Second revised and improved London edition. Edited, with additions, by W. W. KEEN, M. D., Lecturer on Pathological Anatomy in the Jefferson Medical College, Philadelphia. In one handsome royal 12mo. volume of 578 pages, with 247 illustrations. Cloth, $3 50; leather, $4 00. (*Lately Published.*)

Dr. Keen, the American editor of this work, in his preface, says: "In presenting this American edition of 'Heath's Practical Anatomy,' I feel that I have been instrumental in supplying a want long felt for *a real dissector's manual,*" and this assertion of its editor we deem is fully justified, after an examination of its contents, for it is really an excellent work. Indeed, we do not hesitate to say, the best of its class with which we are acquainted; resembling Wilson in terse and clear description, excelling most of the so-called practical anatomical dissectors in the scope of the subject and practical selected matter. . . . In reading this work, one is forcibly impressed with the great pains the author takes to impress the subject upon the mind of the student. He is full of rare and pleasing little devices to aid memory in maintaining its hold upon the slippery slopes of anatomy. —*St. Louis Med. and Surg. Journal,* Mar. 10, 1871.

It appears to us certain that, as a guide in dissection, and as a work containing facts of anatomy in brief and easily understood form, this manual is complete. This work contains, also, very perfect illustrations of parts which can thus be more easily understood and studied; in this respect it compares favorably with works of much greater pretension. Such manuals of anatomy are always favorite works with medical students. We would earnestly recommend this one to their attention; it has excellences which make it valuable as a guide in dissecting, as well as in studying anatomy.—*Buffalo Medical and Surgical Journal,* Jan. 1871.

BELLAMY (E.), F.R.C.S.
THE STUDENT'S GUIDE TO SURGICAL ANATOMY: A Text-Book for Students preparing for their Pass Examination. With engravings on wood. In one handsome royal 12mo. volume. Cloth, $2 25. (*Just Issued.*)

We welcome Mr. Bellamy's work, as a contribution to the study of regional anatomy, of equal value to the student and the surgeon. It is written in a clear and concise style, and its practical suggestions add largely to the interest attaching to its technical details.—*Chicago Med. Examiner,* March 1, 1874.

We cordially congratulate Mr. Bellamy upon having produced it.—*Med. Times and Gaz.*

We cannot too highly recommend it.—*Student's Journal.*

Mr. Bellamy has spared no pains to produce a really reliable student's guide to surgical anatomy—one which all candidates for surgical degrees may consult with advantage, and which possesses much original matter.—*Med. Press and Circular.*

MACLISE (JOSEPH).
SURGICAL ANATOMY. By JOSEPH MACLISE, Surgeon. In one volume, very large imperial quarto; with 68 large and splendid plates, drawn in the best style and beautifully colored, containing 190 figures, many of them the size of life; together with copious explanatory letter-press. Strongly and handsomely bound in cloth. Price $14 00.

We know of no work on surgical anatomy which can compete with it.—*Lancet.*

The work of Maclise on surgical anatomy is of the highest value. In some respects it is the best publication of its kind we have seen, and is worthy of a place in the library of any medical man, while the student could scarcely make a better investment than this.—*The Western Journal of Medicine and Surgery.*

No such lithographic illustrations of surgical regions have hitherto, we think, been given. While the operator is shown every vessel and nerve where an operation is contemplated, the exact anatomist is refreshed by those clear and distinct dissections, which every one must appreciate who has a particle of enthusiasm. The English medical press has quite exhausted the words of praise, in recommending this admirable treatise.—*Boston Med. and Surg. Journ.*

HARTSHORNE (HENRY), M.D.,
Professor of Hygiene, etc, in the Univ. of Penna.
HANDBOOK OF ANATOMY AND PHYSIOLOGY. Second Edition, revised. In one royal 12mo. volume, with 220 wood-cuts; cloth, $1 75. (*Just Issued.*)

MARSHALL (JOHN), F. R. S.,
Professor of Surgery in University College, London, &c.

OUTLINES OF PHYSIOLOGY, HUMAN AND COMPARATIVE.
With Additions by FRANCIS GURNEY SMITH, M. D., Professor of the Institutes of Medicine in the University of Pennsylvania, &c. With numerous illustrations. In one large and handsome octavo volume, of 1026 pages, cloth, $6 50; leather, raised bands, $7 50.

In fact, in every respect, Mr. Marshall has presented us with a most complete, reliable, and scientific work, and we feel that it is worthy our warmest commendation.—*St. Louis Med. Reporter*, Jan. 1869.

We doubt if there is in the English language any compend of physiology more useful to the student than this work.—*St. Louis Med. and Surg. Journal*, Jan. 1869.

It quite fulfils, in our opinion, the author's design of making it truly *educational* in its character—which is, perhaps, the highest commendation that can be asked.—*Am. Journ. Med. Sciences*, Jan. 1869.

We may now congratulate him on having completed the latest as well as the best summary of modern physiological science, both human and comparative, with which we are acquainted. To speak of this work in the terms ordinarily used on such occasions would not be agreeable to ourselves, and would fail to do justice to its author. To write such a book requires a varied and wide range of knowledge, considerable power of analysis, correct judgment, skill in arrangement, and conscientious spirit.—*London Lancet*, Feb. 22, 1868.

There are few, if any, more accomplished anatomists and physiologists than the distinguished professor of surgery at University College; and he has long enjoyed the highest reputation as a teacher of physiology, possessing remarkable powers of clear exposition and graphic illustration. We have rarely the pleasure of being able to recommend a text-book so unreservedly as this.—*British Med. Journal*, Jan. 25, 1868.

CARPENTER (WILLIAM B.), M. D., F. R. S.,
Examiner in Physiology and Comparative Anatomy in the University of London.

PRINCIPLES OF HUMAN PHYSIOLOGY; with their chief applications to Psychology, Pathology, Therapeutics, Hygiene and Forensic Medicine. A new American from the last and revised London edition. With nearly three hundred illustrations. Edited, with additions, by FRANCIS GURNEY SMITH, M. D., Professor of the Institutes of Medicine in the University of Pennsylvania, &c. In one very large and beautiful octavo volume, of about 900 large pages, handsomely printed; cloth, $5 50; leather, raised bands, $6 50.

With Dr. Smith, we confidently believe "that the present will more than sustain the enviable reputation already attained by former editions, of being one of the fullest and most complete treatises on the subject in the English language." We know of none from the pages of which a satisfactory knowledge of the physiology of the human organism can be as well obtained, none better adapted for the use of such as take up the study of physiology in its reference to the institutes and practice of medicine.—*Am. Jour. Med. Sciences.*

We doubt not it is destined to retain a strong hold on public favor, and remain the favorite text-book in our colleges.—*Virginia Medical Journal.*

The above is the title of what is emphatically *the* great work on physiology; and we are conscious that it would be a useless effort to attempt to add anything to the reputation of this invaluable work, and can only say to all with whom our opinion has any influence, that it is our *authority.—Atlanta Med. Journal.*

BY THE SAME AUTHOR.

PRINCIPLES OF COMPARATIVE PHYSIOLOGY. New American, from the Fourth and Revised London Edition. In one large and handsome octavo volume, with over three hundred beautiful illustrations. Pp. 752. Cloth, $5 00.

As a complete and condensed treatise on its extended and important subject, this work becomes a necessity to students of natural science, while the very low price at which it is offered places it within the reach of all.

KIRKES (WILLIAM SENHOUSE), M. D.

A MANUAL OF PHYSIOLOGY. Edited by W. MORRANT BAKER, M. D., F.R.C.S. A new American from the eighth and improved London edition. With about two hundred and fifty illustrations. In one large and handsome royal 12mo. volume. Cloth, $3 25; leather, $3 75. (*Lately Issued.*)

Kirkes' Physiology has long been known as a concise and exceedingly convenient text-book, presenting within a narrow compass all that is important for the student. The rapidity with which successive editions have followed each other in England has enabled the editor to keep it thoroughly on a level with the changes and new discoveries made in the science, and the eighth edition, of which the present is a reprint, has appeared so recently that it may be regarded as the latest accessible exposition of the subject.

On the whole, there is very little in the new book which either the student or practitioner will not find of practical value and consistent with our present knowledge of this rapidly changing science; and we have no hesitation in expressing our opinion that this eighth edition is one of the best handbooks on physiology which we have in our language.—*N. Y. Med. Record*, April 15, 1873.

This volume might well be used to replace many of the physiological text-books in use in this country. It represents more accurately than the works of Dalton or Flint, the present state of our knowledge of most physiological questions, while it is much less bulky and far more readable than the larger text-books of Carpenter or Marshall. The book is admirably adapted to be placed in the hands of students.—*Boston Med. and Surg. Journ.*, April 10, 1873.

In its enlarged form it is, in our opinion, still the best book on physiology, most useful to the student.—*Phila. Med. Times*, Aug. 30, 1873.

This is undoubtedly the best work for students of physiology extant.—*Cincinnati Med. News*, Sept. '73.

It more nearly represents the present condition of physiology than any other text-book on the subject.—*Detroit Rev. of Med. Pharm.*, Nov. 1873.

DALTON (J. C.), M. D.,
Professor of Physiology in the College of Physicians and Surgeons, New York, &c.
A TREATISE ON HUMAN PHYSIOLOGY. Designed for the use of Students and Practitioners of Medicine. Fifth edition, revised, with nearly three hundred illustrations on wood. In one very beautiful octavo volume, of over 700 pages, cloth, $5 25; leather, $6 25.

Preface to the Fifth Edition.

In preparing the present edition of this work, the general plan and arrangement of the previous editions have been retained, so far as they have been found useful and adapted to the purposes of a text-book for students of medicine. The incessant advance of all the natural and physical sciences, never more active than within the last five years, has furnished many valuable aids to the special investigations of the physiologist; and the progress of physiological research, during the same period, has required a careful revision of the entire work, and the modification or rearrangement of many of its parts. At this day, nothing is regarded as of any value in natural science which is not based upon direct and intelligible observation or experiment; and, accordingly, the discussion of doubtful or theoretical questions has been avoided, as a general rule, in the present volume, while new facts, from whatever source, if fully established, have been added and incorporated with the results of previous investigation. A number of new illustrations have been introduced, and a few of the older ones, which seemed to be no longer useful, have been omitted. In all the changes and additions thus made, it has been the aim of the writer to make the book, in its present form, a faithful exponent of the actual conditions of physiological science.

NEW YORK, October, 1871.

In this, the standard text-book on Physiology, all that is needed to maintain the favor with which it is regarded by the profession, is the author's assurance that it has been thoroughly revised and brought up to a level with the advanced science of the day. To accomplish this has required some enlargement of the work, but no advance has been made in the price.

The fifth edition of this truly valuable work on Human Physiology comes to us with many valuable improvements and additions. As a text-book of physiology the work of Prof. Dalton has long been well known as one of the best which could be placed in the hands of student or practitioner. Prof. Dalton has, in the several editions of his work heretofore published, labored to keep step with the advancement in science, and the last edition shows by its improvements on former ones that he is determined to maintain the high standard of his work. We predict for the present edition increased favor, though this work has long been the favorite standard.—*Buffalo Med. and Surg. Journal,* April, 1872.

An extended notice of a work so generally and favorably known as this is unnecessary. It is justly regarded as one of the most valuable text-books on the subject in the English language.—*St. Louis Med. Archives,* May, 1872.

We know no treatise in physiology so clear, complete, well assimilated, and perfectly digested, as Dalton's. He never writes cloudily or dubiously, or in mere quotation. He assimilates all his material, and from it constructs a homogeneous transparent argument, which is always honest and well informed, and hides neither truth, ignorance, nor doubt, so far as either belongs to the subject in hand.—*Brit. Med. Journal,* March 23, 1872.

Dr. Dalton's treatise is well known, and by many highly esteemed in this country. It is, indeed, a good elementary treatise on the subject it professes to teach, and may safely be put into the hands of English students. It has one great merit—it is clear, and, on the whole, admirably illustrated. The part we have always esteemed most highly is that relating to Embryology. The diagrams given of the various stages of development give a clearer view of the subject than do those in general use in this country; and the text may be said to be, upon the whole, equally clear.—*London Med. Times and Gazette,* March 23, 1872.

Dalton's Physiology is already, and deservedly, the favorite text-book of the majority of American medical students. Treating a most interesting department of science in his own peculiarly lively and fascinating style, Dr. Dalton carries his reader along without effort, and at the same time impresses upon his mind the truths taught much more successfully than if they were hurled beneath a multitude of words.—*Kansas City Med. Journal,* April, 1872.

Professor Dalton is regarded justly as *the* authority in this country on physiological subjects, and the fifth edition of his valuable work fully justifies the exalted opinion the medical world has of his labors. This last edition is greatly enlarged.—*Virginia Clinical Record,* April, 1872.

DUNGLISON (ROBLEY), M. D.,
Professor of Institutes of Medicine in Jefferson Medical College, Philadelphia.
HUMAN PHYSIOLOGY. Eighth edition. Thoroughly revised and extensively modified and enlarged, with five hundred and thirty-two illustrations. In two large and handsomely printed octavo volumes of about 1500 pages, cloth, $7 00.

LEHMANN (C. G.).
PHYSIOLOGICAL CHEMISTRY. Translated from the second edition by GEORGE E. DAY, M. D., F. R. S., &c., edited by R. E. ROGERS, M. D., Professor of Chemistry in the Medical Department of the University of Pennsylvania, with illustrations selected from Funke's Atlas of Physiological Chemistry, and an Appendix of plates. Complete in two large and handsome octavo volumes, containing 1200 pages, with nearly two hundred illustrations, cloth, $6 00.

BY THE SAME AUTHOR.
MANUAL OF CHEMICAL PHYSIOLOGY. Translated from the German, with Notes and Additions, by J. CHESTON MORRIS, M. D., with an Introductory Essay on Vital Force, by Professor SAMUEL JACKSON, M. D., of the University of Pennsylvania. With illustrations on wood. In one very handsome octavo volume of 336 pages, cloth, $2 25.

ATTFIELD (JOHN), Ph. D.,
Professor of Practical Chemistry to the Pharmaceutical Society of Great Britain, &c.

CHEMISTRY, GENERAL, MEDICAL, AND PHARMACEUTICAL; including the Chemistry of the U. S. Pharmacopœia. A Manual of the General Principles of the Science, and their Application to Medicine and Pharmacy. Fifth Edition, revised by the author. In one handsome royal 12mo. volume; cloth, $2 75; leather, $3 25. (*Lately Issued.*)

No other American publication with which we are acquainted covers the same ground, or does it so well. In addition to an admirable exposé of the facts and principles of general elementary chemistry, the author has presented us with a condensed mass of practical matter, just such as the medical student and practitioner needs.—*Cincinnati Lancet*, Mar. 1874.

We commend the work heartily as one of the best text-books extant for the medical student.—*Detroit Rev. of Med. and Pharm.*, Feb. 1872.

The best work of the kind in the English language.—*N. Y. Psychological Journal*, Jan. 1872.

The work is constructed with direct reference to the wants of medical and pharmaceutical students; and, although an English work, the points of difference between the British and United States Pharmacopœias are indicated, making it as useful here as in England. Altogether, the book is one we can heartily recommend to practitioners as well as students.—*N. Y. Med. Journal*, Dec. 1871.

It differs from other text-books in the following particulars: first, in the exclusion of matter relating to compounds which, at present, are only of interest to the scientific chemist; secondly, in containing the chemistry of every substance recognized officially or in general, as a remedial agent. It will be found a most valuable book for pupils, assistants, and others engaged in medicine and pharmacy, and we heartily commend it to our readers.—*Canada Lancet*, Oct. 1871.

When the original English edition of this work was published, we had occasion to express our high appreciation of its worth, and also to review, in considerable detail, the main features of the book. As the arrangement of subjects, and the main part of the text of the present edition are similar to the former publication, it will be needless for us to go over the ground a second time; we may, however, call attention to a marked advantage possessed by the American work—we allude to the introduction of the chemistry of the preparations of the United States Pharmacopœia, as well as that relating to the British authority.— *Canadian Pharmaceutical Journal*, Nov. 1871.

Chemistry has borne the name of being a hard subject to master by the student of medicine, and chiefly because so much of it consists of compounds only of interest to the scientific chemist; in this work such portions are modified or altogether left out, and in the arrangement of the subject-matter of the work, the practical utility is sought after, and we think fully attained. We commend it for its clearness and order to both teacher and pupil.—*Oregon Med. and Surg. Reporter*, Oct. 1871.

FOWNES (GEORGE), Ph. D.

A MANUAL OF ELEMENTARY CHEMISTRY; Theoretical and Practical. With one hundred and ninety-seven illustrations. A new American, from the tenth and revised London edition. Edited by ROBERT BRIDGES, M. D. In one large royal 12mo. volume, of about 850 pp., cloth, $2 75; leather, $3 25. (*Lately Issued.*)

This work is so well known that it seems almost superfluous for us to speak about it. It has been a favorite text-book with medical students for years, and its popularity has in no respect diminished. Whenever we have been consulted by medical students, as has frequently occurred, what treatise on chemistry they should procure, we have always recommended Fownes', for we regarded it as the best. There is no work that combines so many excellences. It is of convenient size, not prolix, of plain perspicuous diction, contains all the most recent discoveries, and is of moderate price.—*Cincinnati Med. Repertory*, Aug. 1869.

Large additions have been made, especially in the department of organic chemistry, and we know of no other work that has greater claims on the physician, pharmacentist, or student, than this. We cheerfully recommend it as the best text-book on elementary chemistry, and bespeak for it the careful attention of students of pharmacy.—*Chicago Pharmacist*, Aug. 1869.

Here is a new edition which has been long watched for by eager teachers of chemistry. In its new garb, and under the editorship of Mr. Watts, it has resumed its old place as the most successful of text-books.—*Indian Medical Gazette*, Jan. 1, 1869

It will continue, as heretofore, to hold the first rank as a text-book for students of medicine.—*Chicago Med. Examiner*, Aug. 1869.

ODLING (WILLIAM),
Lecturer on Chemistry at St. Bartholomew's Hospital, &c.

A COURSE OF PRACTICAL CHEMISTRY, arranged for the Use of Medical Students. With Illustrations. From the Fourth and Revised London Edition. In one neat royal 12mo. volume, cloth, $2.

GALLOWAY (ROBERT), F.C.S.,
Prof. of Applied Chemistry in the Royal College of Science for Ireland, &c.

A MANUAL OF QUALITATIVE ANALYSIS. From the Fifth London Edition. In one neat royal 12mo. volume, with illustrations; cloth, $2 50. (*Just Issued.*)

The success which has carried this work through repeated editions in England, and its adoption as a text-book in several of the leading institutions in this country, show that the author has succeeded in the endeavor to produce a sound practical manual and book of reference for the chemical student.

Prof. Galloway's books are deservedly in high esteem, and this American reprint of the fifth edition (1869) of his Manual of Qualitative Analysis, will be acceptable to many American students to whom the English edition is not accessible.—*Am. Jour. of Science and Arts*, Sept. 1872.

We regard this volume as a valuable addition to the chemical text-books, and as particularly calculated to instruct the student in analytical researches of the inorganic compounds, the important vegetable acids, and of compounds and various secretions and excretions of animal origin.—*Am. Journ. of Pharm.*, Sept. 1872.

*B*LOXAM (C. L.),
Professor of Chemistry in King's College, London.
CHEMISTRY, INORGANIC AND ORGANIC. From the Second London Edition. In one very handsome octavo volume, of 700 pages, with about 300 illustrations. Cloth, $4 00; leather, $5 00. (*Lately Issued*.)

It has been the author's endeavor to produce a Treatise on Chemistry sufficiently comprehensive for those studying the science as a branch of general education, and one which a student may use with advantage in pursuing his chemical studies at one of the colleges or medical schools. The special attention devoted to Metallurgy and some other branches of Applied Chemistry renders the work especially useful to those who are being educated for employment in manufacture.

We have in this work a complete and most excellent text-book for the use of schools, and can heartily recommend it as such.—*Boston Med. and Surg. Journ.*, May 28, 1874.

Of all the numerous works upon elementary chemistry that have been published within the last few years, we can point to none that, in fulness, accuracy, and simplicity, can surpass this; while, in the number and detailed descriptions of experiments, as also in the profuseness of its illustrations, we believe it stands above any similar work published in this country. . . . . The statements made are clear and concise, and every step proved by an abundance of experiments, which excite our admiration as much by their simplicity as by their direct conclusiveness.—*Chicago Med. Examiner*, Nov. 15, 1873.

It is seldom that in the same compass so complete and interesting a compendium of the leading facts of chemistry is offered.—*Druggists' Circular*, Nov. '73.

The above is the title of a work which we can most conscientiously recommend to students of chemistry. It is as easy as a work on chemistry could be made, at the same time that it presents a full account of that science as it now stands. We have spoken of the work as admirably adapted to the wants of students; it is quite as well suited to the requirements of practitioners who wish to review their chemistry, or have occasion to refresh their memories on any point relating to it. In a word, it is a book to be read by all who wish to know what is the chemistry of the present day.—*American Practitioner*, Nov. 1873.

Among the various works upon general chemistry issued, we know of none that will supply the average wants of the student or teacher better than this.—*Indiana Journ. of Med.*, Nov. 1873.

We cordially welcome this American reprint of a work which has already won for itself so substantial a reputation in England. Professor Bloxam has condensed into a wonderfully small compass all the important principles and facts of chemical science. Thoroughly imbued with an enthusiastic love for the science he expounds, he has stripped it of all needless technicalities, and rounded out its hard outlines by a fulness of illustration that cannot fail to attract and delight the student. The details of illustrative experiment have been worked up with especial care, and many of the experiments described are both new and striking.—*Detroit Rev. of Med. and Pharm.*, Nov. 1873.

One of the best text-books of chemistry yet published.—*Chicago Med. Journ.*, Nov. 1873.

This is an excellent work, well adapted for the beginner and the advanced student of chemistry.—*Am. Journ. of Pharm.*, Nov. 1873.

Probably the most valuable, and at the same time practical, text-book on general chemistry extant in our language.—*Kansas City Med. Journ.*, Dec. 1873.

Prof. Bloxam possesses pre-eminently the inestimable gift of perspicuity. It is a pleasure to read his books, for he is capable of making very plain what other authors frequently have left very obscure.—*Va. Clinical Record*, Nov. 1873.

It would be difficult for a practical chemist and teacher to find any material fault with this most admirable treatise. The author has given us almost a cyclopedia within the limits of a convenient volume, and has done so without penning the *useless* paragraphs too commonly making up a great part of the bulk of many cumbrous works. The progressive scientist is not disappointed when he looks for the record of new and valuable processes and discoveries, while the cautious conservative does not find its pages monopolized by uncertain theories and speculations. A peculiar point of excellence is the crystallized form of expression in which great truths are expressed in very short paragraphs. One is surprised at the brief space allotted to an important topic, and yet, after reading it, he feels that little, if any more, should have been said. Altogether, it is seldom you see a text-book so nearly faultless.—*Cincinnati Lancet*, Nov. 1873.

Professor Bloxam has given us a most excellent and useful practical treatise. His 666 pages are crowded with facts and experiments, nearly all well chosen, and many quite new, even to scientific men. . . . It is astonishing how much information he often conveys in a few paragraphs. We might quote fifty instances of this.—*Chemical News.*

---

*W*ÖHLER AND FITTIG.
OUTLINES OF ORGANIC CHEMISTRY. Translated with Additions from the Eighth German Edition. By IRA REMSEN, M.D., Ph.D., Professor of Chemistry and Physics in Williams College, Mass. In one handsome volume, royal 12mo. of 550 pp., cloth, $3.

As the numerous editions of the original attest, this work is the leading text-book and standard authority throughout Germany on its important and intricate subject—a position won for it by the clearness and conciseness which are its distinguishing characteristics. The translation has been executed with the approbation of Profs. Wöhler and Fittig, and numerous additions and alterations have been introduced, so as to render it in every respect on a level with the most advanced condition of the science.

---

*B*OWMAN (JOHN E.), M. D.
PRACTICAL HANDBOOK OF MEDICAL CHEMISTRY. Edited by C. L. BLOXAM, Professor of Practical Chemistry in King's College, London. Sixth American, from the fourth and revised English Edition. In one neat volume, royal 12mo., pp. 351, with numerous illustrations, cloth, $2 25.

*B*Y THE SAME AUTHOR. (*Lately Issued*.) ———
INTRODUCTION TO PRACTICAL CHEMISTRY, INCLUDING ANALYSIS. Sixth American, from the sixth and revised London edition. With numerous illustrations. In one neat vol., royal 12mo., cloth, $2 25.

KNAPP'S TECHNOLOGY; or Chemistry Applied to the Arts, and to Manufactures. With American additions, by Prof. WALTER R. JOHNSON. In two very handsome octavo volumes, with 500 wood engravings, cloth, $6 00.

PARRISH (EDWARD),
Late Professor of Materia Medica in the Philadelphia College of Pharmacy.

A TREATISE ON PHARMACY. Designed as a Text-Book for the Student, and as a Guide for the Physician and Pharmaceutist. With many Formulæ and Prescriptions. Fourth Edition, thoroughly revised, by THOMAS S. WIEGAND. In one handsome octavo volume of 977 pages, with 280 illustrations; cloth, $5 50; leather, $6 50. (*Just Issued.*)

The delay in the appearance of the new U. S. Pharmacopœia, and the sudden death of the author, have postponed the preparation of this new edition beyond the period expected. The notes and memoranda left by Mr. Parrish have been placed in the hands of the editor, Mr. Wiegand, who has labored assiduously to embody in the work all the improvements of pharmaceutical science which have been introduced during the last ten years. It is therefore hoped that the new edition will fully maintain the reputation which the volume has heretofore enjoyed as a standard text-book and work of reference for all engaged in the preparation and dispensing of medicines.

Of Dr. Parrish's excellent work on pharmacy it only remains to be said that the editor has accomplished his work so well as to maintain, in this fourth edition, the high standard of excellence which it had attained in previous editions, under the editorship of its accomplished author. This has not been accomplished without much labor, and many additions and improvements, involving changes in the arrangement of the several parts of the work, and the addition of much new matter. With the modifications thus effected it constitutes, as now presented, a compendium of the science and art indispensable to the pharmacist, and of the utmost value to every practitioner of medicine desirous of familiarizing himself with the pharmaceutical preparation of the articles which he prescribes for his patients.—*Chicago Med. Journ.*, July, 1874.

The work is eminently practical, and has the rare merit of being readable and interesting, while it preserves a strictly scientific character. The whole work reflects the greatest credit on author, editor, and publisher. It will convey some idea of the liberality which has been bestowed upon its production when we mention that there are no less than 250 carefully executed illustrations. In conclusion, we heartily recommend the work, not only to pharmacists, but also to the multitude of medical practitioners who are obliged to compound their own medicines. It will ever hold an honored place on our own bookshelves.—*Dublin Medi Press and Circular*, Aug. 12, 1874.

We expressed our opinion of a former edition in terms of unqualified praise, and we are in no mood to detract from that opinion in reference to the present edition, the preparation of which has fallen into competent hands. It is a book with which no pharmacist can dispense, and from which no physician can fail to derive much information of value to him in practice.—*Pacific Med. and Surg. Journ.*, June, '74.

With these few remarks we heartily commend the work, and have no doubt that it will maintain its old reputation as a text-book for the student, and a work of reference for the more experienced physician and pharmacist.—*Chicago Med. Examiner*, June 15, 1874.

Perhaps one, if not the most important book upon pharmacy which has appeared in the English language has emanated from the transatlantic press. "Parrish's Pharmacy" is a well-known work on this side of the water, and the fact shows us that a really useful work never becomes merely local in its fame. Thanks to the judicious editing of Mr. Wiegand, the posthumous edition of "Parrish" has been saved to the public with all the mature experience of its author, and perhaps none the worse for a dash of new blood.—*Lond. Pharm. Journal*, Oct. 17, 1874.

STILLÉ (ALFRED), M. D.,
Professor of Theory and Practice of Medicine in the University of Penna.

THERAPEUTICS AND MATERIA MEDICA; a Systematic Treatise on the Action and Uses of Medicinal Agents, including their Description and History. Fourth edit., revised and enlarged. In two large and handsome 8vo. vols. of about 2000 pages. Cloth, $10; leather, $12. (*Now Ready.*)

The care bestowed by the author on the revision of this edition has kept the work out of the market for nearly two years, and has increased its size about two hundred and fifty pages. Notwithstanding this enlargement, the price has been kept at the former very moderate rate. A few notices of former editions are subjoined.

Dr. Stillé's splendid work on therapeutics and materia medica.—*London Med. Times*, April 8, 1865.

Dr. Stillé stands to-day one of the best and most honored representatives at home and abroad, of American medicine; and these volumes, a library in themselves, a treasure-house for every studious physician, assure his fame even had he done nothing more.—*The Western Journal of Medicine*, Dec. 1868.

We regard this work as the best one on Materia Medica in the English language, and as such it deserves the favor it has received.—*Am. Journ. Medical Sciences*, July 1868.

We need not dwell on the merits of the third edition of this magnificently conceived work. It is the work on Materia Medica, in which Therapeutics are primarily considered—the mere natural history of drugs being briefly disposed of. To medical practitioners this is a very valuable conception. It is wonderful how much of the riches of the literature of Materia Medica has been condensed into this book. The references alone would make it worth possessing. But it is not a mere compilation. The writer exercises a good judgment of his own on the great doctrines and points of Therapeutics. For purposes of practice, Stillé's book is almost unique as a repertory of information, empirical and scientific, on the actions and uses of medicines.—*London Lancet*, Oct. 31, 1868.

Through the former editions, the professional world is well acquainted with this work. At home and abroad its reputation as a standard treatise on Materia Medica is securely established. It is second to no work on the subject in the English tongue, and, indeed, is decidedly superior, in some respects, to any other.—*Pacific Med. and Surg. Journal*, July, 1868.

Stillé's Therapeutics is incomparably the best work on the subject.—*N. Y. Med. Gazette*, Sept. 26, 1868.

Dr. Stillé's work is becoming the best known of any of our treatises on Materia Medica. . . . One of the most valuable works in the language on the subjects of which it treats.—*N. Y. Med. Journal*, Oct. 1868.

The rapid exhaustion of two editions of Prof. Stillé's scholarly work, the consequent necessity for a third edition, is sufficient evidence of the high estimate placed upon it by the profession. It is no exaggeration to say that there is no superior work upon the subject in the English language. The present edition is fully up to the most recent advance in the science and art of therapeutics.—*Leavenworth Medical Herald*, Aug. 1868.

The work of Prof. Stillé has rapidly taken a high place in professional esteem, and to say that a third edition is demanded and now appears before us, sufficiently attests the firm position this treatise has made for itself. As a work of great research, and scholarship, it is safe to say we have nothing superior. It is exceedingly full, and the busy practitioner will find ample suggestions upon almost every important point of therapeutics.—*Cincinnati Lancet*, Aug. 1868.

GRIFFITH (ROBERT E.), M.D.

A UNIVERSAL FORMULARY, Containing the Methods of Preparing and Administering Officinal and other Medicines. The whole adapted to Physicians and Pharmaceutists. Third edition, thoroughly revised, with numerous additions, by JOHN M. MAISCH, Professor of Materia Medica in the Philadelphia College of Pharmacy. In one large and handsome octavo volume of about 800 pages, cloth, $4 50; leather, $5 50. *(Just Issued)*

This work has long been known for the vast amount of information which it presents in a condensed form, arranged for easy reference. The new edition has received the most careful revision at the competent hands of Professor Maisch, who has brought the whole up to the standard of the most recent authorities. More than eighty new headings of remedies have been introduced, the entire work has been thoroughly remodelled, and whatever has seemed to be obsolete has been omitted. As a comparative view of the United States, the British, the German, and the French Pharmacopœias, together with an immense amount of unofficinal formulas, it affords to the practitioner and pharmaceutist an aid in their daily avocations not to be found elsewhere, while three indexes, one of "Diseases and their Remedies," one of Pharmaceutical Names, and a General Index, afford an easy key to the alphabetical arrangement adopted in the text.

The young practitioner will find the work invaluable in suggesting eligible modes of administering many remedies.—*Am. Journ. of Pharm.*, Feb. 1874.

Our copy of Griffith's Formulary, after long use, first in the dispensing shop, and afterwards in our medical practice, had gradually fallen behind in the outward march of materia medica, pharmacy, and therapeutics, until we had ceased to consult it as a daily book of reference. So completely has Prof. Maisch reformed, remodelled, and rejuvenated it in the new edition, we shall gladly welcome it back to our table again beside Dunglison, Webster, and Wood & Bache. The publisher could not have been more fortunate in the selection of an editor. Prof. Maisch is eminently the man for the work, and he has done it thoroughly and ably. To enumerate the alterations, amendments, and additions would be an endless task; everywhere we are greeted with the evidences of his labor. Following the Formulary, is an addendum of useful Recipes, Dietetic Preparations, List of Incompatibles, Posological table, table of Pharmaceutical Names, Officinal Preparations and Directions, Poisons. Antidotes and Treatment, and copious indices, which afford ready access to all parts of the work. We unhesitatingly commend the book as being the best of its kind, within our knowledge.—*Atlanta Med. and Surg. Journ.*, Feb. 1874.

To the druggist a good formulary is simply indispensable, and perhaps no formulary has been more extensively used than the well-known work before us. Many physicians have to officiate, also, as druggists. This is true especially of the country physician, and a work which shall teach him the means by which to administer or combine his remedies in the most efficacious and pleasant manner, will always hold its place upon his shelf. A formulary of this kind is of benefit also to the city physician in largest practice.—*Cincinnati Clinic*, Feb. 21, 1874.

The Formulary has already proved itself acceptable to the medical profession, and we do not hesitate to say that the third edition is much improved, and of greater practical value, in consequence of the careful revision of Prof. Maisch.—*Chicago Med. Examiner*, March 15, 1874.

A more complete formulary than it is in its present form the pharmacist or physician could hardly desire. To the first some such work is indispensable, and it is hardly less essential to the practitioner who compounds his own medicines. Much of what is contained in the introduction ought to be committed to memory by every student of medicine. As a help to physicians it will be found invaluable, and doubtless will make its way into libraries not already supplied with a standard work of the kind.—*The American Practitioner*, Louisville, July, '74.

ELLIS (BENJAMIN), M.D.

THE MEDICAL FORMULARY: being a Collection of Prescriptions derived from the writings and practice of many of the most eminent physicians of America and Europe. Together with the usual Dietetic Preparations and Antidotes for Poisons. The whole accompanied with a few brief Pharmaceutic and Medical Observations. Twelfth edition, carefully revised and much improved by ALBERT H. SMITH, M.D. In one volume 8vo. of 376 pages, cloth, $3 00.

PEREIRA (JONATHAN), M.D., F.R.S. and L.S.

MATERIA MEDICA AND THERAPEUTICS; being an Abridgment of the late Dr. Pereira's Elements of Materia Medica, arranged in conformity with the British Pharmacopœia, and adapted to the use of Medical Practitioners, Chemists and Druggists, Medical and Pharmaceutical Students, &c. By F. J. FARRE, M.D., Senior Physician to St. Bartholomew's Hospital, and London Editor of the British Pharmacopœia; assisted by ROBERT BENTLEY, M.R.C.S., Professor of Materia Medica and Botany to the Pharmaceutical Society of Great Britain; and by ROBERT WARINGTON, F.R.S., Chemical Operator to the Society of Apothecaries. With numerous additions and references to the United States Pharmacopœia, by HORATIO C. WOOD, M.D., Professor of Botany in the University of Pennsylvania. In one large and handsome octavo volume of 1040 closely printed pages, with 236 illustrations, cloth, $7 00; leather, raised bands, $8 00.

DUNGLISON'S NEW REMEDIES, WITH FORMULÆ FOR THEIR PREPARATION AND ADMINISTRATION. Seventh edition, with extensive additions. One vol. 8vo., pp. 770; cloth. $4 00.

ROYLE'S MATERIA MEDICA AND THERAPEUTICS. Edited by JOSEPH CARSON, M.D. With ninety-eight illustrations. 1 vol. 8vo., pp. 700, cloth. $3 00.

CARSON'S SYNOPSIS OF THE LECTURES ON MATERIA MEDICA AND PHARMACY, delivered in the University of Pennsylvania. Fourth and revised edition. Cloth, $2.

CHRISTISON'S DISPENSATORY. With copious additions, and 213 large wood-engravings. By R. EGLESFELD GRIFFITH, M.D. One vol. 8vo., pp. 1000; cloth. $4 00.

CARPENTER'S PRIZE ESSAY ON THE USE OF ALCOHOLIC LIQUORS IN HEALTH AND DISEASE. New edition, with a Preface by D. F. CONDIE, M.D., and explanations of scientific words. In one neat 12mo. volume, pp. 178, cloth. 60 cents.

DE JONGH ON THE THREE KINDS OF COD-LIVER OIL, with their Chemical and Therapeutic Properties. 1 vol. 12mo., cloth. 75 cents.

FENWICK (SAMUEL), M.D.,
Assistant Physician to the London Hospital.

THE STUDENT'S GUIDE TO MEDICAL DIAGNOSIS. From the Third Revised and Enlarged English Edition. With eighty-four illustrations on wood. In one very handsome volume, royal 12mo., cloth, $2 25. (*Just Issued.*)

The very great success which this work has obtained in England, shows that it has supplied an admitted want among elementary books for the guidance of students and junior practitioners. Taking up in order each portion of the body or class of disease, the author has endeavored to present in simple language the value of symptoms, so as to lead the student to a correct appreciation of the pathological changes indicated by them. The latest investigations have been carefully introduced into the present edition, so that it may fairly be considered as on a level with the most advanced condition of medical science.

Of the many guide-books on medical diagnosis, claimed to be written for the special instruction of students, this is the best. The author is evidently a well-read and accomplished physician, and he knows how to teach practical medicine. The charm of simplicity is not the least interesting feature in the manner in which Dr. Fenwick conveys instruction. There are few books of this size on practical medicine that contain so much and convey it so well as the volume before us. It is a book we can sincerely recommend to the student for direct instruction, and to the practitioner as a ready and useful aid to his memory.—*Am. Journ. of Syphilography,* Jan. 1874.

It covers the ground of medical diagnosis in a concise, practical manner, well calculated to assist the student in forming a correct, thorough, and systematic method of examination and diagnosis of disease. The illustrations are numerous, and finely executed. Those illustrative of the microscopic appearance of morbid tissue, &c., are especially clear and distinct.—*Chicago Med. Examiner,* Nov. 1873.

So far superior to any offered to students that the colleges of this country should recommend it to their respective classes.—*N. O. Med. and Surg. Journ.,* March, 1874.

This little book ought to be in the possession of every medical student.—*Boston Medical and Surg. Journ.,* Jan. 15, 1874.

GREEN (T. HENRY), M.D.,
Lecturer on Pathology and Morbid Anatomy at Charing-Cross Hospital Medical School.

PATHOLOGY AND MORBID ANATOMY. With numerous Illustrations on Wood. In one very handsome octavo volume of over 250 pages, cloth, $2 50. (*Lately Published.*)

We have been very much pleased by our perusal of this little volume. It is the only one of the kind with which we are acquainted, and practitioners as well as students will find it a very useful guide; for the information is up to the day, well and compactly arranged, without being at all scanty.—*London Lancet,* Oct. 7, 1871.

It embodies in a comparatively small space a clear statement of the present state of our knowledge of pathology and morbid anatomy. The author shows that he has been not only a student of the teachings of his *confrères* in this branch of science, but a practical and conscientious laborer in the post-mortem chamber. The work will prove useful one to the great mass of students and practitioners whose time for devotion to this class of studies is limited.—*Am. Journ. of Syphilography,* April, 1872.

GLUGE'S ATLAS OF PATHOLOGICAL HISTOLOGY. Translated, with Notes and Additions, by JOSEPH LEIDY, M.D. In one volume, very large imperial quarto, with 320 copper-plate figures, plain and colored, cloth. $4 00.

JONES AND SIEVEKING'S PATHOLOGICAL ANATOMY. With 397 wood-cuts. 1 vol. 8vo., of nearly 750 pages, cloth. $3 50.

HOLLAND'S MEDICAL NOTES AND REFLECTIONS. 1 vol. 8vo., pp. 500, cloth. $3 50

WHAT TO OBSERVE AT THE BEDSIDE AND AFTER DEATH IN MEDICAL CASES. Published under the authority of the London Society for Medical Observation. From the second London edition. 1 vol. royal 12mo., cloth. $1 00.

LA ROCHE ON YELLOW FEVER, considered in its Historical, Pathological, Etiological, and Therapeutical Relations. In two large and handsome octavo volumes of nearly 1500 pages, cloth. $7 00.

LAYCOCK'S LECTURES ON THE PRINCIPLES AND METHODS OF MEDICAL OBSERVATION AND RESEARCH. For the use of advanced students and junior practitioners. In one very neat royal 12mo. volume, cloth. $1 00.

BARLOW'S MANUAL OF THE PRACTICE OF MEDICINE. With Additions by D. F. CONDIE, M.D. 1 vol. 8vo., pp. 600, cloth. $2 50.

TODD'S CLINICAL LECTURES ON CERTAIN ACUTE DISEASES. In one neat octavo volume, of 320 pages, cloth. $2 50.

STURGES (OCTAVIUS), M.D. Cantab.,
Fellow of the Royal College of Physicians, &c. &c.

AN INTRODUCTION TO THE STUDY OF CLINICAL MEDICINE. Being a Guide to the Investigation of Disease, for the Use of Students. In one handsome 12mo. volume, cloth, $1 25. (*Just Issued.*)

DAVIS (NATHAN S.),
Prof. of Principles and Practice of Medicine, etc., in Chicago Med. College.

CLINICAL LECTURES ON VARIOUS IMPORTANT DISEASES; being a collection of the Clinical Lectures delivered in the Medical Wards of Mercy Hospital, Chicago. Edited by FRANK H. DAVIS, M.D. Second edition, enlarged. In one handsome royal 12mo. volume. Cloth, $1 75. (*Now Ready.*)

STOKES (WILLIAM), M.D., D.C.L., F.R.S.,
Regius Professor of Physic in the Univ. of Dublin, &c.

LECTURES ON FEVER, delivered in the Theatre of the Meath Hospital and County of Dublin Infirmary. Edited by JOHN WILLIAM MOORE, M.D , Assistant Physician to the Cork Street Fever Hospital. In one neat octavo volume. (*Preparing.*)
*** To appear in the "MEDICAL NEWS AND LIBRARY" for 1875.

FLINT (AUSTIN), M. D.,
Professor of the Principles and Practice of Medicine in Bellevue Med. College, N. Y.
A TREATISE ON THE PRINCIPLES AND PRACTICE OF MEDICINE; designed for the use of Students and Practitioners of Medicine. Fourth edition, revised and enlarged. In one large and closely printed octavo volume of about 1100 pages; cloth, $6 00; or strongly bound in leather, with raised bands, $7 00. (Just Issued.)

By common consent of the English and American medical press, this work has been assigned to the highest position as a complete and compendious text-book on the most advanced condition of medical science. At the very moderate price at which it is offered it will be found one of the cheapest volumes now before the profession. A few notices of previous editions are subjoined.

Admirable and unequalled.—*Western Journal of Medicine,* Nov. 1869.

Dr. Flint's work, though claiming no higher title than that of a text-book, is really more. He is a man of large clinical experience, and his book is full of such masterly descriptions of disease as can only be drawn by a man intimately acquainted with their various forms. It is not so long since we had the pleasure of reviewing his first edition, and we recognize a great improvement, especially in the general part of the work. It is a work which we can cordially recommend to our readers as fully abreast of the science of the day.—*Edinburgh Med. Journal,* Oct. '69.

One of the best works of the kind for the practitioner, and the most convenient of all for the student.—*Am. Journ. Med. Sciences,* Jan. 1869.

This work, which stands pre-eminently as the advance standard of medical science up to the present time in the practice of medicine, has for its author one who is well and widely known as one of the leading practitioners of this continent. In fact, it is seldom that any work is ever issued from the press more deserving of universal recommendation.—*Dominion Med. Journal,* May, 1869.

The third edition of this most excellent book scarcely needs any commendation from us. The volume, as it stands now, is really a marvel: first of all, it is excellently printed and bound—and we encounter that luxury of America, the ready-cut pages, which the Yankees are 'cute enough to insist upon—nor are these by any means trifles; but the contents of the book are astonishing. Not only is it wonderful that any one man can have grasped in his mind the whole scope of medicine with that vigor which Dr. Flint shows, but the condensed yet clear way in which this is done is a perfect literary triumph. Dr. Flint is pre-eminently one of the strong men, whose right to do this kind of thing is well admitted; and we say no more than the truth when we affirm that he is very nearly the only living man that could do it with such results as the volume before us.—*The London Practitioner,* March, 1869.

This is in some respects the best text-book of medicine in our language, and it is highly appreciated on the other side of the Atlantic, inasmuch as the first edition was exhausted in a few months. The second edition was little more than a reprint, but the present has, as the author says, been thoroughly revised. Much valuable matter has been added, and by making the type smaller, the bulk of the volume is not much increased. The weak point in many American works is pathology, but Dr. Flint has taken peculiar pains on this point, greatly to the value of the book.—*London Med. Times and Gazette,* Feb. 6, 1869.

BY THE SAME AUTHOR.
ESSAYS ON CONSERVATIVE MEDICINE AND KINDRED TOPICS. In one very handsome royal 12mo. volume. Cloth, $1 38. (Just Issued.)

CONTENTS.

I. Conservative Medicine. II. Conservative Medicine as applied to Therapeutics. III. Conservative Medicine as applied to Hygiene. IV. Medicine in the Past, the Present, and the Future. V. Alimentation in D'sease. VI. Tolerance of Disease. VII. On the Agency of the Mind in Etiology, Prophylaxis, and Therapeutics. VIII. Divine design as exemplified in the Natural History of Disease.

WATSON (THOMAS), M. D., &c.
LECTURES ON THE PRINCIPLES AND PRACTICE OF PHYSIC. Delivered at King's College, London. A new American, from the Fifth revised and enlarged English edition. Edited, with additions, and several hundred illustrations, by HENRY HARTSHORNE, M.D., Professor of Hygiene in the University of Pennsylvania. In two large and handsome 8vo. vols. Cloth, $9 00; leather, $11 00. (Lately Published.)

It is a subject for congratulation and for thankfulness that Sir Thomas Watson, during a period of comparative leisure, after a long, laborious, and most honorable professional career, while retaining full possession of his high mental faculties, should have employed the opportunity to submit his Lectures to a more thorough revision than was possible during the earlier and busier period of his life. Carefully passing in review some of the most intricate and important pathological and practical questions, the results of his clear insight and his calm judgment are now recorded for the benefit of mankind, in language which, for precision, vigor, and classical elegance, has rarely been equalled, and never surpassed. The revision has evidently been most carefully done, and the results appear in almost every page.—*Brit. Med. Journ.,* Oct. 14, 1871.

The lectures are so well known and so justly appreciated, that it is scarcely necessary to do more than call attention to the special advantages of the last over previous editions. The author's rare combination of great scientific attainments combined with wonderful forensic eloquence has exerted extraordinary influence over the last two generations of physicians. His clinical descriptions of most diseases have never been equalled; and on this score at least his work will live long in the future. The work will be sought by all who appreciate a great book.—*Amer. Journ. of Syphilography,* July, 1872.

We are exceedingly gratified at the reception of this new edition of Watson, pre-eminently the prince of English authors, on "Practice." We, who read the first edition shall never forget the great pleasure and profit we derived from its graphic delineations of disease, its vigorous style and splendid English. Maturity of years, extensive observation, profound research, and yet continuous enthusiasm, have combined to give us in this latest edition a model of professional excellence in teaching with rare hearty in the mode of communication. But this *classic* needs no eulogium of ours.—*Chicago Med. Journ.,* July, 1872.

DUNGLISON, FORBES, TWEEDIE, AND CONOLLY.
THE CYCLOPÆDIA OF PRACTICAL MEDICINE: comprising Treatises on the Nature and Treatment of Diseases, Materia Medica and Therapeutics, Diseases of Women and Children, Medical Jurisprudence, &c. &c. In four large super-royal octavo volumes, of 3254 double-columned pages, strongly and handsomely bound in leather, $15; cloth, $11.

16 · HENRY C. LEA'S PUBLICATIONS—(*Practice of Medicine*).

HARTSHORNE (HENRY), M.D.,
Professor of Hygiene in the University of Pennsylvania.
ESSENTIALS OF THE PRINCIPLES AND PRACTICE OF MEDI-
CINE. A handy-book for Students and Practitioners. Fourth edition, revised and improved. With about one hundred illustrations. In one handsome royal 12mo volume, of about 550 pages, cloth, $2 63; half bound, $2 88. (*Just Ready.*)
The thorough manner in which the author has labored to fully represent in this favorite handbook the most advanced condition of practical medicine is shown by the fact that the present edition contains more than 250 additions, representing the investigations of 172 authors not referred to in previous editions. Notwithstanding an enlargement of the page, the size has been increased by sixty pages. A number of illustrations have been introduced which it is hoped will facilitate the comprehension of details by the reader, and no effort has been spared to make the volume worthy a continuance of the very great favor with which it has hitherto been received.

The work is brought fully up with all the recent advances in medicine, is admirably condensed, and yet sufficiently explicit for all the purposes intended, thus making it by far the best work of its character ever published.—*Cincinnati Clinic*, Oct. 24, 1874.

We have already had occasion to notice the previous editions of this work. It is excellent of its kind. The author has given a very careful revision, in view of the rapid progress of medical science.—*N. Y. Med. Journ.*, Nov. 1874.

The present edition of Dr. Hartshorne's work is a very decided improvement upon the former ones in

many particulars, and is fully up to the most advanced state of the science.—*Leavenworth Medical Herald*, Nov. 1874.

Without doubt the best book of the kind published in the English language.—*St. Louis Med. and Surg. Journ.*, Nov. 1874.

As a handbook, which clearly sets forth the ESSENTIALS of the PRINCIPLES AND PRACTICE OF MEDICINE, we do not know of its equal.— *Va. Med. Monthly.*

As a brief, condensed, but comprehensive handbook, it cannot be improved upon.—*Chicago Med. Examiner*, Nov. 15, 1874.

PAVY (F. W.), M.D., F.R.S.,
Senior Asst. Physician to and Lecturer on Physiology, at Guy's Hospital, &c.
A TREATISE ON THE FUNCTION OF DIGESTION; its Disor-
ders and their Treatment. From the second London edition. In one handsome volume, small octavo, cloth, $2 00.

BY THE SAME AUTHOR. (Just Ready.)
A TREATISE ON FOOD AND DIETETICS, PHYSIOLOGI-
CALLY AND THERAPEUTICALLY CONSIDERED. In one handsome octavo volume of nearly 600 pages, cloth, $4 75.

SUMMARY OF CONTENTS.

Introductory Remarks on the Dynamic Relations of Food—On the Origination of Food—The Constituent Relations of Food—Alimentary Principles, their Classification, Chemical Relations, Digestion, Assimilation, and Physiological Uses—Nitrogenous Alimentary Principles—Non-Nitrogenous Alimentary Principles—The Carbo-Hydrates—The Inorganic Alimentary Principles—Alimentary Substances—Animal Alimentary Substances—Vegetable Alimentary Substances—Beverages—Condiments—The Preservation of Food—Principles of Dietetics—Practical Dietetics—Diet of Infants—Diet for Training—Therapeutic Dietetics—Dietetic Preparations for the Invalid—Hospital Dietaries.

CHAMBERS (T. K.), M.D. (Lately Published.)
Consulting Physician to St. Mary's Hospital, London, &c.
THE INDIGESTIONS; or, Diseases of the Digestive Organs Functionally
Treated. Third and revised Edition. In one handsome 8vo. vol. of 333 pages, cloth, $3 00.

BY THE SAME AUTHOR. (Lately Published.)
RESTORATIVE MEDICINE. An Harveian Annual Oration. With
Two Sequels. In one very handsome volume, small 12mo., cloth, $1 00.

BY THE SAME AUTHOR. (Now Ready.)
A MANUAL OF DIET AND REGIMEN IN HEALTH AND SICK-
NESS. In one handsome octavo volume. Cloth, $2 75.
The aims of this handbook are purely practical, and therefore it has not been thought right to increase its size by the addition of the chemical, botanical, and industrial learning which rapidly collects round the nucleus of every article interesting as an eatable. Space has been thus gained for a full discussion of many matters connecting food and drink with the daily current of social life, which the position of the author as a practising physician has led him to believe highly important to the present and future of our race.—*Preface.*

SUMMARY OF CONTENTS.

PART I. *General Dietetics.* CHAP. I. Theories of Dietetics. II. On the Choice of Food. III. On the Preparation of Food. IV. On Digestion and Nutrition.
PART II. *Special Dietetics of Health.* CHAP. I. Regimen of Infancy and Motherhood. II. Regimen of Childhood and Youth. III. Commercial Life. IV. Literary and Professional Life. V. Noxious Trades. VI. Athletic Training. VII. Hints for Healthy Travellers. VIII. Effects of Climate. IX. Starvation, Poverty, and Fasting. X. The Decline of Life. XI. Alcohol.
PART III. *Dietetics in Sickness.* CHAP. I. Dietetics and Regimen in Acute Fevers. II. The Diet and Regimen of certain other Inflammatory States. III. The Diet and Regimen of Weak Digestion. IV. Gout and Rheumatism. V. Gravel, Stone, Albuminuria, and Diabetes. VI. Deficient Evacuation. VII. Nerve Disorders. VIII. Scrofula, Rickets, and Consumption. IX. Diseases of Heart and Arteries.

FLINT (AUSTIN), M. D.,
Professor of the Principles and Practice of Medicine in Bellevue Hospital Med. College, N. Y.

A PRACTICAL TREATISE ON THE DIAGNOSIS, PATHOLOGY, AND TREATMENT OF DISEASES OF THE HEART. Second revised and enlarged edition. In one octavo volume of 550 pages, with a plate, cloth, $4.

Dr. Flint chose a difficult subject for his researches, and has shown remarkable powers of observation and reflection, as well as great industry, in his treatment of it. His book must be considered the fullest and clearest practical treatise on those subjects, and should be in the hands of all practitioners and students. It is a credit to American medical literature.—*Amer. Journ. of the Med. Sciences*, July, 1860.

We question the fact of any recent American author in our profession being more extensively known, or more deservedly esteemed in this country than Dr. Flint. We willingly acknowledge his success, more particularly in the volume on diseases of the heart, in making an extended personal clinical study available for purposes of illustration, in connection with cases which have been reported by other trustworthy observers.—*Brit. and For. Med.-Chirurg. Review.*

In regard to the merits of the work, we have no hesitation in pronouncing it full, accurate, and judicious. Considering the present state of science, such a work was much needed. It should be in the hands of every practitioner.—*Chicago Med. Journ.*

With more than pleasure do we hail the advent of this work, for it fills a wide gap on the list of textbooks for our schools, and is, for the practitioner, the most valuable practical work of its kind.—*N. O. Med. News.*

BY THE SAME AUTHOR.

A PRACTICAL TREATISE ON THE PHYSICAL EXPLORATION OF THE CHEST AND THE DIAGNOSIS OF DISEASES AFFECTING THE RESPIRATORY ORGANS. Second and revised edition. In one handsome octavo volume of 595 pages, cloth, $4 50.

Dr. Flint's treatise is one of the most trustworthy guides which we can consult. The style is clear and distinct, and is also concise, being free from that tendency to over-refinement and unnecessary minuteness which characterizes many works on the same subject.—*Dublin Medical Press*, Feb. 6, 1867.

The chapter on Phthisis is replete with interest; and his remarks on the diagnosis, especially in the early stages, are remarkable for their acumen and great practical value. Dr. Flint's style is clear and elegant, and the tone of freshness and originality which pervades his whole work lend an additional force to its thoroughly practical character, which cannot fail to obtain for it a place as a standard work on diseases of the respiratory system.—*London Lancet*, Jan. 19, 1867.

This is an admirable book. Excellent in detail and execution, nothing better could be desired by the practitioner. Dr. Flint enriches his subject with much solid and not a little original observation.—*Ranking's Abstract*, Jan. 1867.

FULLER (HENRY WILLIAM), M. D.,
Physician to St. George's Hospital, London.

ON DISEASES OF THE LUNGS AND AIR-PASSAGES. Their Pathology, Physical Diagnosis, Symptoms, and Treatment. From the second and revised English edition. In one handsome octavo volume of about 500 pages, cloth, $3 50.

WILLIAMS (C. J. B.), M.D.,
Senior Consulting Physician to the Hospital for Consumption, Brompton, and
WILLIAMS (CHARLES T.), M.D.,
Physician to the Hospital for Consumption.

PULMONARY CONSUMPTION; Its Nature, Varieties, and Treatment. With an Analysis of One Thousand Cases to exemplify its duration. In one neat octavo volume of about 350 pages, cloth, $2 50. (*Lately Published.*)

He can still speak from a more enormous experience, and a closer study of the morbid processes involved in tuberculosis, than most living men. He owed it to himself, and to the importance of the subject, to embody his views in a separate work, and we are glad that he has accomplished this duty.

After all, the grand teaching which Dr Williams has for the profession is to be found in his therapeutical chapters, and in the history of individual cases extended, by dint of care, over ten, twenty, thirty, and even forty years.—*London Lancet*, Oct. 21, 1871.

LA ROCHE ON PNEUMONIA. 1 vol. 8vo., cloth, of 500 pages. Price $3 00.
SMITH ON CONSUMPTION; ITS EARLY AND REMEDIABLE STAGES. 1 vol. 8vo., pp. 254. $2 25.

WALSHE ON THE DISEASES OF THE HEART AND GREAT VESSELS. Third American edition. In 1 vol. 8vo., 420 pp., cloth. $3 00.

FOX (WILSON), M.D.,
Holme Prof. of Clinical Med., University Coll., London.

THE DISEASES OF THE STOMACH: Being the Third Edition of the "Diagnosis and Treatment of the Varieties of Dyspepsia." Revised and Enlarged. With illustrations. In one handsome octavo volume, cloth, $2 00. (*Now Ready.*)

Dr. Fox has put forth a volume of uncommon excellence, which we feel very sure will take a high rank among works that treat of the stomach.—*Am. Practitioner*, March, 1873.

BRINTON (WILLIAM), M.D., F.R.S.

LECTURES ON THE DISEASES OF THE STOMACH; with an Introduction on its Anatomy and Physiology. From the second and enlarged London edition. With illustrations on wood. In one handsome octavo volume of about 300 pages, cloth, $3 25.

ROBERTS (WILLIAM), M. D.,
Lecturer on Medicine in the Manchester School of Medicine, &c.

A PRACTICAL TREATISE ON URINARY AND RENAL DIS-
EASES, including Urinary Deposits. Illustrated by numerous cases and engravings. Second American, from the Second Revised and Enlarged London Edition. In one large and handsome octavo volume of 616 pages, with a colored plate; cloth, $4 50. (*Lately Published.*)

The author has subjected this work to a very thorough revision, and has sought to embody in it the results of the latest experience and investigations. Although every effort has been made to keep it within the limits of its former size, it has been enlarged by a hundred pages, many new wood-cuts have been introduced, and also a colored plate representing the appearance of the different varieties of urine, while the price has been retained at the former very moderate rate.

The plan, it will thus be seen, is very complete, and the manner in which it has been carried out is in the highest degree satisfactory. The characters of the different deposits are very well described, and the microscopic appearances they present are illustrated by numerous well executed engravings. It only remains to us to strongly recommend to our readers Dr. Roberts's work, as containing an admirable *résumé* of the present state of knowledge of urinary diseases, and as a safe and reliable guide to the clinical observer.—*Edin. Med. Jour.*

The most complete and practical treatise upon renal diseases we have examined. It is peculiarly adapted to the wants of the majority of American practitioners from its clearness and simple announcement of the facts in relation to diagnosis and treatment of urinary disorders, and contains in condensed form the investigations of Bence Jones, Bird, Beale, Hassall, Prout, and a host of other well-known writers upon this subject. The characters of urine, physiological and pathological, as indicated to the naked eye as well as by microscopical and chemical investigations, are concisely represented both by description and by well executed engravings.—*Cincinnati Journ. of Med.*

BASHAM (W. R.), M.D.,
Senior Physician to the Westminster Hospital, &c.

RENAL DISEASES: a Clinical Guide to their Diagnosis and Treatment.
With illustrations. In one neat royal 12mo. volume of 304 pages, cloth, $2 00.

The chapters on diagnosis and treatment are very good, and the student and young practitioner will find them full of valuable practical hints. The third part, on the urine, is excellent, and we cordially recommend its perusal. The author has arranged his matter in a somewhat novel, and, we think, useful form. Here everything can be easily found, and, what is more important, easily read, for all the dry details of larger books here acquire a new interest from the author's arrangement. This part of the book is full of good work.—*Brit. and For. Medico-Chirurgical Review,* July, 1870.

The easy descriptions and compact modes of statement render the book pleasing and convenient.—*Am. Journ. Med. Sciences,* July, 1870.

LINCOLN (D. F.) M.D.,
Physician to the Department of Nervous Diseases, Boston Dispensary.

ELECTRO-THERAPEUTICS; A Concise Manual of Medical Electri-
city. In one very neat royal 12mo. volume, cloth, with illustrations, $1 50. (*Just Ready.*)

The work is convenient in size, its descriptions of methods and appliances are sufficiently complete for the general practitioner, and the chapters on Electro-physiology and diagnosis are well written and readable. For those who wish a handy-book of directions for the employment of galvanism in medicine, this will serve as a very good and reliable guide.—*New Remedies,* Oct. 1874.

It is a well written work, and calculated to meet the demands of the busy practitioner. It contains the latest researches in this important branch of medicine.—*Peninsular Journ. of Med.,* Oct. 1874.

Eminently practical in character. It will amply repay any one for a careful perusal.—*Leavenworth Med. Herald,* Oct. 1874.

This little book is, considering its size, one of the very best of the English treatises on its subject that has come to our notice, possessing, among others, the rare merit of dealing avowedly and actually with principles, mainly, rather than with practical details, thereby supplying a real want, instead of helping merely to flood the literary market. Dr. Lincoln's style is usually remarkably clear, and the whole book is readable and interesting.—*Boston Med. and Surg. Journ.,* July 23, 1874.

We have here in a small compass a great deal of valuable information upon the subject of Medical Electricity.—*Canada Med. and Surg. Journ.,* Nov. 1874.

LEE (HENRY),
Prof. of Surgery at the Royal College of Surgeons of England, etc.

LECTURES ON SYPHILIS AND ON SOME FORMS OF LOCAL
DISEASE AFFECTING PRINCIPALLY THE ORGANS OF GENERATION. In one handsome octavo volume.

CONTENTS.

LECTURES I., II., III. General.—IV. Treatment of Syphilis—V. Treatment of Particular and Modified Syphilitic Affections—VI. Second Stage of Lues Venerea; Treatment—VII. Local Suppurating Venereal Sore; Syphilization; Lymphatic Absorption; Physiological Absorption; Twofold Inoculation—VIII. Urethral Discharges: different kinds; Treatment; Conclusions of Hunter and Ricord—IX. Prostatic Discharges—X. Lymphatic Absorption continued; Local Affections; Warts and Excrescences.

DIPHTHERIA; its Nature and Treatment, with an account of the History of its Prevalence in various Countries. By D. D. SLADE, M.D. Second and revised edition. In one neat royal 12mo. volume, cloth, $1 25.

LECTURES ON THE STUDY OF FEVER. By A. HUDSON, M.D., M.R.I.A., Physician to the Meath Hospital. In one vol. 8vo., cloth, $2 50.

A TREATISE ON FEVER. By ROBERT D LYONS, K C C. In one octavo volume of 362 pages, cloth, $2 25.

CLINICAL OBSERVATIONS ON FUNCTIONAL NERVOUS DISORDERS By C. HANDFIELD JONES, M.D., Physician to St. Mary's Hospital, &c. Second American Edition. In one handsome octavo volume of 348 pages, cloth, $3 25.

BUMSTEAD (FREEMAN J.), M.D.,
Professor of Venereal Diseases at the Col. of Phys. and Surg., New York, &c.

THE PATHOLOGY AND TREATMENT OF VENEREAL DISEASES.
Including the results of recent investigations upon the subject. Third edition, revised and enlarged, with illustrations. In one large and handsome octavo volume of over 700 pages, cloth, $5 00; leather, $6 00.

In preparing this standard work again for the press, the author has subjected it to a very thorough revision. Many portions have been rewritten, and much new matter added, in order to bring it completely on a level with the most advanced condition of syphilography, but by careful compression of the text of previous editions, the work has been increased by only sixty-four pages. The labor thus bestowed upon it, it is hoped, will insure for it a continuance of its position as a complete and trustworthy guide for the practitioner.

It is the most complete book with which we are acquainted in the language. The latest views of the best authorities are put forward, and the information is well arranged—a great point for the student, and still more for the practitioner. The subjects of visceral syphilis, syphilitic affections of the eyes, and the treatment of syphilis by repeated inoculations, are very fully discussed.—*London Lancet,* Jan. 7, 1871.

Dr. Bumstead's work is already so universally known as the best treatise in the English language on venereal diseases, that it may seem almost superfluous to say more of it than that a new edition has been issued. But the author's industry has rendered this new edition virtually a new work, and so merits as much special commendation as if its predecessors had not been published. As a thoroughly practical book on a class of diseases which form a large share of nearly every physician's practice, the volume before us is by far the best of which we have knowledge.—*N. Y. Medical Gazette,* Jan. 28, 1871.

It is rare in the history of medicine to find any one book which contains all that a practitioner needs to know; while the possessor of "Bumstead on Venereal" has no occasion to look outside of its covers for anything practical connected with the diagnosis, history, or treatment of these affections.—*N. Y. Medical Journal,* March, 1871.

CULLERIER (A.), and BUMSTEAD (FREEMAN J.),
Surgeon to the Hôpital du Midi. *Professor of Venereal Diseases in the College of Physicians and Surgeons, N. Y.*

AN ATLAS OF VENEREAL DISEASES.
Translated and Edited by FREEMAN J. BUMSTEAD. In one large imperial 4to. volume of 328 pages, double-columns, with 26 plates, containing about 150 figures, beautifully colored, many of them the size of life; strongly bound in cloth, $17 00; also, in five parts, stout wrappers for mailing, at $3 per part.

Anticipating a very large sale for this work, it is offered at the very low price of THREE DOLLARS a Part, thus placing it within the reach of all who are interested in this department of practice. Gentlemen desiring early impressions of the plates would do well to order it without delay. A specimen of the plates and text sent free by mail, on receipt of 25 cents.

We wish for once that our province was not restricted to methods of treatment, that we might say something of the exquisite colored plates in this volume.—*London Practitioner,* May, 1869.

As a whole, it teaches all that can be taught by means of plates and print.—*London Lancet,* March 13, 1869.

Superior to anything of the kind ever before issued on this continent.—*Canada Med. Journal,* March, '69.

The practitioner who desires to understand this branch of medicine thoroughly should obtain this, the most complete and best work ever published.—*Dominion Med. Journal,* May, 1869.

This is a work of master hands on both sides. M. Cullerier is scarcely second to, we think we may truly say is a peer of the illustrious and venerable Ricord, while in this country we do not hesitate to say that Dr. Bumstead, as an authority, is without a rival. Assuring our readers that these illustrations tell the whole history of venereal disease, from its inception to its end, we do not know a single medical work, which for its kind is more *necessary* for them to have.—*California Med. Gazette,* March, 1869.

The most splendidly illustrated work in the language, and in our opinion far more useful than the French original.—*Am. Journ. Med. Sciences,* Jan.'69.

The fifth and concluding number of this magnificent work has reached us, and we have no hesitation in saying that its illustrations surpass those of previous numbers.—*Boston Med. and Surg. Journal,* Jan. 14, 1869.

Other writers besides M. Cullerier have given us a good account of the diseases of which he treats, but no one has furnished us with such a complete series of illustrations of the venereal diseases. There is, however, an additional interest and value possessed by the volume before us; for it is an American reprint and translation of M. Cullerier's work, with incidental remarks by one of the most eminent American syphilographers, Mr. Bumstead.—*Brit. and For. Medico-Chir. Review,* July, 1869.

HILL (BERKELEY),
Surgeon to the Lock Hospital, London.

ON SYPHILIS AND LOCAL CONTAGIOUS DISORDERS.
In one handsome octavo volume; cloth, $3 25.

Bringing, as it does, the entire literature of the disease down to the present day, and giving with great ability the results of modern research, it is in every respect a most desirable work, and one which should find a place in the library of every surgeon.—*California Med. Gazette,* June, 1869.

Considering the scope of the book and the careful attention to the manifold aspects and details of its subject, it is wonderfully concise. All these qualities render it an especially valuable book to the beginner, to whom we would most earnestly recommend its study; while it is no less useful to the practitioner.—*St. Louis Med. and Surg. Journal,* May, 1869.

The most convenient and ready book of reference we have met with.—*N. Y. Med. Record,* May 1, 1869.

Most admirably arranged for both student and practitioner, no other work on the subject equals it; it is more simple, more easily studied.—*Buffalo Med. and Surg. Journal,* March, 1869.

ZEISSL (H.), M.D.
A COMPLETE TREATISE ON VENEREAL DISEASES.
Translated from the Second Enlarged German Edition, by FREDERIC R. STURGIS, M.D In one octavo volume, with illustrations. (*Preparing.*)

WILSON (ERASMUS), F.R.S.
ON DISEASES OF THE SKIN. With Illustrations on wood. Seventh American, from the sixth and enlarged English edition. In one large octavo volume of over 800 pages, $5.

A SERIES OF PLATES ILLUSTRATING "WILSON ON DISEASES OF THE SKIN;" consisting of twenty beautifully executed plates, of which thirteen are exquisitely colored, presenting the Normal Anatomy and Pathology of the Skin, and embracing accurate representations of about one hundred varieties of disease, most of them the size of nature. Price, in extra cloth, $5 50. Also, the Text and Plates, bound in one handsome volume. Cloth, $10.

No one treating skin diseases should be without a copy of this standard work.—*Canada Lancet.*

We can safely recommend it to the profession as the best work on the subject now in existence in the English language.—*Medical Times and Gazette.*

Mr. Wilson's volume is an excellent digest of the actual amount of knowledge of cutaneous diseases; it includes almost every fact or opinion of importance connected with the anatomy and pathology of the skin.—*British and Foreign Medical Review.*

Such a work as the one before us is a most capital and acceptable help. Mr. Wilson has long been held as high authority in this department of medicine, and his book on diseases of the skin has long been regarded as one of the best text-books extant on the subject. The present edition is carefully prepared, and brought up in its revision to the present time. In this edition we have also included the beautiful series of plates illustrative of the text, and in the last edition published separately. There are twenty of these plates, nearly all of them colored to nature, and exhibiting with great fidelity the various groups of diseases.—*Cincinnati Lancet.*

BY THE SAME AUTHOR.
THE STUDENT'S BOOK OF CUTANEOUS MEDICINE and DISEASES OF THE SKIN. In one very handsome royal 12mo. volume. $3 50.

NELIGAN (J. MOORE), M.D., M.R.I.A.
A PRACTICAL TREATISE ON DISEASES OF THE SKIN. Fifth American, from the second and enlarged Dublin edition by T. W. Belcher, M.D. In one neat royal 12mo. volume of 462 pages, cloth, $2 25.

Fully equal to all the requirements of students and young practitioners.—*Dublin Med. Press.*

Of the remainder of the work we have nothing beyond unqualified commendation to offer. It is so far the most complete one of its size that has appeared, and for the student there can be none which can compare with it in practical value. All the late discoveries in Dermatology have been duly noticed, and their value justly estimated; in a word, the work is fully up to the times, and is thoroughly stocked with most valuable information.—*New York Med. Record,* Jan. 15, 1867.

The most convenient manual of diseases of the skin that can be procured by the student.—*Chicago Med. Journal,* Dec. 1866.

BY THE SAME AUTHOR.
ATLAS OF CUTANEOUS DISEASES. In one beautiful quarto volume, with exquisitely colored plates, &c., presenting about one hundred varieties of disease. Cloth, $5 50.

The diagnosis of eruptive disease, however, under all circumstances, is very difficult. Nevertheless, Dr. Neligan has certainly, "as far as possible," given a faithful and accurate representation of this class of diseases, and there can be no doubt that these plates will be of great use to the student and practitioner in drawing a diagnosis as to the class, order, and species to which the particular case may belong. While looking over the "Atlas" we have been induced to examine also the "Practical Treatise," and we are inclined to consider it a very superior work, combining accurate verbal description with sound views of the pathology and treatment of eruptive diseases.—*Glasgow Med. Journal.*

A compend which will very much aid the practitioner in this difficult branch of diagnosis. Taken with the beautiful plates of the Atlas, which are remarkable for their accuracy and beauty of coloring, it constitutes a very valuable addition to the library of a practical man.—*Buffalo Med. Journal.*

HILLIER (THOMAS), M.D.,
Physician to the Skin Department of University College Hospital, &c.
HAND-BOOK OF SKIN DISEASES, for Students and Practitioners. Second American Edition. In one royal 12mo. volume of 358 pp. With Illustrations. Cloth, $2 25.

We can conscientiously recommend it to the student; the style is clear and pleasant to read, the matter is good, and the descriptions of disease, with the modes of treatment recommended, are frequently illustrated with well-recorded cases.—*London Med. Times and Gazette,* April 1, 1865.

It is a concise, plain, practical treatise on the various diseases of the skin; just such a work, indeed, as was much needed, both by medical students and practitioners.—*Chicago Medical Examiner,* May, 1865.

ANDERSON (McCALL), M.D.,
Physician to the Dispensary for Skin Diseases, Glasgow, &c.
ON THE TREATMENT OF DISEASES OF THE SKIN. With an Analysis of Eleven Thousand Consecutive Cases. In one vol. 8vo. $1. (*Lately Published.*)

GUERSANT'S SURGICAL DISEASES OF INFANTS AND CHILDREN. Translated by R. J. Dunglison, M.D. 1 vol. 8vo. Cloth, $2 50.

DEWEES ON THE PHYSICAL AND MEDICAL TREATMENT OF CHILDREN. Eleventh edition. 1 vol. 8vo. of 548 pages. Cloth, $2 80.

SMITH (J. LEWIS), M. D.,
Professor of Morbid Anatomy in the Bellevue Hospital Med. College, N. Y.
A COMPLETE PRACTICAL TREATISE ON THE DISEASES OF
CHILDREN. Second Edition, revised and greatly enlarged. In one handsome octavo volume of 742 pages, cloth, $5; leather, $6. (*Lately Published.*)

FROM THE PREFACE TO THE SECOND EDITION.

In presenting to the profession the second edition of his work, the author gratefully acknowledges the favorable reception accorded to the first. He has endeavored to merit a continuance of this approbation by rendering the volume much more complete than before. Nearly twenty additional diseases have been treated of, among which may be named Diseases Incidental to Birth, Rachitis, Tuberculosis, Scrofula, Intermittent, Remittent, and Typhoid Fevers, Chorea, and the various forms of Paralysis. Many new formulæ, which experience has shown to be useful, have been introduced, portions of the text of a less practical nature have been condensed, and other portions, especially those relating to pathological histology, have been rewritten to correspond with recent discoveries. Every effort has been made, however, to avoid an undue enlargement of the volume, but, notwithstanding this, and an increase in the size of the page, the number of pages has been enlarged by more than one hundred.

227 WEST 49TH STREET, NEW YORK, April, 1872.

The work will be found to contain nearly one-third more matter than the previous edition, and it is confidently presented as in every respect worthy to be received as the standard American text-book on the subject.

Eminently practical as well as judicious in its teachings.—*Cincinnati Lancet and Obs.*, July, 1872.

A standard work that leaves little to be desired.—*Indiana Journal of Medicine*, July, 1872.

We know of no book on this subject that we can more cordially recommend to the medical student and the practitioner.—*Cincinnati Clinic*, June 29, '72.

We regard it as superior to any other single work on the diseases of infancy and childhood.—*Detroit Rev. of Med. and Pharmacy*, Aug. 1872.

We confess to increased enthusiasm in recommending this second edition.—*St. Louis Med. and Surg. Journal*, Aug. 1872.

CONDIE (D. FRANCIS), M. D.
A PRACTICAL TREATISE ON THE DISEASES OF CHILDREN.
Sixth edition, revised and augmented. In one large octavo volume of nearly 800 closely-printed pages, cloth, $5 25; leather, $6 25.

The present edition, which is the sixth, is fully up to the times in the discussion of all those points in the pathology and treatment of infantile diseases which have been brought forward by the German and French teachers. As a whole, however, the work is the best American one that we have, and in its special adaptation to American practitioners it certainly has no equal.— *New York Med. Record*, March 2, 1868.

WEST (CHARLES), M. D.,
Physician to the Hospital for Sick Children, &c.
LECTURES ON THE DISEASES OF INFANCY AND CHILD-
HOOD. Fifth American from the sixth revised and enlarged English edition. In one large and handsome octavo volume of 678 pages. Cloth, $4 50; leather, $5 50. (*Just Issued.*)

The continued demand for this work on both sides of the Atlantic, and its translation into German, French, Italian, Danish, Dutch, and Russian, show that it fills satisfactorily a want extensively felt by the profession. There is probably no man living who can speak with the authority derived from a more extended experience than Dr. West, and his work now presents the results of nearly 2000 recorded cases, and 600 post-mortem examinations selected from among nearly 40,000 cases which have passed under his care. In the preparation of the present edition he has omitted much that appeared of minor importance, in order to find room for the introduction of additional matter, and the volume, while thoroughly revised, is therefore not increased materially in size.

Of all the English writers on the diseases of children, there is no one so entirely satisfactory to us as Dr. West. For years we have held his opinion as judicial, and have regarded him as one of the highest living authorities in the difficult department of medical science in which he is most widely known.— *Boston Med. and Surg. Journal*.

BY THE SAME AUTHOR. (*Lately Issued.*)
ON SOME DISORDERS OF THE NERVOUS SYSTEM IN CHILD-
HOOD; being the Lumleian Lectures delivered at the Royal College of Physicians of London, in March, 1871. In one volume, small 12mo., cloth, $1 00.

SMITH (EUSTACE), M. D.,
Physician to the Northwest London Free Dispensary for Sick Children.
A PRACTICAL TREATISE ON THE WASTING DISEASES OF
INFANCY AND CHILDHOOD. Second American, from the second revised and enlarged English edition. In one handsome octavo volume, cloth, $2 50. (*Lately Issued.*)

This is in every way an admirable book. The modest title which the author has chosen for it scarcely conveys an adequate idea of the many subjects upon which it treats. Wasting is so constant an attendant upon the maladies of childhood, that a treatise upon the wasting diseases of children must necessarily embrace the consideration of many affections of which it is a symptom; and this is excellently well done by Dr. Smith. The book might fairly be described as a practical handbook of the common diseases of children, so numerous are the affections considered either collaterally or directly. We are acquainted with no safer guide to the treatment of children's diseases, and few works give the insight into the physiological and other peculiarities of children that Dr. Smith's book does.—*Brit. Med. Journ.*, April 8, 1871.

ish and Foreign Contributors; Transactions of the Obstetrical Societies in England and abroad; Reports of Hospital Practice; Reviews and Bibliographical Notices; Articles and Notes, Editorial, Historical, Forensic, and Miscellaneous; Selections from Journals; Correspondence, &c. Collecting together the vast amount of material daily accumulating in this important and rapidly improving department of medical science, the value of the information which it presents to the subscriber may be estimated from the character of the gentlemen who have already promised their support, including such names as those of Drs. ATTHILL, ROBERT BARNES, HENRY BENNET, THOMAS CHAMBERS, FLEETWOOD CHURCHILL, MATTHEWS DUNCAN, GRAILY HEWITT, BRAXTON HICKS, ALFRED MEADOWS, W. LEISHMAN, ALEX. SIMPSON, TYLER SMITH, EDWARD J. TILT, SPENCER WELLS, &c. &c.; in short, the representative men of British Obstetrics and Gynæcology.

In order to render the OBSTETRICAL JOURNAL fully adequate to the wants of the American profession, each number contains a Supplement devoted to the advances made in Obstetrics and Gynæcology on this side of the Atlantic. This portion of the Journal is under the editorial charge of Dr. WILLIAM F. JENKS, to whom editorial communications, exchanges, books for review, &c., may be addressed, to the care of the publisher.

*** Complete sets from the beginning can no longer be furnished, but subscriptions can commence with January, 1875, or with Vol. II., April, 1874.

THOMAS (T. GAILLARD), M. D.,
Professor of Obstetrics, &c., in the College of Physicians and Surgeons, N. Y., &c.

A PRACTICAL TREATISE ON THE DISEASES OF WOMEN. Fourth edition, enlarged and thoroughly revised. In one large and handsome octavo volume of 800 pages, with 191 illustrations. Cloth, $5 00; leather, $6 00. (*Now Ready*.)

The author has taken advantage of the opportunity afforded by the call for another edition of this work to render it worthy a continuance of the very remarkable favor with which it has been received. Every portion has been subjected to a conscientious revision, and no labor has been spared to make it a complete treatise on the most advanced condition of its important subject.

A few notices of the previous editions are subjoined:—

Professor Thomas fairly took the Profession of the United States by storm when his book first made its appearance early in 1868. Its reception was simply enthusiastic, notwithstanding a few adverse criticisms from our transatlantic brethren, the first large edition was rapidly exhausted, and in six months a second one was issued, and in two years a third one was announced and published, and we are now promised the fourth. The popularity of this work was not ephemeral, and its success was unprecedented in the annals of American medical literature. Six years is a long period in medical scientific research, but Thomas's work on "Diseases of Women" is still the leading native production of the United States. The order, the matter, the absence of theoretical disputativeness, the fairness of statement, and the elegance of diction, preserved throughout the entire range of the book, indicate that Professor Thomas did not overestimate his powers when he conceived the idea and executed the work of producing a new treatise upon diseases of women.—PROF. PALLEN, in *Louisville Med. Journal*, Sept. 1874.

Briefly, we may say that we know of no book which so completely and concisely represents the present state of gynæcology; none so full of well-digested and reliable teaching; none which bespeaks an author more apt in research and abundant in resources.—*N. Y. Med. Record*, May 1, 1872.

We should not be doing our duty to the profession did we not tell those who are unacquainted with the book, how much it is valued by gynæcologists, and how it is in many respects one of the best text-books on the subject we possess in our language. We have no hesitation in recommending Dr. Thomas's work as one of the most complete of its kind ever published. It should be in the possession of every practitioner for reference and for study.—*London Lancet*, April 27, 1872.

We are free to say that we regard Dr. Thomas the best American authority on diseases of women.— *Cincinnati Lancet and Observer*, May, 1872.

No general practitioner can afford to be without it.—*St. Louis Med. and Surg. Journal*, May, 1872.

Its able author need not fear comparison between it and any similar work in the English language; nay more, as a text-book for students and as a guide for practitioners, we believe it is unequalled. If either student or practitioner can get but one book on diseases of women, that book should be "Thomas." —*Amer. Jour. Med. Sciences*, April, 1872.

To students we unhesitatingly recommend it as the best text-book on diseases of females extant.— *St. Louis Med. Reporter*, June, 1869.

Of all the army of books that have appeared of late years, on the diseases of the uterus and its appendages, we know of none that is so clear, comprehensive, and practical as this of Dr. Thomas', or one that we should more emphatically recommend to the young practitioner, as his guide.—*California Med. Gazette*, June, 1869.

It would be superfluous to give an extended review of what is now firmly established as *the* American text-book of Gynæcology.—*N. Y. Med. Gazette*, July 17, 1869.

This is a new and revised edition of a work which we recently noticed at some length, and earnestly commended to the favorable attention of our readers. The fact that, in the short space of one year, this second edition makes its appearance, shows that the general judgment of the profession has largely confirmed the opinion we gave at that time.—*Cincinnati Lancet*, Aug. 1869.

It is so short a time since we gave a full review of the first edition of this book, that we deem it only necessary now to call attention to the second appearance of the work. Its success has been remarkable, and we can only congratulate the author on the brilliant reception his book has received.—*N. Y. Med. Journal*, April, 1869.

HODGE (HUGH L.), M.D.,
Emeritus Professor of Obstetrics, &c., in the University of Pennsylvania.

ON DISEASES PECULIAR TO WOMEN; including Displacements of the Uterus. With original illustrations. Second edition, revised and enlarged. In one beautifully printed octavo volume of 531 pages, cloth, $4 50.

From PROF. W. H. BYFORD, *of the Rush Medical College, Chicago.*

The book bears the impress of a master hand, and must, as its predecessor, prove acceptable to the profession. In diseases of women Dr. Hodge has established a school of treatment that has become worldwide in fame.

Professor Hodge's work is truly an original one from beginning to end, consequently no one can peruse its pages without learning something new. The book, which is by no means a large one, is divided into two grand sections, so to speak: first, that treating of the nervous sympathies of the uterus, and, secondly,

that which speaks of the mechanical treatment of displacements of that organ. He is disposed, as a nonbeliever in the frequency of inflammations of the uterus, to take strong ground against many of the highest authorities in this branch of medicine, and the arguments which he offers in support of his position are, to say the least, well put. Numerous woodcuts adorn this portion of the work, and add incalculably to the proper appreciation of the variously shaped instruments referred to by our author. As a contribution to the study of women's diseases, it is of great value, and is abundantly able to stand on its own merits.—*N. Y. Medical Record*, Sept. 15, 1868.

WEST (CHARLES), M.D.

LECTURES ON THE DISEASES OF WOMEN. Third American, from the Third London edition. In one neat octavo volume of about 550 pages, cloth, $3 75; leather, $4 75.

As a writer, Dr. West stands, in our opinion, second only to Watson, the "Macaulay of Medicine;" he possesses that happy faculty of clothing instruction in easy garments; combining pleasure with profit, he leads his pupils, in spite of the ancient proverb, along a royal road to learning. His work is one which will not satisfy the extreme on either side, but it is one that will please the great majority who are

seeking truth, and one that will convince the student that he has committed himself to a candid, safe, and valuable guide.—*N. A. Med.-Chirurg Review.*

We have to say of it, briefly and decidedly, that it is the best work on the subject in any language, and that it stamps Dr. West as the *facile princeps* of British obstetric authors.—*Edinburgh Med. Journal.*

BARNES (ROBERT), M.D., F.R.C.P.,
Obstetric Physician to St. Thomas's Hospital, &c.

A CLINICAL EXPOSITION OF THE MEDICAL AND SURGICAL DISEASES OF WOMEN. In one handsome octavo volume of about 800 pages, with 169 illustrations. Cloth. $5 00; leather, $6 00. (*Just Issued.*)

The very complete scope of this volume and the manner in which it has been filled out, may be seen by the subjoined Summary of Contents.

INTRODUCTION. CHAPTER I. Ovaries; Corpus Luteum. II. Fallopian Tubes. III. Shape of Uterine Cavity. IV. Structure of Uterus. V. The Vagina. VI. Examinations and Diagnosis. VII. Significance of Leucorrhœa. VIII. Discharges of Air. IX. Watery Discharges. X. Purulent Discharges. XI. Hemorrhagic Discharges. XII. Significance of Pain. XIII. Significance of Dyspareunia. XIV. Significance of Sterility. XV. Instrumental Diagnosis and Treatment. XVI. Diagnosis by the Touch, the Sound, the Speculum. XVII. Menstruation and its Disorders. XVIII. Amenorrhœa. XIX. Amenorrhœa (continued). XX. Dysmenorrhœa. XXI. Ovarian Dysmenorrhœa, &c. XXII. Inflammatory Dysmenorrhœa. XXIII. Irregularities of Change of Life. XXIV. Relations between Menstruation and Diseases. XXV. Disorders of Old Age. XXVI. Ovary, Absence and Hernia of. XXVII. Ovary, Hemorrhage, &c., of. XXVIII. Ovary, Tubercle, Cancer, &c., of. XXIX. Ovarian Cystic Tumors. XXX. Dermoid Cysts of Ovary. XXXI. Ovarian Tumors, Prognosis of. XXXII. Diagnosis of Ovarian Tumors. XXXIII. Ovarian Cysts, Treatment of. XXXIV. Fallopian Tubes, Diseases of. XXXV. Broad Ligaments, Diseases of. XXXVI. Extra-uterine Gestation. XXXVII. Special Pathology of Uterus. XXXVIII. General Uterine Pathology. XXXIX. Alterations of Blood Supply. XL. Metritis, Endometritis, &c. XLI. Pelvic Cellulitis and Peritonitis, &c. XLII. Hæmatocele, &c. XLIII. Displacements of Uterus. XLIV. Displacements (continued). XLV. Retroversion and Retroflexion. XLVI. Inversion. XLVII. Uterine Tumors. XLVIII. Polypus Uteri. XLIX. Polypus Uteri (continued). L. Cancer. LI. Diseases of Vagina. LII. Diseases of the Vulva.

Embodying the long experience and personal observation of one of the greatest of living teachers in diseases of women, it seems pervaded by the presence of the author, who speaks directly to the reader, and speaks, too, as one having authority. And yet, notwithstanding this distinct personality, there is nothing narrow as to time, place, or individuals, in the views presented, and in the instructions given; Dr. Barnes has been an attentive student, not only of European, but also of American literature, pertaining to diseases of females, and enriched his own experience by treasures thence gathered; he seems as familiar, for example, with the writings of Sims, Emmet, Tho-

mas, and Peaslee, as if these eminent men were his countrymen and colleagues, and gives them a credit which must be gratifying to every American physician.—*Am. Journ. Med. Sci.*, April, 1874.

Throughout the whole book it is impossible not to feel that the author has a spontaneously, conscientiously, and fearlessly performed his task. He goes direct to the point, and does not loiter on the way to gossip or quarrel with other authors. Dr. Barnes's book will he eagerly read all over the world, and will everywhere be admired for its comprehensiveness, honesty of purpose, and ability.—*The Obstet. Journ. of Great Britain and Ireland*, March, 1874.

CHURCHILL ON THE PUERPERAL FEVER AND OTHER DISEASES PECULIAR TO WOMEN. 1 vol. 8vo., pp. 450, cloth. $2 50.

MEIGS ON WOMAN: HER DISEASES AND THEIR REMEDIES. A Series of Lectures to his Class. Fourth and Improved Edition. 1 vol. 8vo., over 700 pages, cloth, $5 00; leather, $6 00.

MEIGS ON THE NATURE, SIGNS, AND TREATMENT OF CHILDBED FEVER. 1 vol. 8vo., pp. 365, cloth. $2 00.

ASHWELL'S PRACTICAL TREATISE ON THE DISEASES PECULIAR TO WOMEN. Third American, from the Third and revised London edition. 1 vol. 8vo., pp. 528, cloth. $3 50.

DEWEES'S TREATISE ON THE DISEASES OF FEMALES. With Illustrations. Eleventh Edition, with the Author's last Improvements and corrections. In one octavo volume of 536 pages, with plates, cloth. $3 00.

HODGE (HUGH L.), M.D.,
Emeritus Professor of Midwifery, &c., in the University of Pennsylvania, &c.

THE PRINCIPLES AND PRACTICE OF OBSTETRICS. Illustrated with large lithographic plates containing one hundred and fifty-nine figures from original photographs, and with numerous wood-cuts. In one large and beautifully printed quarto volume of 550 double-columned pages, strongly bound in cloth, $14.

The work of Dr. Hodge is something more than a simple presentation of his particular views in the department of Obstetrics; it is something more than an ordinary treatise on midwifery; it is, in fact, a cyclopædia of midwifery. He has aimed to embody in a single volume the whole science and art of Obstetrics. An elaborate text is combined with accurate and varied pictorial illustrations, so that no fact or principle is left unstated or unexplained.—*Am. Med. Times*, Sept. 3, 1864.

We should like to analyze the remainder of this excellent work, but already has this review extended beyond our limited space. We cannot conclude this notice without referring to the excellent finish of the work. In typography it is not to be excelled; the paper is superior to what is usually afforded by our American cousins, quite equal to the best of English books. The engravings and lithographs are most beautifully executed. The work recommends itself for its originality, and is in every way a most valuable addition to those on the subject of obstetrics.—*Canada Med. Journal*, Oct. 1864.

It is very large, profusely and elegantly illustrated, and is fitted to take its place near the works of great obstetricians. Of the American works on the subject it is decidedly the best.—*Edinb. Med. Jour.*, Dec. '64.

We have examined Professor Hodge's work with great satisfaction; every topic is elaborated most fully. The views of the author are comprehensive, and concisely stated. The rules of practice are judicious, and will enable the practitioner to meet every emergency of obstetric complication with confidence.—*Chicago Med. Journal*, Aug. 1864.

More time than we have had at our disposal since we received the great work of Dr. Hodge is necessary to do it justice. It is undoubtedly by far the most original, complete, and carefully composed treatise on the principles and practice of Obstetrics which has ever been issued from the American press.—*Pacific Med. and Surg. Journal*, July, 1864.

We have read Dr. Hodge's book with great pleasure, and have much satisfaction in expressing our commendation of it as a whole. It is certainly highly instructive, and in the main, we believe, correct. The great attention which the author has devoted to the mechanism of parturition, taken along with the conclusions at which he has arrived, point, we think, conclusively to the fact that, in Britain at least, the doctrines of Naegele have been too blindly received.—*Glasgow Med. Journal*, Oct. 1864.

*** Specimens of the plates and letter-press will be forwarded to any address, free by mail, on receipt of six cents in postage stamps.

TANNER (THOMAS H.), M.D.

ON THE SIGNS AND DISEASES OF PREGNANCY. First American from the Second and Enlarged English Edition. With four colored plates and illustrations on wood. In one handsome octavo volume of about 500 pages, cloth, $4 25.

The very thorough revision the work has undergone has added greatly to its practical value, and increased materially its efficiency as a guide to the student and to the young practitioner.—*Am. Journ. Med. Sci.*, April, 1868.

With the immense variety of subjects treated of and the ground which they are made to cover, the impossibility of giving an extended review of this truly remarkable work must be apparent. We have not a single fault to find with it, and most heartily commend it to the careful study of every physician who would not only always be sure of his diagnosis of pregnancy, but always ready to treat all the numerous ailments that are, unfortunately for the civilized women of to-day, so commonly associated with the function.—*N. Y. Med. Record*, March 16, 1868.

We recommend obstetrical students, young and old, to have this volume in their collections. It contains not only a fair statement of the signs, symptoms, and diseases of pregnancy, but comprises in addition much interesting relative matter that is not to be found in any other work that we can name.—*Edinburgh Med. Journal*, Jan. 1868.

SWAYNE (JOSEPH GRIFFITHS), M.D.,
Physician-Accoucheur to the British General Hospital, &c.

OBSTETRIC APHORISMS FOR THE USE OF STUDENTS COMMENCING MIDWIFERY PRACTICE. Second American, from the Fifth and Revised London Edition, with Additions by E. R. HUTCHINS, M.D. With Illustrations. In one neat 12mo. volume. Cloth, $1 25. (*Lately Issued.*)

*** See p. 3 of this Catalogue for the terms on which this work is offered as a premium to subscribers to the "AMERICAN JOURNAL OF THE MEDICAL SCIENCES."

It is really a capital little compendium of the subject, and we recommend young practitioners to buy it and carry it with them when called to attend cases of labor. They can while away the otherwise tedious hours of waiting, and thoroughly fix in their memories the most important practical suggestions it contains. The American editor has materially added by his notes and the concluding chapters to the completeness and general value of the book.—*Chicago Med. Journal*, Feb. 1870.

The manual before us contains in exceedingly small compass—small enough to carry in the pocket—about all there is of obstetrics, condensed into a nutshell of Aphorisms. The illustrations are well selected, and serve as excellent reminders of the conduct of labor—regular and difficult.—*Cincinnati Lancet*, April, '70.

This is a most admirable little work, and completely answers the purpose. It is not only valuable for young beginners, but no one who is not a proficient in the art of obstetrics should be without it, because it condenses all that is necessary to know for ordinary midwifery practice. We commend the book most favorably.—*St. Louis Med. and Surg. Journal*, Sept. 10, 1870.

A studied perusal of this little book has satisfied us of its eminently practical value. The object of the work, the author says, in his preface, is to give the student a few brief and practical directions respecting the management of ordinary cases of labor; and also to point out to him in extraordinary cases when and how he may act upon his own responsibility, and when he ought to send for assistance.—*N. Y. Medical Journal*, May, 1870.

WINCKEL (F.),
Professor and Director of the Gynæcological Clinic in the University of Rostock.

A COMPLETE TREATISE ON THE PATHOLOGY AND TREATMENT OF CHILDBED, for Students and Practitioners. Translated, with the consent of the author, from the Second German Edition, by JAMES READ CHADWICK, M.D. In one octavo volume. (*Preparing.*)

*L*EISHMAN (WILLIAM), M.D.,
Regius Professor of Midwifery in the University of Glasgow, &c.

A SYSTEM OF MIDWIFERY, INCLUDING THE DISEASES OF PREGNANCY AND THE PUERPERAL STATE. In one large and very handsome octavo volume of over 700 pages, with one hundred and eighty-two illustrations. Cloth, $5 00; leather, $6 00. (*Lately Published.*)

This is one of a most complete and exhaustive character. We have gone carefully through it, and there is no subject in Obstetrics which has not been considered well and fully. The result is a work, not only admirable as a text-book, but valuable as a work of reference to the practitioner in the various emergencies of obstetric practice. Take it all in all, we have no hesitation in saying that it is in our judgment the best English work on the subject.—*London Lancet*, Aug. 23, 1873.

The work of Leishman gives an excellent view of modern midwifery, and evinces its author's extensive acquaintance with British and foreign literature; and not only acquaintance with it, but wholesome digestion and sound judgment of it. He has, withal, a manly, free style, and can state a difficult and complicated matter with remarkable clearness and brevity.—*Edin. Med. Journ.*, Sept. 1873.

The author has succeeded in presenting to the profession an admirable treatise, especially in its practical aspects; one which is, in general, clearly written, and sound in doctrine, and one which cannot fail to add to his already high reputation. In concluding our examination of this work, we cannot avoid again saying that Dr. Leishman has fully accomplished that difficult task of presenting a good text-book upon obstetrics. We know none better for the use of the student or junior practitioner.—*Am. Practitioner*, Mar. 1874.

It proposes to offer to practitioners and students

"A Complete System of the Midwifery of the Present Day," and well redeems the promise. In all that relates to the subject of labor, the teaching is admirably clear, concise, and practical, representing not alone British practice, but the contributions of Continental and American schools.—*N. Y. Med. Record*, March 2, 1874.

The work of Dr. Leishman is, in many respects, not only the best treatise on midwifery that we have seen, but one of the best treatises on any medical subject that has been published of late years.—*Lond. Practitioner*, Feb. 1874.

It was written to supply a desideratum, and we will be much surprised if it does not fulfil the purpose of its author. Taking it as a whole, we know of no work on obstetrics by an English author in which the student and the practitioner will find the information so clear and so completely abreast of the present state of our knowledge on the subject.—*Glasgow Med. Journ.*, Aug. 1873.

Dr. Leishman's System of Midwifery, which has only just been published, will go far to supply the want which has so long been felt, of a really good modern English text-book. Although large, as is inevitable in a work on so extensive a subject, it is so well and clearly written, that it is never wearisome to read. Dr. Leishman's work may be confidently recommended as an admirable text-book, and is sure to be largely used.—*Lond. Med. Record*, Sept. 1873.

---

*R*AMSBOTHAM (FRANCIS H.), M.D.

THE PRINCIPLES AND PRACTICE OF OBSTETRIC MEDICINE AND SURGERY, in reference to the Process of Parturition. A new and enlarged edition, thoroughly revised by the author. With additions by W. V. KEATING, M. D., Professor of Obstetrics, &c., in the Jefferson Medical College, Philadelphia. In one large and handsome imperial octavo volume of 650 pages, strongly bound in leather, with raised bands; with sixty-four beautiful plates, and numerous wood-cuts in the text, containing in all nearly 200 large and beautiful figures. $7 00.

We will only add that the student will learn from it all he need to know, and the practitioner will find it, as a book of reference, surpassed by none other.—*Stethoscope*.

The character and merits of Dr. Ramsbotham's work are so well known and thoroughly established, that comment is unnecessary and praise superfluous. The illustrations, which are numerous and accurate, are executed in the highest style of art. We cannot too highly recommend the work to our readers.—*St. Louis Med. and Surg. Journal*.

To the physician's library it is indispensable, while to the student, as a text-book, from which to extract the material for laying the foundation of an education on obstetrical science, it has no superior.—*Ohio Med. and Surg. Journal*.

When we call to mind the toil we underwent in acquiring a knowledge of this subject, we cannot but envy the student of the present day the aid which this work will afford him.—*Am. Jour. of the Med. Sciences*.

---

*C*HURCHILL (FLEETWOOD), M.D., M.R.I.A.

ON THE THEORY AND PRACTICE OF MIDWIFERY. A new American from the fourth revised and enlarged London edition. With notes and additions by D. FRANCIS CONDIE, M. D., author of a "Practical Treatise on the Diseases of Children," &c. With one hundred and ninety-four illustrations. In one very handsome octavo volume of nearly 700 large pages. Cloth, $4 00; leather, $5 00.

These additions render the work still more complete and acceptable than ever; and we can commend it to the profession with great cordiality and pleasure.—*Cincinnati Lancet*.

Few works on this branch of medical science are equal to it, certainly none excel it, whether in regard to theory or practice.—*Brit. Am. Journal*.

No treatise on obstetrics with which we are ac-

quainted can compare favorably with this, in respect to the amount of material which has been gathered from every source.—*Boston Med. and Surg. Journal*.

There is no better text-book for students, or work of reference and study for the practising physician than this. It should adorn and enrich every medical library.—*Chicago Med. Journal*.

MONTGOMERY'S EXPOSITION OF THE SIGNS AND SYMPTOMS OF PREGNANCY. With two exquisite colored plates, and numerous wood-cuts. In 1 vol. 8vo., of nearly 600 pp., cloth. $3 75.

RIGBY'S SYSTEM OF MIDWIFERY. With Notes and Additional Illustrations. Second American edition. One volume octavo, cloth, 422 pages. $2 50.

GROSS (SAMUEL D.), M.D.,
Professor of Surgery in the Jefferson Medical College of Philadelphia.

A SYSTEM OF SURGERY: Pathological, Diagnostic, Therapeutic, and Operative. Illustrated by upwards of Fourteen Hundred Engravings. Fifth edition, carefully revised, and improved. In two large and beautifully printed imperial octavo volumes of about 2300 pages, strongly bound in leather, with raised bands, $15. (*Just Issued.*)

The continued favor, shown by the exhaustion of successive large editions of this great work, proves that it has successfully supplied a want felt by American practitioners and students. In the present revision no pains have been spared by the author to bring it in every respect fully up to the day. To effect this a large part of the work has been rewritten, and the whole enlarged by nearly one-fourth, notwithstanding which the price has been kept at its former very moderate rate. By the use of a close, though very legible type, an unusually large amount of matter is condensed in its pages, the two volumes containing as much as four or five ordinary octavos. This, combined with the most careful mechanical execution, and its very durable binding, renders it one of the cheapest works accessible to the profession. Every subject properly belonging to the domain of surgery is treated in detail, so that the student who possesses this work may be said to have in it a surgical library. A few notices of the previous edition are subjoined :—

It must long remain the most comprehensive work on this important part of medicine.—*Boston Medical and Surgical Journal*, March 23, 1865.

We have compared it with most of our standard works, such as those of Erichsen, Miller, Fergusson, Syme, and others, and we must, in justice to our author, award it the pre-eminence. As a work, complete in almost every detail, no matter how minute or trifling, and embracing every subject known in the principles and practice of surgery, we believe it stands without a rival. Dr. Gross, in his preface, remarks "my aim has been to embrace the whole domain of surgery, and to allot to every subject its legitimate claim to notice;" and, we assure our readers, he has kept his word. It is a work which we can most confidently recommend to our brethren, for its utility is becoming the more evident the longer it is upon the shelves of our library.—*Canada Med. Journal*, September, 1865.

The first two editions of Professor Gross' System of Surgery are so well known to the profession, and so highly prized, that it would be idle for us to speak in praise of this work.—*Chicago Medical Journal*, September, 1865.

We gladly indorse the favorable recommendation of the work, both as regards matter and style, which we made when noticing its first appearance.—*British and Foreign Medico-Chirurgical Review*, Oct. 1865.

The most complete work that has yet issued from the press on the science and practice of surgery.—*London Lancet.*

This system of surgery is, we predict, destined to take a commanding position in our surgical literature, and be the crowning glory of the author's well earned fame. As an authority on general surgical subjects, this work is long to occupy a pre-eminent place, not only at home, but abroad. We have no

hesitation in pronouncing it without a rival in our language, and equal to the best systems of surgery in any language.—*N. Y. Med. Journal.*

Not only by far the best text-book on the subject, as a whole, within the reach of American students, but one which will be much more than ever likely to be resorted to and regarded as a high authority abroad.—*Am. Journal Med. Sciences*, Jan. 1865.

The work contains everything, minor and major, operative and diagnostic, including mensuration and examination, venereal diseases, and uterine manipulations and operations. It is a complete Thesaurus of modern surgery, where the student and practitioner shall not seek in vain for what they desire.—*San Francisco Med. Press*, Jan. 1865.

Open it where we may, we find sound practical information conveyed in plain language. This book is no mere provincial or even national system of surgery, but a work which, while very largely indebted to the past, has a strong claim on the gratitude of the future of surgical science.—*Edinburgh Med. Journal*, Jan. 1865.

A glance at the work is sufficient to show that the author and publisher have spared no labor in making it the most complete "System of Surgery" ever published in any country.—*St. Louis Med. and Surg. Journal*, April, 1865.

A system of surgery which we think unrivalled in our language, and which will indelibly associate his name with surgical science. And what, in our opinion, enhances the value of the work is that, while the practising surgeon will find all that he requires in it, it is at the same time one of the most valuable treatises which can be put into the hands of the student seeking to know the principles and practice of this branch of the profession which he designs subsequently to follow.—*The Brit. Am. Journ., Montreal.*

BY THE SAME AUTHOR.

A PRACTICAL TREATISE ON FOREIGN BODIES IN THE AIR-PASSAGES. In 1 vol. 8vo., with illustrations, pp. 468, cloth, $2 75.

SKEY'S OPERATIVE SURGERY. In 1 vol. 8vo. cloth, of over 650 pages; with about 100 wood-cuts. $3 25.
COOPER'S LECTURES ON THE PRINCIPLES AND PRACTICE OF SURGERY. In 1 vol. 8vo. cloth, 750 p. $2.

GIBSON'S INSTITUTES AND PRACTICE OF SURGERY. Eighth edition, improved and altered. With thirty-four plates. In two handsome octavo volumes, about 1000 pp., leather, raised bands. $6 50.

MILLER (JAMES),
Late Professor of Surgery in the University of Edinburgh, &c.

PRINCIPLES OF SURGERY. Fourth American, from the third and revised Edinburgh edition. In one large and very beautiful volume of 700 pages, with two hundred and forty illustrations on wood, cloth, $3 75.

BY THE SAME AUTHOR.

THE PRACTICE OF SURGERY. Fourth American, from the last Edinburgh edition. Revised by the American editor. Illustrated by three hundred and sixty-four engravings on wood. In one large octavo volume of nearly 700 pages, cloth, $3 75.

SARGENT (F. W.), M.D.
ON BANDAGING AND OTHER OPERATIONS OF MINOR SURGERY. New edition, with an additional chapter on Military Surgery. One handsome royal 12mo. volume, of nearly 400 pages, with 184 wood-cuts. Cloth, $1 75.

ASHHURST (JOHN, Jr.), M.D.,
Surgeon to the Episcopal Hospital, Philadelphia.
THE PRINCIPLES AND PRACTICE OF SURGERY. In one very large and handsome octavo volume of about 1000 pages, with nearly 550 illustrations, cloth, $6 50; leather, raised bands, $7 50. (Lately Published.)

The object of the author has been to present, within as condensed a compass as possible, a complete treatise on Surgery in all its branches, suitable both as a text-book for the student and a work of reference for the practitioner. So much has of late years been done for the advancement of Surgical Art and Science, that there seemed to be a want of a work which should present the latest aspects of every subject, and which, by its American character, should render accessible to the profession at large the experience of the practitioners of both hemispheres. This has been the aim of the author, and it is hoped that the volume will be found to fulfil its purpose satisfactorily. The plan and general outline of the work will be seen by the annexed

CONDENSED SUMMARY OF CONTENTS.

CHAPTER I. Inflammation. II. Treatment of Inflammation. III. Operations in general: Anæsthetics. IV. Minor Surgery. V. Amputations. VI. Special Amputations. VII. Effects of Injuries in General: Wounds. VIII. Gunshot Wounds. IX. Injuries of Bloodvessels. X. Injuries of Nerves, Muscles and Tendons, Lymphatics, Bursæ, Bones, and Joints. XI. Fractures. XII. Special Fractures. XIII. Dislocations. XIV. Effects of Heat and Cold. XV. Injuries of the Head. XVI. Injuries of the Back. XVII. Injuries of the Face and Neck. XVIII. Injuries of the Chest. XIX. Injuries of the Abdomen and Pelvis. XX. Diseases resulting from Inflammation. XXI. Erysipelas. XXII. Pyæmia. XXIII. Diathetic Diseases: Struma (including Tubercle and Scrofula); Rickets. XXIV. Venereal Diseases; Gonorrhœa and Chancroid. XXV. Venereal Diseases continued: Syphilis. XXVI. Tumors. XXVII. Surgical Diseases of Skin, Areolar Tissue, Lymphatics, Muscles, Tendons, and Bursæ. XXVIII. Surgical Disease of Nervous System (including Tetanus). XXIX. Surgical Diseases of Vascular System (including Aneurism). XXX. Diseases of Bone. XXXI. Diseases of Joints. XXXII. Excisions. XXXIII. Orthopædic Surgery. XXXIV. Diseases of Head and Spine. XXXV. Diseases of the Eye. XXXVI. Diseases of the Ear. XXXVII. Diseases of the Face and Neck. XXXVIII. Diseases of the Mouth, Jaws, and Throat. XXXIX. Diseases of the Breast. XL. Hernia. XLI. Special Herniæ. XLII. Diseases of Intestinal Canal. XLIII. Diseases of Abdominal Organs, and various operations on the Abdomen. XLIV. Urinary Calculus. XLV. Diseases of Bladder and Prostate. XLVI. Diseases of Urethra. XLVII. Diseases of Generative Organs. INDEX.

Its author has evidently tested the writings and experiences of the past and present in the crucible of a careful, analytic, and honorable mind, and faithfully endeavored to bring his work up to the level of the highest standard of practical surgery. He is frank and definite, and gives us opinions, and generally sound ones, instead of a mere *résumé* of the opinions of others. He is conservative, but not hidebound by authority. His style is clear, elegant, and scholarly. The work is an admirable text-book, and a useful book of reference It is a credit to American professional literature, and one of the first ripe fruits of the soil fertilized by the blood of our late unhappy war.—*N. Y. Med. Record*, Feb. 1, 1872.

Indeed, the work as a whole must be regarded as an excellent and concise exponent of modern surgery, and as such it will be found a valuable textbook for the student, and a useful book of reference for the general practitioner.—*N. Y. Med. Journal*, Feb. 1872.

It gives us great pleasure to call the attention of the profession to this excellent work. Our knowledge of its talented and accomplished author led us to expect from him a very valuable treatise upon subjects to which he has repeatedly given evidence of having probably devoted much time and labor, and we are in no way disappointed.—*Phila. Med. Times*, Feb. 1, 1872.

PIRRIE (WILLIAM), F. R. S. E.,
Professor of Surgery in the University of Aberdeen.
THE PRINCIPLES AND PRACTICE OF SURGERY. Edited by JOHN NEILL, M. D., Professor of Surgery in the Penna. Medical College, Surgeon to the Pennsylvania Hospital, &c. In one very handsome octavo volume of 780 pages, with 316 illustrations, cloth, $3 75.

HAMILTON (FRANK H.), M.D.,
Professor of Fractures and Dislocations, &c., in Bellevue Hosp. Med. College, New York.
A PRACTICAL TREATISE ON FRACTURES AND DISLOCATIONS. Fourth edition, thoroughly revised. In one large and handsome octavo volume of nearly 800 pages, with several hundred illustrations. Cloth, $5 75; leather, $6 75.

It is not, of course, our intention to review *in extenso*, Hamilton on "Fractures and Dislocations." Eleven years ago such review might not have been out of place; to-day the work is an authority, so well, so generally, and so favorably known, that it only remains for the reviewer to say that a new edition is just out, and it is better than either of its predecessors.—*Cincinnati Clinic*, Oct. 14, 1871.

Undoubtedly the best work on Fractures and Dislocations in the English language.—*Cincinnati Med. Repertory*, Oct. 1871.

We have once more before us Dr. Hamilton's admi-

rable treatise, which we have always considered the most complete and reliable work on the subject. As a whole, the work is without an equal in the literature of the profession.—*Boston Med. and Surg. Journ.*, Oct. 12, 1871.

It is unnecessary at this time to commend the book, except to such as are beginners in the study of this particular branch of surgery. Every practical surgeon in this country and abroad knows of it as a most trustworthy guide, and one which they, in common with us, would unqualifiedly recommend as the highest authority in any language.—*N. Y. Med. Record*, Oct. 16, 1871.

ERICHSEN (JOHN E.),
Professor of Surgery in University College, London, etc.

THE SCIENCE AND ART OF SURGERY; being a Treatise on Surgical Injuries, Diseases, and Operations. Revised by the author from the Sixth and enlarged English Edition. Illustrated by over seven hundred engravings on wood. In two large and beautiful octavo volumes of over 1700 pages, cloth, $9 00; leather, $11 00. (*Lately Issued.*)

Author's Preface to the New American Edition.

"The favorable reception with which the 'Science and Art of Surgery' has been honored by the Surgical Profession in the United States of America has been not only a source of deep gratification and of just pride to me, but has laid the foundation of many professional friendships that are amongst the agreeable and valued recollections of my life.

"I have endeavored to make the present edition of this work more deserving than its predecessors of the favor that has been accorded to them. In consequence of delays that have unavoidably occurred in the publication of the Sixth British Edition, time has been afforded to me to add to this one several paragraphs which I trust will be found to increase the practical value of the work."
LONDON, Oct. 1872.

On no former edition of this work has the author bestowed more pains to render it a complete and satisfactory exposition of British Surgery in its modern aspects. Every portion has been sedulously revised, and a large number of new illustrations have been introduced. In addition to the material thus added to the English edition, the author has furnished for the American edition such material as has accumulated since the passage of the sheets through the press in London, so that the work as now presented to the American profession, contains his latest views and experience. The increase in the size of the work has seemed to render necessary its division into two volumes. Great care has been exercised in its typographical execution, and it is confidently presented as in every respect worthy to maintain the high reputation which has rendered it a standard authority on this department of medical science.

These are only a few of the points in which the present edition of Mr. Erichsen's work surpasses its predecessors. Throughout there is evidence of a laborious care and solicitude in seizing the passing knowledge of the day, which reflects the greatest credit on the author, and much enhances the value of his work. We can only admire the industry which has enabled Mr. Erichsen thus to succeed, amid the distractions of active practice, in producing emphatically THE book of reference and study for British practitioners of surgery.—*London Lancet*, Oct. 26, 1872.

Considerable changes have been made in this edition, and nearly a hundred new illustrations have been added. It is difficult in a small compass to point out the alterations and additions; for, as the author states in his preface, they are not confined to any one portion, but are distributed generally through the subjects of which the work treats. Certainly one of the most valuable sections of the book seems to us to be that which treats of the diseases of the arteries and the operative proceedings which they necessitate. In few text-books is so much carefully arranged information collected.—*London Med. Times and Gas.*, Oct. 26, 1872.

The entire work, complete, as the great English treatise on Surgery of our own time, is, we can assure our readers, equally well adapted for the most junior student, and, as a book of reference, for the advanced practitioner.—*Dublin Quarterly Journal.*

DRUITT (ROBERT), M.R.C.S., &c.

THE PRINCIPLES AND PRACTICE OF MODERN SURGERY. A new and revised American, from the eighth enlarged and improved London edition. Illustrated with four hundred and thirty-two wood engravings. In one very handsome octavo volume, of nearly 700 large and closely printed pages, cloth, $4 00; leather, $5 00.

All that the surgical student or practitioner could desire.—*Dublin Quarterly Journal.*

It is a most admirable book. We do not know when we have examined one with more pleasure.—*Boston Med. and Surg. Journal.*

In Mr. Druitt's book, though containing only some seven hundred pages, both the principles and the practice of surgery are treated, and so clearly and perspicuously, as to elucidate every important topic. We have examined the book most thoroughly, and can say that this success is well merited. His book, moreover, possesses the inestimable advantages of having the subjects perfectly well arranged and classified, and of being written in a style at once clear and succinct.—*Am. Journal of Med. Sciences.*

ASHTON (T. J.).

ON THE DISEASES, INJURIES, AND MALFORMATIONS OF THE RECTUM AND ANUS; with remarks on Habitual Constipation. Second American, from the fourth and enlarged London edition. With handsome illustrations. In one very beautifully printed octavo volume of about 300 pages, cloth, $3 25.

BIGELOW (HENRY J.), M. D.,
Professor of Surgery in the Massachusetts Med. College.

ON THE MECHANISM OF DISLOCATION AND FRACTURE OF THE HIP. With the Reduction of the Dislocation by the Flexion Method. With numerous original illustrations. In one very handsome octavo volume. Cloth, $2 50.

LAWSON (GEORGE), F. R. C. S., Engl.,
Assistant Surgeon to the Royal London Ophthalmic Hospital, Moorfields, &c.

INJURIES OF THE EYE, ORBIT, AND EYELIDS: their Immediate and Remote Effects. With about one hundred illustrations. In one very handsome octavo volume, cloth, $3 50

It is an admirable practical book in the highest and best sense of the phrase.—*London Medical Times and Gazette*, May 18, 1867.

BRYANT (THOMAS), F.R.C.S.,
Surgeon to Guy's Hospital.

THE PRACTICE OF SURGERY. With over Five Hundred Engravings on Wood. In one large and very handsome octavo volume of nearly 1000 pages, cloth, $6 25; leather, raised bands, $7 25. (*Lately Published.*)

Again, the author gives us his own practice, his own beliefs, and illustrates by his own cases, or those treated in Guy's Hospital. This feature adds joint emphasis, and a solidity to his statements that inspire confidence. One feels himself almost by the side of the surgeon, seeing his work and hearing his living words. The views, etc., of other surgeons are considered calmly and fairly, but Mr. Bryant's are adopted. Thus the work is not a compilation of other writings; it is not an encyclopædia, but the plain statements, on practical points, of a man who has lived and breathed and had his being in the richest surgical experience. The whole profession owe a debt of gratitude to Mr. Bryant, for his work in their behalf. We are confident that the American profession will give substantial testimonial of their feelings towards both author and publisher, by speedily exhausting this edition. We cordially and heartily commend it to our friends, and think that no live surgeon can afford to be without it.—*Detroit Review of Med. and Pharmacy*, August, 1873.

As a manual of the practice of surgery for the use of the student, we do not hesitate to pronounce Mr. Bryant's book a first-rate work. Mr. Bryant has a good deal of the dogmatic energy which goes with the clear, pronounced opinions of a man whose reflections and experience have moulded a character not wanting in firmness and decision. At the same time he teaches with the enthusiasm of one who has faith in his teaching; he speaks as one having authority, and herein lies the charm and excellence of his work. He states the opinions of others freely

and fairly, yet it is no mere compilation. The book combines much of the merit of the manual with the merit of the monograph. One may recognize in almost every chapter of the ninety-four of which the work is made up the acuteness of a surgeon who has seen much, and observed closely, and who gives forth the results of actual experience. In conclusion we repeat what we stated at first, that Mr. Bryant's book is one which we can conscientiously recommend both to practitioners and students as an admirable work. —*Dublin Journ. of Med. Science*, August, 1873.

Mr. Bryant has long been known to the reading portion of the profession as an able, clear, and graphic writer upon surgical subjects. The volume before us is one eminently upon the practice of surgery and not one which treats at length on surgical pathology, though the views that are entertained upon this subject are sufficiently interspersed through the work for all practical purposes. As a text-book we cheerfully recommend it, feeling convinced that, from the subject-matter, and the concise and true way Mr. Bryant deals with his subject, it will prove a formidable rival among the numerous surgical text-books which are offered to the student.—*N. Y. Med. Record*, June, 1873.

This is, as the preface states, an entirely new book, and contains in a moderately condensed form all the surgical information necessary to a general practitioner. It is written in a spirit consistent with the present improved standard of medical and surgical science.—*American Journal of Obstetrics*, August, 1873.

WELLS (J. SOELBERG),
Professor of Ophthalmology in King's College Hospital, &c.

A TREATISE ON DISEASES OF THE EYE. Second American, from the Third and Revised London Edition, with additions; illustrated with numerous engravings on wood, and six colored plates. Together with selections from the Test-types of Jaeger and Snellen. In one large and very handsome octavo volume of nearly 800 pages; cloth, $5 00; leather, $6 00. (*Lately Published.*)

The continued demand for this work, both in England and this country, is sufficient evidence that the author has succeeded in his effort to supply within a reasonable compass a full practical digest of ophthalmology in its most modern aspects, while the call for repeated editions has enabled him in his revisions to maintain its position abreast of the most recent investigations and improvements. In again reprinting it, every effort has been made to adapt it thoroughly to the wants of the American practitioner. Such additions as seemed desirable have been introduced by the editor, Dr. I. Minis Hays, and the number of illustrations has been largely increased. The importance of test-types as an aid to diagnosis is so universally acknowledged at the present day that it seemed essential to the completeness of the work that they should be added, and as the author recommends the use of those both of Jaeger and of Snellen for different purposes, selections have been made from each, so that the practitioner may have at command all the assistance necessary. Although enlarged by one hundred pages, it has been retained at the former very moderate price, rendering it one of the cheapest volumes before the profession.

A few notices of the previous edition are subjoined.

On examining it carefully, one is not at all surprised that it should meet with universal favor. It is, in fact, a comprehensive and thoroughly practical treatise on diseases of the eye, setting forth the practice of the leading oculists of Europe and America, and giving the author's own opinions and preferences, which are quite decided and worthy of high consideration. The third English edition, from which this is taken, having been revised by the author, comprises a notice of all the more recent advances made in ophthalmic science. The style of the writer is

lucid and flowing, therein differing materially from some of the translations of Continental writers on this subject that are in the market. Special pains are taken to explain, at length, those subjects which are particularly difficult of comprehension to the beginner, as the use of the ophthalmoscope, the interpretation of its images, etc. The book is profusely and ably illustrated, and at the end are to be found 16 excellent colored ophthalmoscopic figures, which are copies of some of the plates of Liebreich's admirable atlas.—*Kansas City Med. Journ.*, June, 1874.

LAURENCE (JOHN Z.), F. R. C. S.,
Editor of the Ophthalmic Review, &c.

A HANDY-BOOK OF OPHTHALMIC SURGERY, for the use of Practitioners. Second Edition, revised and enlarged. With numerous illustrations. In one very handsome octavo volume, cloth, $3 00.

For those, however, who must assume the care of diseases and injuries of the eye, and who are too much pressed for time to study the classic works on the subject, or those recently published by Stellwag, Wells, Bader, and others, Mr. Laurence will prove a safe and trustworthy guide. He has described in this

edition those novelties which have secured the confidence of the profession since the appearance of his last. The volume has been considerably enlarged and improved by the revision and additions of its author, expressly for the American edition.—*Am. Journ. Med. Sciences*, Jan. 1870.

THOMPSON (SIR HENRY),
Surgeon and Professor of Clinical Surgery to University College Hospital.
LECTURES ON DISEASES OF THE URINARY ORGANS. With
illustrations on wood. Second American from the Third English Edition. In one neat octavo volume. Cloth, $2 25. (*Now Ready*.)

My aim has been to produce in the smallest possible compass an epitome of practical knowledge concerning the nature and treatment of the diseases which form the subject of the work; and I venture to believe that my intention has been more fully realized in this volume than in either of its predecessors.—*Author's Preface*.

BY THE SAME AUTHOR.

ON THE PATHOLOGY AND TREATMENT OF STRICTURE OF
THE URETHRA AND URINARY FISTULÆ. With plates and wood-cuts. From the third and revised English edition. In one very-handsome octavo volume, cloth, $3 50. (*Lately Published*.)

BY THE SAME AUTHOR. (Just Issued.)

THE DISEASES OF THE PROSTATE, THEIR PATHOLOGY
AND TREATMENT. Fourth Edition, Revised. In one very handsome octavo volume of 355 pages, with thirteen plates, plain and colored, and illustrations on wood. Cloth, $3 75.

TAYLOR (ALFRED S.), M. D.,
Lecturer on Med. Jurisp. and Chemistry in Guy's Hospital
MEDICAL JURISPRUDENCE. Seventh American Edition. Edited
by JOHN J. REESE, M.D., Prof. of Med. Jurisp. in the Univ. of Penn. In one large octavo volume of nearly 900 pages. Cloth, $5 00; leather, $6 00. (*Just Issued.*)

In preparing for the press this *seventh* American edition of the "Manual of Medical Jurisprudence" the editor has, through the courtesy of Dr. Taylor, enjoyed the very great advantage of consulting the sheets of the new edition of the author's larger work, "The Principles and Practice of Medical Jurisprudence," which is now ready for publication in London. This has enabled him to introduce the author's latest views upon the topics discussed, which are believed to bring the work fully up to the present time.

The notes of the former editor, Dr. Hartshorne, as also the numerous valuable references to American practice and decisions by his successor, Mr. Penrose, have been retained, with but few slight exceptions; they will be found inclosed in brackets, distinguished by the letters (H.) and (P.). The additions made by the present editor, from the material at his command, amount to about one hundred pages; and his own notes are designated by the letter (R.).

Several subjects, not treated of in the former edition, have been noticed in the present one, and the work, it is hoped, will be found to merit a continuance of the confidence which it has so long enjoyed as a standard authority.

BY THE SAME AUTHOR. (Now Ready.)

THE PRINCIPLES AND PRACTICE OF MEDICAL JURISPRU-
DENCE. Second Edition, Revised, with numerous Illustrations. In two large octavo volumes, cloth, $10 00; leather, $12 00.

This great work is now recognized in England as the fullest and most authoritative treatise on every department of its important subject. In laying it, in its improved form, before the American profession, the publisher trusts that it will assume the same position in this country.

BY THE SAME AUTHOR. New Edition—Nearly Ready.

POISONS IN RELATION TO MEDICAL JURISPRUDENCE AND
MEDICINE. Third American, from the Third and Revised English Edition. In one large octavo volume of 850 pages.

This work, which has been so long recognized as a leading authority on its important subject, has received a very thorough revision at the hands of the author, and may be regarded as a new book rather than as a mere revision. He has sought to bring it on all points to a level with the advanced science of the day; many portions have been rewritten, much that was of minor importance has been omitted, and every effort made to condense a complete view of the subject within the limits of a single volume. Dr. Taylor's position as an expert has brought him into connection with nearly all important cases in England for many years. He thus speaks with an authority that few other living men possess, while his intimate acquaintance with the literature of toxicology on both sides of the Atlantic, renders his work equally adapted as a text-book in this country as in Great Britain.

CONTENTS.

Poisons.—Absorption and Elimination—Detection—Action—Influence of Habit—Classification of Poisons—Evidence of Poisoning—Diseases resembling Poisoning—Inspection of the Dead Body—Objects of Chemical Analysis—Moral and Circumstantial Evidence in Poisoning, &c. &c. *Irritant Poisons.*—Mineral Irritants—Acid Poisons—Alkaline Poisons—Non-Metallic Irritants—Metallic Irritants—Vegetable Irritants—Animal Irritants.

Neurotic Poisons.—Cerebral or Narcotic Poisons—Spinal Poisons—Cerebro-Spinal Poisons—Cerebro-Cardiac Poisons.

TUKE (DANIEL HACK), M.D.,
Joint author of "The Manual of Psychological Medicine," &c.

ILLUSTRATIONS OF THE INFLUENCE OF THE MIND UPON THE BODY IN HEALTH AND DISEASE. Designed to illustrate the Action of the Imagination. In one handsome octavo volume of 416 pages, cloth, $3 25. (*Just Issued.*)

The object of the author in this work has been to show not only the effect of the mind in causing and intensifying disease, but also its curative influence, and the use which may be made of the imagination and the emotions as therapeutic agents. Scattered facts bearing upon this subject have long been familiar to the profession, but no attempt has hitherto been made to collect and systematize them so as to render them available to the practitioner, by establishing the several phenomena upon a scientific basis. In the endeavor thus to convert to the use of legitimate medicine the means which have been employed so successfully in many systems of quackery, the author has produced a work of the highest freshness and interest as well as of permanent value.

BLANDFORD (G. FIELDING), M. D., F. R. C P.,
Lecturer on Psychological Medicine at the School of St. George's Hospital, &c.

INSANITY AND ITS TREATMENT: Lectures on the Treatment, Medical and Legal, of Insane Patients. With a Summary of the Laws in force in the United States on the Confinement of the Insane. By ISAAC RAY, M. D. In one very handsome octavo volume of 471 pages; cloth, $3 25.

This volume is presented to meet the want, so frequently expressed, of a comprehensive treatise, in moderate compass, on the pathology, diagnosis, and treatment of insanity. To render it of more value to the practitioner in this country, Dr. Ray has added an appendix which affords information, not elsewhere to be found in so accessible a form, to physicians who may at any moment be called upon to take action in relation to patients.

It satisfies a want which must have been sorely felt by the busy general practitioners of this country. It takes the form of a manual of clinical description of the various forms of insanity, with a description of the mode of examining persons suspected of Insanity. We call particular attention to this feature of the book, as giving it a unique value to the general practitioner. If we pass from theoretical considerations to descriptions of the varieties of insanity as actually seen in practice and the appropriate treatment for them, we find in Dr. Blandford's work a considerable advance over previous writings on the subject. His pictures of the various forms of mental disease are so clear and good that no reader can fail to be struck with their superiority to those given in ordinary manuals in the English language or (so far as our own reading extends) in any other.—*London Practitioner*, Feb. 1871.

WINSLOW (FORBES), M.D., D.C.L., &c.

ON OBSCURE DISEASES OF THE BRAIN AND DISORDERS OF THE MIND; their incipient Symptoms, Pathology, Diagnosis, Treatment, and Prophylaxis. Second American, from the third and revised English edition. In one handsome octavo volume of nearly 600 pages, cloth, $4 25.

LEA (HENRY C.).

SUPERSTITION AND FORCE: ESSAYS ON THE WAGER OF LAW, THE WAGER OF BATTLE, THE ORDEAL, AND TORTURE. Second Edition, Enlarged. In one handsome volume royal 12mo. of nearly 500 pages; cloth, $2 75. (*Lately Published.*)

We know of no single work which contains, in so small a compass, so much illustrative of the strangest operations of the human mind. Foot-notes give the authority for each statement, showing vast research and wonderful industry. We advise our *confrères* to read this book and ponder its teachings.—*Chicago Med. Journal*, Aug. 1870.

As a work of curious inquiry on certain outlying points of obsolete law, "Superstition and Force" is one of the most remarkable books we have met with.—*London Athenæum*, Nov. 3, 1866.

He has thrown a great deal of light upon what must be regarded as one of the most instructive as well as interesting phases of human society and progress. . . The fulness and breadth with which he has carried out his comparative survey of this repulsive field of history [Torture], are such as to preclude our doing justice to the work within our present limits. But here, as throughout the volume, there will be found a wealth of illustration and a critical grasp of the philosophical import of facts which will render Mr. Lea's labors of sterling value to the historical student.—*London Saturday Review*, Oct. 8, 1870.

As a book of ready reference on the subject, it is of the highest value.—*Westminster Review*, Oct. 1867.

BY THE SAME AUTHOR. (Lately Published.)

STUDIES IN CHURCH HISTORY—THE RISE OF THE TEMPORAL POWER—BENEFIT OF CLERGY—EXCOMMUNICATION. In one large royal 12mo. volume of 516 pp. cloth, $2 75.

The story was never told more calmly or with greater learning or wiser thought. We doubt, indeed, if any other study of this field can be compared with this for clearness, accuracy, and power.—*Chicago Examiner*, Dec. 1870.

Mr. Lea's latest work, "Studies in Church History," fully sustains the promise of the first. It deals with three subjects—the Temporal Power, Benefit of Clergy, and Excommunication, the record of which has a peculiar importance for the English student, and is a chapter on Ancient Law likely to be regarded as final. We can hardly pass from our mention of such works as these—with which that on "Sacerdotal Celibacy" should be included—without noting the literary phenomenon that the head of one of the first American houses is also the writer of some of its most original books.—*London Athenæum*, Jan. 7, 1871.

Mr. Lea has done great honor to himself and this country by the admirable works he has written on ecclesiological and cognate subjects. We have already had occasion to commend his "Superstition and Force" and his "History of Sacerdotal Celibacy." The present volume is fully as admirable in its method of dealing with topics and in the thoroughness—a quality so frequently lacking in American authors—with which they are investigated.—*N. Y. Journal of Psychol. Medicine*, July, 1870.

INDEX TO CATALOGUE.

	PAGE
American Journal of the Medical Sciences	1
Abstract, Half-Yearly, of the Med. Sciences	3
Anatomical Atlas, by Smith and Horner	6
Anderson on Diseases of the Skin	20
Ashton on the Rectum and Anus	28
Attfield's Chemistry	10
Ashwell on Diseases of Females	23
Ashhurst's Surgery	27
Barnes on Diseases of Women	23
Bellamy's Surgical Anatomy	7
Bryant's Practical Surgery	29
Bloxam's Chemistry	11
Blandford on Insanity	31
Basham on Renal Diseases	18
Brinton on the Stomach	17
Bigelow on the Hip	28
Barlow's Practice of Medicine	14
Bowman's (John E.) Practical Chemistry	11
Bowman's (John E.) Medical Chemistry	11
Bumstead on Venereal	19
Bumstead and Cullerier's Atlas of Venereal	19
Carpenter's Human Physiology	8
Carpenter's Comparative Physiology	9
Carpenter on the Use and Abuse of Alcohol	13
Carson's Synopsis of Materia Medica	13
Chambers on Diet and Regimen	16
Chambers on the Indigestions	16
Chambers's Restorative Medicine	16
Christison and Griffith's Dispensatory	13
Churchill's System of Midwifery	25
Churchill on Puerperal Fever	23
Condie on Diseases of Children	21
Cooper's (B. B.) Lectures on Surgery	26
Cullerier's Atlas of Venereal Diseases	19
Cyclopedia of Practical Medicine	15
Dalton's Human Physiology	9
Davis' Clinical Lectures	14
De Jongh on Cod-Liver Oil	13
Dewees on Diseases of Females	23
Dewees on Diseases of Children	20
Druitt's Modern Surgery	28
Dunglison's Medical Dictionary	4
Dunglison's Human Physiology	9
Dunglison on New Remedies	13
Ellis's Medical Formulary, by Smith	13
Erichsen's System of Surgery	28
Fenwick's Diagnosis	14
Flint on Respiratory Organs	17
Flint on the Heart	17
Flint's Practice of Medicine	15
Flint's Essays	15
Fownes's Elementary Chemistry	10
Fox on Diseases of the Stomach	17
Fuller on the Lungs, &c.	17
Green's Pathology and Morbid Anatomy	14
Gibson's Surgery	26
Ginge's Pathological Histology, by Leidy	14
Galloway's Qualitative Analysis	10
Gray's Anatomy	6
Griffith's (R. E.) Universal Formulary	13
Gross on Foreign Bodies in Air-Passages	26
Gross's Principles and Practice of Surgery	26
Guersant on Surgical Diseases of Children	20
Hamilton on Dislocations and Fractures	27
Hartshorne's Essentials of Medicine	16
Hartshorne's Conspectus of the Medical Sciences	5
Hartshorne's Anatomy and Physiology	7
Heath's Practical Anatomy	7
Hoblyn's Medical Dictionary	4
Hodge on Women	23
Hodge's Obstetrics	24
Hodges' Practical Dissections	6
Holland's Medical Notes and Reflections	14
Horner's Anatomy and Histology	6
Hudson on Fevers	18
Hill on Venereal Diseases	19
Hillier's Handbook of Skin Diseases	20
Jones and Sieveking's Pathological Anatomy	14
Jones (C. Handfield) on Nervous Disorders	18

	PAGE
Kirkes' Physiology	8
Knapp's Chemical Technology	11
Lea's Superstition and Force	31
Lea's Studies in Church History	31
Lee on Syphilis	18
Lincoln on Electro-Therapeutics	18
Leishman's Midwifery	25
La Roche on Yellow Fever	14
La Roche on Pneumonia, &c.	17
Laurence and Moon's Ophthalmic Surgery	29
Lawson on the Eye	28
Laycock on Medical Observation	14
Lehmann's Physiological Chemistry, 2 vols.	9
Lehmann's Chemical Physiology	5
Ludlow's Manual of Examinations	5
Lyons on Fever	18
Maclise's Surgical Anatomy	7
Marshall's Physiology	8
Medical News and Library	2
Meigs's Lectures on Diseases of Women	23
Meigs on Puerperal Fever	23
Miller's Practice of Surgery	26
Miller's Principles of Surgery	26
Montgomery on Pregnancy	25
Neill and Smith's Compendium of Med. Science	5
Neligan's Atlas of Diseases of the Skin	20
Neligan on Diseases of the Skin	20
Obstetrical Journal	22
Odling's Practical Chemistry	10
Pavy on Digestion	16
Pavy on Food	16
Parrish's Practical Pharmacy	12
Pirrie's System of Surgery	27
Pereira's Mat. Medica and Therapeutics, abridged	13
Quain and Sharpey's Anatomy, by Leidy	6
Roberts on Urinary Diseases	18
Ramsbotham on Parturition	25
Rigby's Midwifery	25
Royle's Materia Medica and Therapeutics	13
Swayne's Obstetric Aphorisms	24
Sargent's Minor Surgery	26
Sharpey and Quain's Anatomy, by Leidy	6
Skey's Operative Surgery	26
Slade on Diphtheria	19
Smith (J. L.) on Children	21
Smith (H. H.) and Horner's Anatomical Atlas	6
Smith (Edward) on Consumption	17
Smith on Wasting Diseases c. Children	21
Stillé's Therapeutics	12
Sturges on Clinical Medicine	14
Stokes on Fever	14
Tanner's Manual of Clinical Medicine	5
Tanner on Pregnancy	24
Taylor's Medical Jurisprudence	30
Taylor's Principles and Practice of Med Jurisp	30
Taylor on Poisons	39
Tuke on the Influence of the Mind	31
Thomas on Diseases of Females	22
Thompson on Urinary Organs	30
Thompson on Stricture	30
Thompson on the Prostate	30
Todd on Acute Diseases	14
Walshe on the Heart	17
Watson's Practice of Physic	15
Wells on the Eye	29
West on Diseases of Females	23
West on Diseases of Children	21
West on Nervous Disorders of Children	21
What to Observe in Medical Cases	14
Williams on Consumption	17
Wilson s Human Anatomy	7
Wilson on Diseases of the Skin	20
Wilson's Plates on Diseases of the Skin	20
Wilson's Handbook of Cutaneous Medicine	20
Winslow on Brain and Mind	31
Wöhler's Organic Chemistry	11
Winckel on Childbed	4
Zeissl on Venereal	19

For "THE OBSTETRICAL JOURNAL," FIVE DOLLARS a year, see p. 22.

www.ingramcontent.com/pod-product-compliance
Lightning Source LLC
Chambersburg PA
CBHW021201230426
43667CB00006B/499